AA003496

13th Process Plant Safety Symposium 2011

(PPSS)

Topical Conference at the 2011 AIChE Spring Meeting and 7th Global Congress on Process Safety

Chicago, Illinois, USA
13 – 17 March 2011

ISBN: 978-1-61782-887-4

Printed from e-media with permission by:

Curran Associates, Inc.
57 Morehouse Lane
Red Hook, NY 12571

proceedings.com

Some format issues inherent in the e-media version may also appear in this print version.

Copyright© (2011) by AIChE
All rights reserved.

Printed by Curran Associates, Inc. (2011)

For permission requests, please contact AIChE
at the address below.

AIChE
3 Park Avenue
New York, NY 10016-5991

Phone: (203) 702-7660
Fax: (203) 775-5177

www.aiche.org

Additional copies of this publication are available from:

Curran Associates, Inc.
57 Morehouse Lane
Red Hook, NY 12571 USA
Phone: 845-758-0400
Fax: 845-758-2634
Email: curran@proceedings.com
Web: www.proceedings.com

TABLE OF CONTENTS

Asset-Based MOC Approval

William R. Mathews
Lyondell Chemical Company,
A LyondellBasell Company
Bayport Plant
10801 Choate Rd
Pasadena, Tx 77520
William.Matthews@Lyondellbasell.com

Keywords: Management of Change, Approval, Review

Abstract

One of the more difficult tasks involved in the Management of Change process is determining how to obtain the most effective Safety, Environmental and Technical review and approval of the proposed change. A novel approach based on the type of asset involved with the change has been successfully implemented at this Houston-based chemical plant. The MOC process uses a series of questions and logic branches that can be used by even the most inexperienced employee to determine who needs to review and approve the change. This has proven to be an effective and efficient method of obtaining an effective review and authorization of all MOCs.

1. Management of Change Approval

The Management of Change process is a very complex and broad system that can cover changes with varying size and complexity, ranging from changing an instrument range to building an entire chemical plant. One of the problems this creates is the issue of determining how to get the most effective Safety, Environmental and Technical review along with the proper approval of the proposed change. There are many industry approaches that try to achieve this goal. They range from all Management and Engineering staff at the site reviewing and approving the change, to a single "MOC Expert" performing the review.

If a proposed change is not properly reviewed and approved, it can cause unsafe, inefficient, unreliable or non-compliant conditions in the facility. If the review and approval process is excessive, it can increase costs and timing of projects (and the economic well-being of the company.

In this paper, we examine a method of selecting the personnel required to obtain proper review and approval of changes. This determination can be conducted with the least experienced personnel and can facilitate consistent review of approvals across the company.

2. MOC Approval and Review Definition

First, we need to look at what is meant by approval and review of a change. While related, they are different functions. They are also linked to the concept of someone being notified of the change. Note that each MOC (no matter how large or small) is a "project" to be handled and routed just like every major capital project. There may be some steps that are simplified, but they must follow this pattern to be successful. So we will start with the RACI model of management of a project presented by the Project Management Institute [1,2].

RACI Model of Project Management Institute (PMI.ORG)			
Symbol	**Meaning**	**Duty**	**Typical Function**
R	**R**esponsible Person who does the task	Coordinate	Project Engineer
A	**A**ccountable Person who is ultimately responsible	Approval	Area Manager
C	**C**onsult Person who provides input, but cannot veto	Review	Engineering Expert
I	**I**nformed Person who must be notified	Respond	Change Implementer

Every MOC must have a sponsor, champion, Lead Engineer or other person who is "Responsible" for the change. They coordinate all of its activities and requirements. They may not do all the work, but they make sure that it is done properly and in the correct sequence. They are an "unofficial" approver and reviewer, in that they help select the scope and timing of the change.

Every project has one or more "Approvers" who are ultimately responsible for the success of the project. The larger and more complex the change, the more Approvers required. Note that Approvers are the only ones who can veto or cancel a change. Also note that often an Approver is not an expert on the process with detailed knowledge of the Safety, Environmental, and Technical. Their job is to make sure that all of this review is done properly.

Every change has one or more "Reviewers." These are the experts in the "technical" aspects of the change. They make sure that the change follows any applicable regulations, standards or practices. Note that the Reviewers cannot veto or cancel the change, but they can affect the design of the change or how the change is implemented.

Finally, every change has one or more persons who must be Notified. Note that there are people who must be notified at different stages of the change. Typically, there are people who need to be notified of "change approval" (Interestingly, these are the people who will implement the change. They need to know that the proverbial check has been signed and it is OK to cash it). Secondly, there are people who need to know that the change has been implemented (per OSHA, the technicians and Operators who are affected by the change). An Environmental representative may need to be notified that a change is in effect so that certain monitoring requirements can be started.

Note that there are many times when someone wants to "approve" an MOC, but when looking at these definitions, all they really need to receive notification of the approval or implementation of the change.

3. Industry Methods of MOC Approval and Review

We will examine how to select the appropriate personnel or functions required to properly approve and review and MOC.

Currently there are many different methods to determine who approves and/or reviews an MOC. Some of the more frequently used methods include:

Cost of MOC –This is a model based on the Capital Project systems present in chemical companies today. The higher the cost of a project, the more approvers required, and the higher in the Company organization the approvers need to be. Also, the higher the cost, the more likely Corporate Engineering and Staff are required to review and approve it. Note that this model works well for the financial control aspects of a project, but it adds little value to the safety aspects, industrial hygiene aspects, and more generally, the technical aspects of the MOC. It also adds to the duration of the project while waiting for the higher management personnel to become available to review the details of the MOC. Involvement of Corporate Engineering and Staff is often helpful in providing a better design; however, approvals which solely use corporate groups have some inherent risk since the corporate engineering groups may not always grasp the nuances of the local site processes.

All Managers Approve – This model is often found at smaller sites where Managers serve as Technical Resources. To avoid someone being left out of the loop, it is "better" to have every Manager approve every MOC, even if it has no effect on them. This model slows down the MOC process and increases costs. It can also lead to the "Approval Bypass," a situation when a Manager is not available and someone assumes that it does not affect their area – and the change must be done immediately. Managers are usually assumed to know the Technical reviews required, and they assign or infer who needs to review the change.

MOC Expert Approve- This model calls for a single MOC Expert to approve and determine who needs to review the change. This model is common in chemical plants with unique or esoteric processes that are not normally found in the rest of the Industry. This person is an expert on the process so they are relied upon to make sure all aspects are properly reviewed. Depending on the experience level of this person, they can select most of the proper reviewers, but they can overlook some disciplines or assume that they know enough about that discipline to be the reviewer (when they are not). So despite their wealth of knowledge of their process, they can sometimes miss some of the required reviewers of other disciplines.

Number of Documents Affected – This model calls for more and higher level Approvers based on the number of PSI documents that are affected by the change. Again, the Approvers would determine what disciplines need to "review" the change. The issue with this system is who determines what and how many documents are affected by a change?

Change Location – This model calls for the Area Manager to be the Approver and to determine who needs to review the change. Like the MOC Expert model, it can be successful depending on the experience and insight of the Area Manager. It can also lead to inconsistent review and approval depending on how soon the change is "needed."

While most of these systems will work and provide a proper Safety, Health and Technical review of most of the changes involved, they all can result in inconsistent review and approval of changes. They are dependent on the experience and ability of the people in the positions to make the right call.

4. Asset-Based MOC Review and Approval

About 12 years ago, a multi-discipline team at the LyondellBasell Bayport plant implemented a different system to determine who needs to review and approve changes at our site. This system uses a series of branched checklists to make this determination. It has proven to be very effective. We have modified it through the years as we have learned what kinds of changes require review – and we will probably revise it in the future as we learn more. The point is that we capture these learnings. They are not sent to another site as someone who has learned it is transferred.

We call this system…

Asset-Based MOC Review and Approval – On all MOC changes, we determine all of the types of assets - Instruments, Pipe, Structures, Pumps, Vessel, Electrical, DCS, etc, - that are involved in the change to determine who must review and approve. We use the "location" of Assets to determine which Area Manager should approve. If there is more than one location, then multiple Area Managers are required to approve.

Next we use a rule-based Checklist for each type of Asset that looks deeper into the details of the Asset change to see what other disciplines or topic specialists should review the MOC. The questions identify "special cases" that may not be immediately obvious to inexperienced personnel along with the "obvious" functional reviewers to provide a thorough comprehensive review of the change.

For example, a change to a pump would obviously require review by a Rotating Equipment Engineer. If the pump was a Fire-Water supply pump, then a review by the Safety Specialist in charge of the site Firewater system would be required to review the change. If the change to the pump involved adding new thermocouples, then an Instrument Engineer would be required to review the change. If the change involved a larger Electrical Motor, then the Electrical Engineer would be involved. There are several other disciplines involved. While many are "obvious," they can be missed by an inexperienced Engineer.

Our Asset-Based Review and Approval System consists of a series of checklists that cover the major groups of Assets

1. Piping and Valves
2. Rotating Equipment
3. Pressure Vessels
4. Heat Exchangers
5. Reactors
6. Relief Valves
7. Instrumentation and Control Valves.
8. Structures
9. Distributed Control Systems (DCS)

If there are some specialty types of assets at a location, they can be added. Note that if a Checklist is not involved with the change, it is not included in the MOC package. Each Checklist has a series of questions that are used to determine which discipline should review the MOC as well as provide a basis for a Safety, Health and Technical review for the change. Some of the questions are intended to determine if an unsafe practice or design is being used. They are also "reminders" of key standards or practices.

One of the checklists we use covers any change to instruments at our site.

Instrument Checklist

This checklist must be filled out for all Full MOCs that involve changes to instruments, online analyzers, control valves, PLC's or field controllers Not Applicable ☒

Instrument #	Instrument Description

Item	Reqd Review	Does this change involve:	Yes	No
I-1	**P, IR**	Fail safe position of control valves?		
I-2	**P, IR, PS**	Changes to Annunciator panel alarms and pushbuttons?		
I-3	**P, IR**	Electrical power failure reaction of transmitters / thermocouples?		
I-4	**P, IR**	Electrical power failure reaction of solenoid valves?		
I-5	**P, IR**	Tracing or insulating instrument lead lines or transmitters?		
I-6	**P, CS**	Control sensors that also act as interlock sensors? (or vice versa?)		
I-7	**P, IR**	Adding, Removing or Changing Safety Critical Equipment?		
I-8	**IR**	Electrically powered equipment or instruments in electrically classified areas?		
I-9	**IR, P**	Change of process conditions on instruments and control valves?		
I-10	**IR, P**	Materials of construction of instrument parts contacting process fluid or parts that could be contacted if a instrument seal leaks?		
I-11	**P, IR**	Change of elevation of transmitters, leads or other pressure indicating devices.		
I-12	**P, IR**	Change to or addition of sensing instruments?		
I-13	**P, IR**	Change to or addition of Control valves?		
I-14	**P, AS, IR**	Change of online analyzer including sample takeoff, sample conditioning, sample return and analyzer.		
I-15	**CS, P, IR**	Addition, change or deletion of instruments reading to the DCS or Interlock system or control valves receiving output from the DCS or Interlock System?		
I-16	**IR, P**	Changes to PLC logic, input or output?		
I-17	**IR, P**	Limit switches or position switches on valves?		

P – Process/Operations Eng **PS**-Process Specialist **S**- Area Safety Rep **IH**-Industrial Hygiene **IR**-Instrument Rel Eng **MR**- Mechanical Rel Eng **CS** – Control Systems Rep **AS**- Analyzer Engineer

Each question answered with a "yes" must be reviewed by the discipline indicated in the Required Review column. This checklist is filled out independently for each Instrument Loop because some may have different answers to the same question. Note that these questions also trigger Process Safety concerns that need to be resolved.

We also ask that the MOC preparer list comments on all questions answered "yes" to address any potential Process Safety issues that were addressed during MOC preparation and review.

Checklist Answers

Item	Add comments to questions above answered "YES"

Another example is the Checklist used to change Heat Exchangers

Exchanger Checklist

This checklist must be filled out for all Full MOCs that involve changes to all heat exchangers. One checklist per equipment no. Not Applicable ☒				
	Equip No.	**Service**		
Item	**Reqd Review**	**Does this change involve:**	**Yes**	**No**
EX-1	**P, MR, RS**	Design pressure, design temperature or materials of construction change to exchanger?		
EX-2	**P, MR, PS**	New Heat Exchanger?		
EX-3	**MR**	Exchanger support structure?		
EX-4	**P, MR**	Changes to exchanger nozzles?		
EX-5	**P, MR**	Changes to exchanger tubes, baffles or other internals?		
EX-6	**P**	Performance criteria or process conditions for exchangers?		
EX-7	**PS, MR,S**	Access to fans or motors on fin-fans?		
EX-8	**ER**	Modifications to fin-fan motors? (Not swapping of like motors)		
EX-9	**MR**	Changes to fans or drive equipment?		
EX-10	**P, MR,PS**	Hot (>140°F) vessel surfaces that would require insulation for personnel protection?		
P – Process/Operations Eng **PS**-Process Specialist **S**- Area Safety Rep **IH**-Industrial Hygiene **IR**-Instrument Rel Eng **RS** – Reliability Supt **MR**- Mechanical Rel Eng **CS** – Control Systems Rep **ER**-Electrical Rel Eng				

Again, any "yes" answer to a question requires review by the discipline identified and a comment recorded on any potential Process Safety concerns or issues addressed by the MOC.

This concept is carried over into the site's list of "Short MOCs". These are MOCs for small or well understood changes that have been implemented frequently at the Site. The normal checklists are not used for Short MOCs, but there is a list of cases where the Short MOC process can be applied. This list also includes who must review and approve the MOC. Again, the Short MOC list is based on the type of Asset involved and matches the requirements of the normal MOC checklists.

The table below shows examples of some our "Short MOCs" and how the applicable reviewers and approvers are determined.

No.	SHORTENED MOC CASES Description	Area Supt	Tech Svcs Supt	Env Rep	Mech Rel Eng	Proc Eng	Contr Sys Eng	Inst Rel Eng	Other
1.3	Changes in physical dimensions or materials of construction of thermowells or thermocouples	X	X	X		X		X	
1.4	Change, add or delete smoke or cabinet purge alarm per company standards	X	X	X				X	
1.5a	Instrument range changes for local instruments that do not go to DCS	X	X	X		X		X	
1.5b	Instrument range changes for local instruments that **DO** go to DCS	X	X	X		X	X	X	
2.7	Addition, change or deletion of hydrocarbon monitoring or smoke detection sensors.	X	X	X		X		X	Procs Special, Ind Hyg
3.1	Permanent addition, change or deletion of insulation to piping per company Piping Specification. (Not tracing)	X	X	X	X	X			
4.3	Pump Coupling design changes per company standards	X	X	X	X				Rotat Equip Special

These two processes have proven to be very effective at our site. New Engineers can come to the site and – with a small amount of training – quickly become productive at preparing and implementing MOCs. If there are issues with "interpretation" of the checklists with regard to who has to review and approve a change, they are handled by other more experienced Engineers who are more readily available than the site MOC Coordinator.

Lastly, this process has proven to be very adaptable. If an unusual situation arises that is not often dealt with (Turnarounds, pipeline outages, etc.), then the determination can be recorded and maintained for the next crew of Engineering staff. If a situation or concept had been missed and an incident occurred, it can be corrected in the future by changes or additions to the checklists.

5. Conclusion

Use of an Asset-Based MOC Review and Approval system has proven very effective in the past 12 years at the LyondellBasell Bayport site. It has allowed new Engineers to become effective and consistent in a short period of time while leading an MOC project. The system has proven to be adaptable to newly discovered issues, and permits the incorporation of lessons learned from incident investigations. The system has allowed more consistent review of changes at the site. A huge side benefit of this system is the determination of which PSI must be modified. The correct drawings and applications can be easily identified to make sure that all appropriate drawings are kept up-to-date.

6. References

[1] "Responsibility Assignment Matrix", http://en.wikipedia.org/wiki/Responsibility_assignment_matrix (accessed Jan 11, 2011)

[2] "Responsibility Assignment Matrix" http://www.pmi.org, (accessed Jan 11, 2011)

Quantitative Models for Assessing MOC Performance

Rainer Hoff
Gateway Consulting Group, Inc.
8610 Transit Road, Suite 100, East Amherst, NY, 14051
rhoff@gatewaygroup.com

Keywords: Management of Change, MOC, performance modeling

Abstract

The safety community, bolstered by the PSM regulations, has emphasized the importance of proper management of change to ensure safe, reliable and economic operation of chemical plants and refineries. Yet, at many sites, the MOC process is often harshly criticized as being too burdensome. In order to assess the validity of these criticisms, it is necessary to quantify and measure characteristics of the MOC business process.

This paper describes conceptual and computational models of the MOC business process, which provide a platform for quantitative assessments of various parameters such as effort, duration, cost, and so on. An MOC Standard Model is developed and validated. This provides a benchmark against which MOC alternatives can be measured.

The characteristics of the MOC Standard Model provide guidance for policy decisions that are immediately applicable by practitioners wishing to create or improve their MOC processes.

1. Introduction

Managing change effectively has always been important at chemical plants and refineries. Since the promulgation of OSHA 1910.119 [1]—the Process Safety Management regulations—proper Management of Change is a regulatory requirement, for Covered Processes (as defined in the regulation).

The Process Safety Management regulations cover 14 elements, much more than just Management of Change. As a result, when the regulations were first promulgated in 1992, substantial effort was needed to bring most plants into compliance. In this environment, with deadlines and potential penalties, Management of Change business processes were either hastily created, if they did not previously exist at the site, or quickly reworked and formalized from an existing business process.

Since then, some sites have made occasional improvements to the MOC business process, often using CCPS guidance [2], although many have not. Personnel who are required to use the MOC process, the "users", often have a negative view of the MOC process. User dissatisfaction may be due to a poorly-designed or poorly-implemented process.

What constitutes good MOC practice, and how would one optimize a given MOC process? This paper begins with the representation of an MOC using a lifecycle concept. The lifecycle model is expanded into a detailed task-level model of an MOC. The task-level model is used as the basis for a computer simulation model using the Petri net method [3]. The Petri net model is used to determine realistic values of important MOC performance characteristics.

2. Models of the MOC Process

2.1 Lifecycle Model

A high-level model of MOC, referred to as a "lifecycle" model, is shown in Figure 1. The diagram depicts a permanent, non-emergency MOC process.

A lifecycle is composed of "states": Initiation, Scoping, Change Design, etc. While the MOC is active, it could be considered to be "in" a state: e.g. the MOC is in the Approvals state.

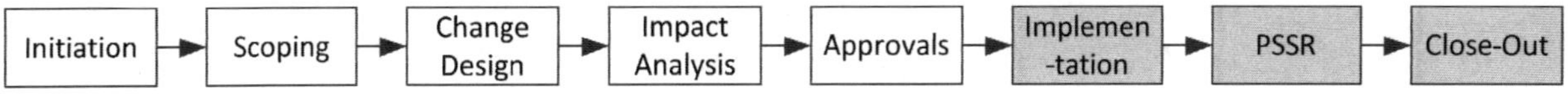

Figure 1. Lifecycle model of an MOC process.

The notion of a business process represented as a collection of states is not new:

- NASA, instituted a "phased review process" in the 1960's,
- various project management methodologies use the concept of a "stage-gate" methodology [4],
- although much more formalized, mathematicians and electrical engineers have used finite automata and finite state machine concepts for decades.

The more common tasks performed during each lifecycle state are listed in Table 1.

State	Major Activities
Initiation	• The change is formally started. • Risk pre-screening is conducted
Scoping	• The change is scoped, which identifies, at a minimum: o What tasks need to be done o Who reviews the MOC o Who approves the MOC o Who is notified
Change Design	• Documents that describe the change are created or redlined
Impact Analysis	• The potential impacts of the change are analyzed. These include: o Process safety impacts (typically PHA's) o Environmental impacts (e.g. permitting impacts) o Financial impacts (i.e. cost estimates) o Safety impacts (e.g. personal protective equipment, etc.) o Industrial hygiene impacts o Product and quality impacts
Approvals	• Previously-identified reviewers and approvers accept, reject or cancel the proposed change. • Rejection implies that further work is needed to detail the physical or financial aspects of the change proposal.
Installation	• The physical change is made in the plant. • Inspections and mechanical integrity activities are conducted. • Procedures are updated. • Training is conducted.
PSSR	• In the case of plant-based MOCs, the Pre-Startup Safety Review validates that the change has been made safely. Discrepancies are noted and resolved. PSSR ends with an authorization to start up. • For non-plant-based MOCs (e.g. lab), or for changes that don't end in a "start-up", this is usually called the "Validation" state, rather than PSSR. The logic is the same inasmuch as a final check is made before the change is put into production.
Close-out	• The change is formally closed. o Metrics are gathered. o Redlines are incorporated into the base documents and released. o Quality assurance steps may be conducted.

Table 1: Typical tasks performed during each state of an MOC.

2.2 *Task-Level Model*

While an MOC is in a given state, a number of tasks must be completed in order to accomplish the work of the task. Once all the tasks for a given state are completed, the gating criteria are satisfied, and the MOC is "promoted" to the next state.

The task-level model shows the individual tasks that take place during a state. For instance, Figure 2 shows the first few tasks that take place during the Change Design state. The logic of this model fragment is as follows:

Assumptions:

- The MOC is small
- Small MOCs, by definition, require that 3 documents are redlined
- Small MOCs, also by definition, require that only 1 person do the redlines. That person is the MOC "owner"

The tasks are:

- CD-01 The Initiator decides whether s/he has the expertise to redline the drawings
- CD-02 If the Initiator does not have the expertise to redline the drawings, then someone, perhaps his/her manager, reassigns ownership of the MOC
- CD-03 The new owner of the MOC must spend some time understanding the requirements of the MOC; i.e. the new owner rises the learning curve for this MOC
- CD-04 Once the new owner understands what is needed, s/he redlines the 3 documents needed to define the proposed change. If ownership didn't change, as a result of a "No" decision in CD-01, then the Initiator redlines the 3 documents.

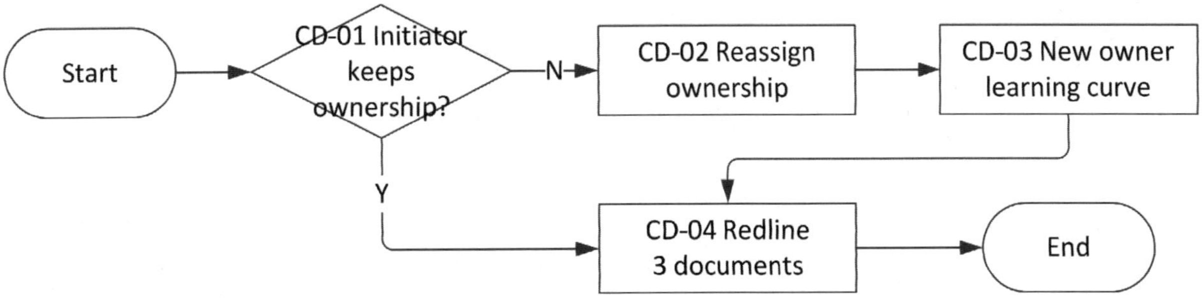

Figure 2. Task-level model fragment, for part of Change Design state.

Only 3 tasks and 1 branch are shown in Figure 2. The entire MOC task-level model has over 300 tasks and over 180 branches.

Note that the lifecycle model, Figure 1, only has ordering at the state level: i.e. all the tasks for one state have to be completed prior to the next state beginning. Within a state, the lifecycle model does not recognize task ordering. The task-level model, Figure 2, is highly ordered, although parallel tasks are certainly permitted. This poses a challenge when creating a model for the typical MOC since the exact order that tasks are performed during the actual execution of an MOC varies from one MOC to another. For the purposes of MOC performance assessment, any reasonable task-level model is likely to be adequate. Certainly the tasks and branches in Figure 2 look reasonable.

2.3 Computational Model

The task-level model describes the MOC process in sufficient detail that interested individuals can discuss the details of what happens in a given state. In order to quantitatively analyze the performance characteristics of MOC, it's necessary to develop a computational model. There are three common approaches to business process quantification:

1. Spreadsheets: while possible to create an MOC computational model using a spreadsheet, there are formidable difficulties dealing with complexity and the demands of stochastic processes,
2. Packaged business process simulation software: a number of vendors offer software to model business processes, and this is certainly a viable approach,
3. Petri nets: Petri nets were developed by C.A. Petri, in 1962 [3], and have been popularized for business process use by vanderAalst [5] and Reijers [6]. The strength of Petri nets is that they can represent business processes at a very granular level, and the size and complexity of the model is only limited by the analyst's time in assembling it. Timed Petri nets can easily handle stochastic processes.

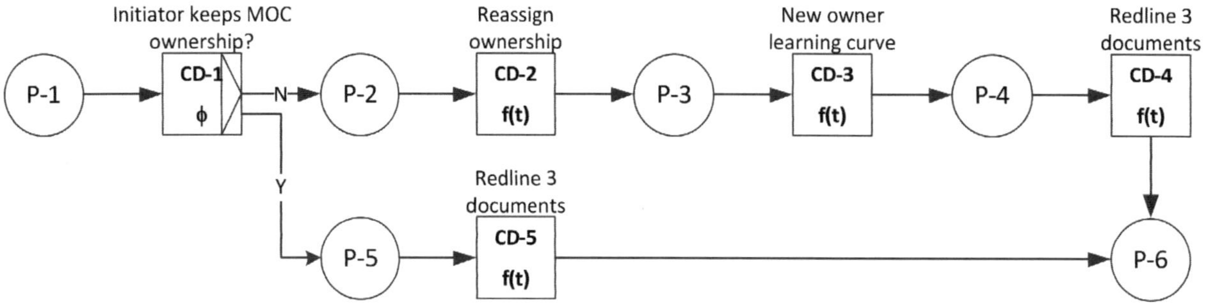

Figure 3. Computational model fragment for part of Change Design state.

While a comprehensive discussion of Petri nets is beyond the scope of this paper, the use of Petri nets is very analogous to the use of finite element methods for structural analysis problems. Specifically, a finite element model consists of nodes, which describe the model's topology, and elements which have prescribed behavior. In a Petri net, as shown in Figure 3, the topology of the business process is described by "places", indicated by the circular symbols, and the behavior of the model is governed by "transitions", indicated by the square symbols.

The places in Figure 3 are:

- P-1 is the start place.
- P-2 thru P-5 are places
- P-6 is the end place.

Place and transition labeling is arbitrary; these places could just as easily have been called Tom, Richard, Harriett, etc. The only requirement is that the names are unique.

The transitions in Figure 3 are:

CD-1 The ϕ signifies that this is a "no operation" transition. Nothing happens, except that this output has an OR-split. Process flow goes from CD-1 to P-2 some percentage of the time, say, x%. Process flow goes from CD-1 to P-5 (1-x)% of the time. Whether the process flow goes to P-2 or P-5, for a specific MOC, is unknowable in advance. However, over many iterations, the percentage of times that P-2 is activated will tend to x%.

CD-2 The f(t) signifies that a resource performs a task during this transition, and that performing the task takes time. In this case, since the Initiator has not kept ownership of the MOC, the Initiator's manager reassigns ownership to someone else.

CD-3 Like CD-2, this is a transition where work takes place. The new owner must rise the learning curve on the MOC just assigned to him/her.

CD-4 The new owner redlines 3 documents, to document the proposed changes for a small MOC.

CD-5 The Initiator has kept ownership of the MOC and redlines the 3 documents.

The transitions CD-2, 3, 4 and 5 model two different aspects of time:

- Start Time: the time before the person starts to work on the task. Just because someone has been assigned a task, doesn't mean that they are available to perform the task—they may be busy with other work. So each transition is given a start time of, say, 8 hr, whereupon the resource begins the work.
- Duration: the time it actually takes the resource to perform the task, say, 1 hr. Multiplying the duration of a task by the hourly rate of the resource, say $50/hr, gives the labor cost for the task.

The MOC performance model is intended to be as realistic as possible. Assigning a start time of exactly 8 hr and a duration of exactly 1 hr is not realistic, since these times are distributed according to some probability density function. Possible probability density functions include:

- **Fixed time:** t = 8, means that each event occurs at exactly the 8 hr point.
- **Uniform distribution:** events are evenly distributed in the interval $0 \le t \le 16$.
- **Gaussian distribution:** the "bell curve" distribution cannot be used for this work because the left tail of the distribution extends to $-\infty$, which would result in negative time.
- **Weibull distribution:** events conform to the probability distribution function shown below. In this example, the shape parameter, k = 2.0425, and the scale parameter, $\lambda = 9.5724$, produces a distribution with the median time of 8 hr.

$$p(t) = \left(\frac{k}{\lambda}\right)\left(\frac{t}{\lambda}\right)^{(k-1)} e^{-\left(\frac{t}{\lambda}\right)^k}, \text{ for } t, k, \lambda > 0$$

This is what is meant by the term "stochastic process", i.e. that many of the parameters are not given fixed values, but are specified as a statistical distribution, with the appropriate parameters.

2.4 The Complete Model

A Petri net model was created of the entire MOC process. It consists of 1,410 places and 1,273 transitions. In addition, 152 parameters were given realistic values: for example, average wait time before someone begins his/her task is 8 hr; percent of time that the Initiator retains ownership of the MOC and creates the redlines is 50%; the hourly rate for a professional person is $75/hr, and so on for the remaining parameters.

The Petri net model determines two kinds of costs:

1. Labor costs: this is the effort involved in each task times the hourly rate for the resource,
2. Scrap costs: material costs are not included in the analysis, because they are determined by the physics of the change, and not impacted by the MOC business process. Scrap costs, however, are related to the business process: a poor MOC process, with inadequate documentation, will lead to higher scrap costs. Due to the variability of the true cost of scrap, an estimate of $1,000 per scrap event is used.

2.5 Small, Medium and Large MOCs

A review of MOC data from Gateway clients reveals that there is no one "typical" MOC, at least not to at the level of detail of this study. There is a large variation in size and complexity of MOCs. One way to deal with this is to define certain MOC configurations that are representative of the universe of MOCs. The author's team acquired MOC data for thousands of completed changes. Only permanent and non-cancelled MOCs were analyzed. The MOCs were assigned 2 values: (1) the number of documents redlined during the Change Design state, and (2) the number of approvers that affixed their signatures during the Approvals state. Each MOC, thus scored, was then sorted according to the scores. The 10th-percentile, 50th-percentile and 90th-percentile points were then identified, and labeled "small", "medium" and "large".

MOC Label	**Small**	**Medium**	**Large**
Size	10th percentile	50th percentile	90th percentile
Number of documents redlined	3	5	12
Number of approvers	3	5	8
Process safety impact analysis	Checklist	What-if	HAZOP
Number of crafts involved in implementation	2	5	8
Number of inspections	2	5	8
Shutdowns during Implementation	No shutdown	Isolated equipment shutdown, in 20% of cases	Unit shutdown, in 20% of cases

Table 2. Parameters describing small, medium and large MOCs.

3. Analysis

A Petri net model was constructed that models the behavior of small, medium and large MOCs. The collection of Petri net model plus the associated set of parameters has been labeled the "MOC Standard Model". This is the first attempt at defining such a model, and should therefore be considered preliminary.

3.1 Validation

A numerical model, such as a finite element or computational fluid dynamics model is only useful to the extent that the results can be trusted. Analysts using either of these methods typically validate their models by comparing results for steady-state cases for simple geometries, with known theoretical solutions.

For the MOC Petri net model, a similarly careful model validation exercise was undertaken. The validation approach was as follows:

- The model was constructed one state at a time. So the first model part was created for Initiation. The second part was created for Scoping, and so on.
- A parallel exercise was undertaken to create a spreadsheet model of each MOC state. The spreadsheet model could only represent the MOC process using the Fixed Time distribution for the time-dependent tasks, and with very restricted branching. More complex task time distributions or more complex branching made the spreadsheet calculations intractable.
- Results were calculated using the Petri net model, which was set to use the Fixed Time distribution. These results were compared to the spreadsheet results. Discrepancies were addressed until these 2 independent approaches yielded identical results.
- This process was repeated for each lifecycle state.
- As the complete model was being assembled, these validation tests were rerun. First the process was rerun with Initiation + Scoping, then Initiation thru Change Design, and so on until the entire process Initiation thru Close-out was validated.
- Once the entire model was thoroughly validated, all analyses were rerun with time-dependent parameters conforming to the Uniform distribution, and then again with the Weibull distribution, by setting the appropriate flag.

4. Results

The results in sections 4.1 through 4.5 are taken from the MOC Standard Model. Section 4.5 describes the results of using the Standard Model to assess other conditions, specifically the impact of errors.

4.1 Basic Results

The basic results for the entire MOC are shown in Table 3a (unless otherwise indicated, the results henceforth assume that time-dependent parameters follow a Weibull distribution). While the specific values in Table 3(a) are applicable to a specific set of parameters used in the Standard Model, the ratios between large, medium and small, from Table 3(b), are more generally instructive.

The duration calculation is based on the notion of an 8 hr workday, with no activity presumed for off-shift hours. Also, the Standard Model presumes the existence of an electronic MOC system, and electronic access to supporting documentation. It is highly unlikely that the durations shown in Table 3(a) could be achieved with a paper-based MOC system.

The effort and cost are closely related. Greater effort implies a greater cost. While cost and effort are linearly related for individual tasks, the entire MOC requires a mixture of resources at different hourly rates. That explains why the Effort Ratio and the Cost Ratio for medium and large MOCs are similar, but not identical.

MOC Size	Effort [person-hr]	Duration [d]	Cost [$]
Small	37.51	22.95	2,603
Medium	112.58	34.18	7,201
Large	438.14	76.60	28,582

(a)

MOC Size	Effort Ratio	Duration Ratio	Cost Ratio
Small	1	1	1
Medium	3.00	1.49	2.77
Large	11.68	3.34	10.98

(b)

Table 3. Basic results for entire MOC process: permanent, non-emergency.
(a) Time and cost values. (b) Values normalized to small MOCs.

4.2 Cost and Duration Distributions

The 3 cost values, from Table 3(a), are fit to a curve in illustrate the probability distribution of MOCs with respect to cost, shown in Figure 4(a). It's clear that most of the MOCs are at the low cost end of the scale. A similar analysis is provided for MOC durations, in Figure 4(b).

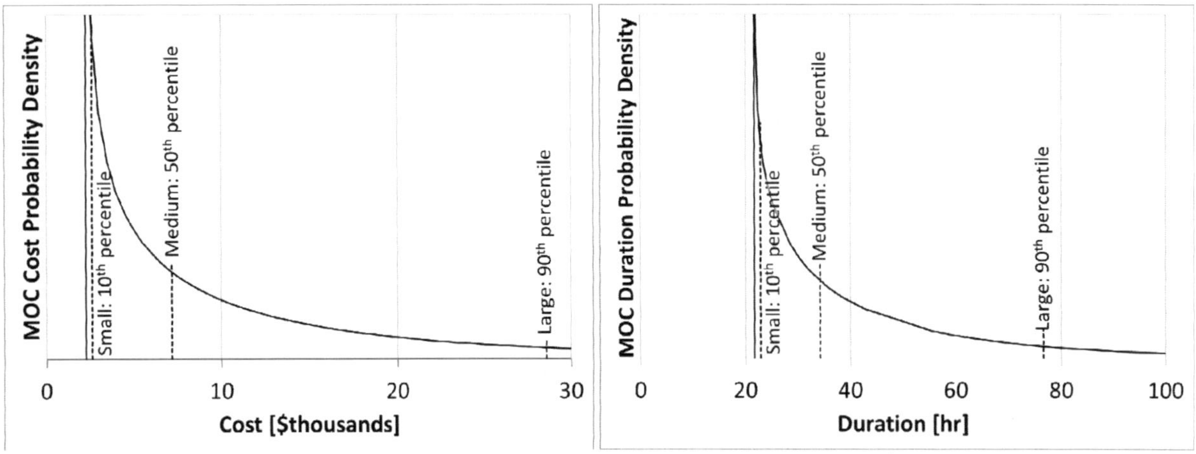

(a) (b)

Figure 4. (a) Distribution of MOC costs. (b) Distribution of MOC durations.

These diagrams point to an ongoing problem that faces MOC practitioners. When an MOC process is designed (i.e. the MOC procedure is written) at a site, the process must work for all MOCs. Therefore, the MOC process must accommodate the needs of the largest MOCs. But, at least 50% of the time, the MOC process is used for MOCs that are 4 to 11 times smaller than what the process was designed for. This partially explains why MOC users often find the process burdensome.

One solution to this problem is to make the MOC adaptive. One common technique, for paper MOCs, is for the MOC procedure to be comprehensive (i.e. covers even the largest MOCs), but many of the supporting documents or checklists are optional (and not needed from small MOCs). Implementing adaptive MOCs using a paper MOC process is challenging:

- if all the optional documents are provided as part of the MOC procedure, then the MOC package quickly becomes unwieldy (one site had 48 pages, including the optional ones!),
- if the optional documents are not provided as part of the MOC procedure, then the user must search for them in order to complete the MOC.

Electronic MOC systems are better suited to implementing adaptive MOCs.

4.3 State Breakdown

Figure 5 presents the cost and duration breakdown by state.

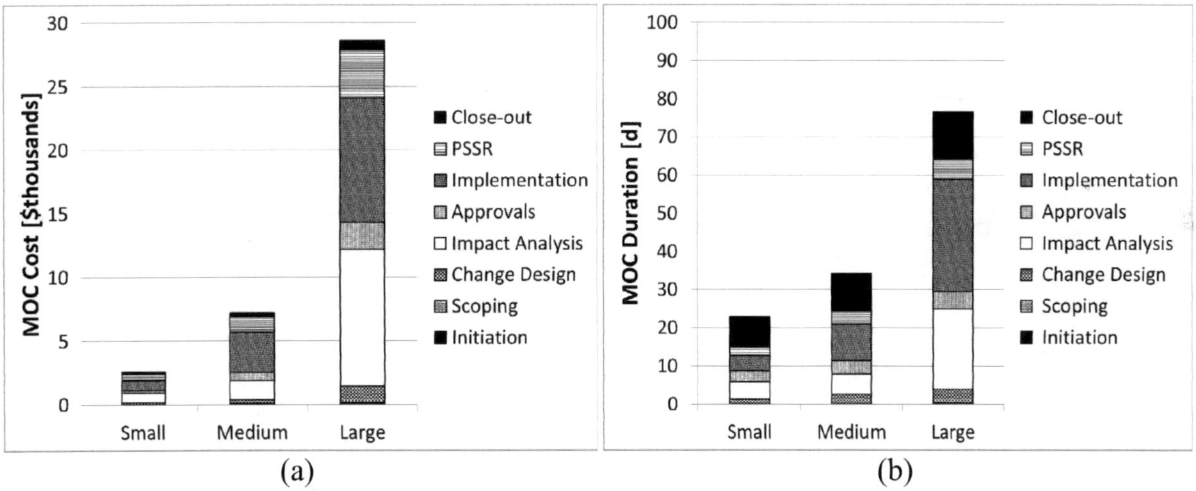

Figure 5. (a) Cost breakdown by state. (b) Duration breakdown by state.

The relative cost and duration of Initiation and Scoping are so small that they are not visible at the scale of the bar graphs.

The relative ease of MOC creation has policy implications. In the interests of safety and encouraging employee participation, most sites permit almost anyone to initiate an MOC. But MOC initiation has huge leverage: for every dollar spent on initiation (i.e. the initiator's salary for the time involved), the company will spend another ~$300 on pursuing the change (and this doesn't even include material cost). Therefore, it appears reasonable that the manager who is responsible for the asset should be required to accept the MOC after it has been scoped. In the absence of this prioritization function, the MOC system at the site will become clogged with dozens of hundreds of in-process MOCs.

Scoping also has huge leverage, like Initiation. Typically, scoping represents about 1/500th the cost of the overall MOC. Yet, the "work" of the MOC is determined during the scoping activity. This suggests that adequate attention must be paid to effectively scope the change.

4.4 Implementation: (Typically) The Largest Cost

The Petri net model attempts to realistically model the implementation process. Key features include:

- unit or equipment shutdown may be required (see Table 2)
- development of detailed change specifications, in addition to the redlines created during Change Design
- purchase requisition generation, ordering, material receipt, work order generation and issuing
- activities of representative crafts
- procedure updates
- training material development and training activities.

With this scope of work, it is not surprising that the Implementation state has the greatest cost.

4.5 Impact Analysis: (Typically) The Second Largest Cost

The MOC Standard Model considers three kinds of impacts:

- process safety impacts, usually addressed using a process hazards analysis
- environmental impacts
- financial impacts

It is assumed that the risks associated with small changes require less detailed analysis. So, the process safety impacts are analyzed using a checklist for small changes. Similarly rapid analysis techniques are used for environmental and financial analyses. Large MOCs are assumed to have risks that require more detailed analysis. So, the process safety risks are analyzed using the HAZOP methodology. Similarly detailed techniques are used for the environmental and financial analyses.

The detailed analysis techniques have two common features:

1. They require multiple people to assemble for a meeting to discuss the specific impacts of the MOC, thereby increasing the resource costs significantly.
2. The meetings must be scheduled in advance, thereby significantly increasing the duration of the impact analyses.

Clearly, when a detailed impact analysis is needed for risk mitigation purposes, its expense and time are unavoidable. However, if there is no reason for a detailed impact analysis, then there are significant cost and time benefits to performing only the necessary analysis.

How and when would one determine what kind of impact analysis is necessary? This has been done using pre-established rules, applicable to the site and its inherent hazards. The best time to identify the impact analysis methodologies is during scoping, since this allows the necessary resources to be arranged ahead of time.

4.6 The Effects of Errors

The Standard model uses a zero-error assumption. Of course, that is not realistic—errors occur all the time, and error mitigation creates extra cost. The kinds of errors represented in the Standard Model are:

1. **Errors discovered during Impact Analysis:** If the change proposal, as documented in the redlines, does not provide sufficient information, the team/person conducting the impact analysis may stop and await additional information.
2. **Approvers rejecting the change:** Approvers may reject a change if it is improperly or inadequately described. The process stops awaiting additional information.
3. **Errors discovered during Implementation:** Prior to implementation, the change could be considered somewhat abstract—during implementation, the change is most definitely concrete. During implementation four kinds of problems are commonly encountered:
    a) **Design deficiencies:** These are errors in the design of the proposed change. "Design" is to be interpreted as any error in documentation, which makes it difficult or impossible to continue work on the change.
    b) **Physical discrepancies:** This is a common problem whereby it's discovered that the documentation does not correspond to what is actually installed in the plant already. When the documentation does not accurately reflect the current plant configuration, it brings into the question the adequacy of the change proposal as well as the adequacy of the various impact analyses.
    c) **Execution deficiencies:** These are mistakes made by the resources implementing the change.
    d) **Material deficiencies:** These are inadequate parts or materials, whose inadequacy may be due to the incorrect materials being installed or damaged during installation.

In all cases, the errors are further classified as follows:

- **Minor errors:** Minor errors are those that can be remedied quickly, and will not change any of the impact analyses in any way. The MOC remains in the same state, and the process continues after the error is remediated.
- **Major errors:** Major errors are those where documentation is significantly altered, and the changed information causes one or more of the impact analyses to be modified. When a major error is encountered, the MOC is usually demoted to the Change Design state. Once the documents are modified, the change must work its way through its lifecycle again, beginning at Change Design.

Reasonable values were selected for the probability of major or minor errors of types 1 thru 3(d), from the previous list. The minor errors have little effect on the already expensive MOC process. The charts in Figure 6 compare results with and without errors. The minority of

MOCs that do have major errors (~10%), have their costs and durations almost double. But, when the doubling of costs and durations is averaged over all MOCs, the average results are still significant, but not that startling.

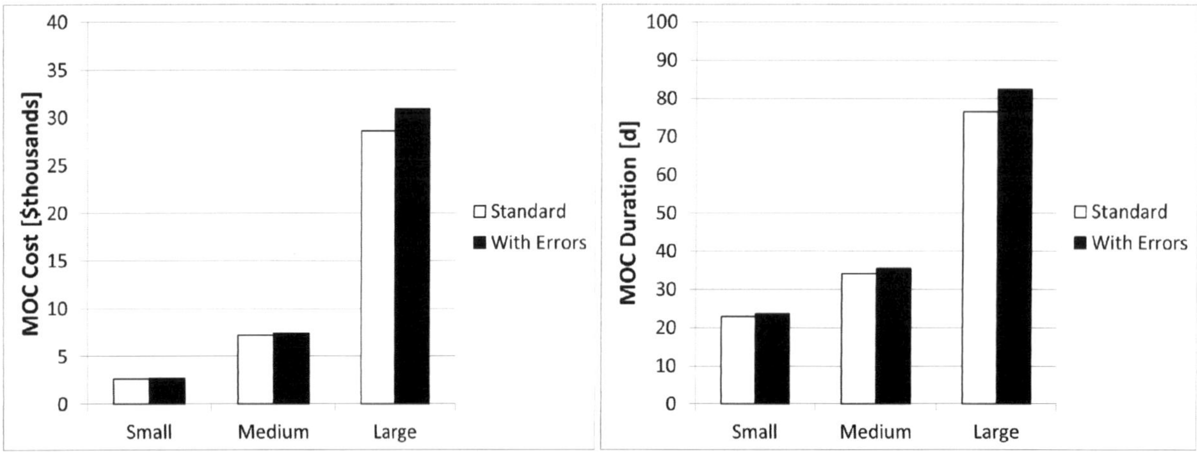

(a) (b)

Figure 6. Impact of errors on (a) costs, and, (b) duration.

Anything that can be done to reduce the frequency of errors in MOCs is clearly a benefit: perhaps better scoping, perhaps greater care in redlining the proposed change?

5. Conclusions

The lifecycle or "stage-gate" model is useful for describing the high level MOC business process, and the lifecycle is further detailed into a computational model. The computational model uses the Petri net method which permits very granular representation of business processes including branching and stochastic behavior of resources. The model was extensively validated.

A "Standard MOC Model" is proposed which can be used as a benchmark in business process improvement studies. The standard model reveals several things about MOCs:

- large MOCs are ~11 times as costly as small ones
- large MOCs have a duration of about 3 times as long as small ones
- Implementation and Impact Analysis are the most costly states, and require the greatest time
- When major errors occur in the execution of an MOC, the cost, as well as the duration, of the MOC can easily double

As discussed in the text, the following ideas may make MOCs more efficient:

- Managerial review of MOCs just after initiation and scoping
- Better scoping tools, to ensure that the work of the MOC is properly scoped
- Adequate time and care to ensure that the proposed change is properly documented on the redlines.

6. References

[1] OSHA, 1992, "Process safety management of highly hazardous chemicals," 29CFR1910.119, OSHA, Washington.

[2] American Institute of Chemical Engineers. Center for Chemical Process Safety., 2008, Guidelines for the Management of Change for Process Safety, Wiley-Interscience, Hoboken, N.J.

[3] Petri, C. A., 1962, "Kommunikation mit Automaten," Ph.D. Dissertation, University of Bonn, Bonn.

[4] Belliveau, P., Griffin, A., and Somermeyer, S., 2002, The PDMS toolbook for new product development, John Wiley & Sons, inc., New York.

[5] vanderAalst, W., and vanHee, K. M., 2002, Workflow management: models, methods, and systems, MIT Press, Cambridge, MA.

[6] Reijers, H. A., 2003, Design and control of workflow processes: business process management for the service industry, Springer, Berlin, New York.

Management of Change (MOC) Auditing System

R Wayne Garland,
Eastman Chemical Company
P O Box 511 Bldg 54D
Kingsport, TN 37662
423-229-5518
wgarland@eastman

Abstract

Our company has discovered that regular auditing of the MOC process is essential to enforcing its proper use. We have developed an audit that can be performed by plant personnel to identify the effectiveness of our MOC process in 2 critical categories: (1) identifying changes and initiating MOC for them, and (2) after MOC is initiated, following the process correctly and completely. In addition, we have a set of measures associated with these audits that we track to give us an overall picture of how effectively our MOC process is operating and identify any systemic issues. The audit form, process data collection and analysis system will be explained.

Background

MOC is one of 14 elements included in the Process Safety Management (PSM) regulation 29 CFR 1910.119. The Process Plant Safety Symposium of the American Institute of Chemical Engineers (AIChE) has called Management of Change "the most difficult PSM challenge". There certainly is ample justification for this characterization:

1. Changes to manufacturing processes can involve coordination of many different persons and organizations.
2. Changes can have various and sometimes unintended consequences.
3. There are different types of changes, which may require different handling.
4. Updates may be required to numerous documents.
5. A high level of training on the MOC process is necessary for it to be successful.
6. Changes must be truly managed and not just documented.
7. Expectations for following the process must be clearly set by management personnel at a sufficiently high level.

Section (o)(1) of the PSM regulation contains a requirement for a PSM audit at least every three years which would include MOC. The requirement reads as follows: "Employers shall certify that they have evaluated compliance with the provisions of this section at least every three years to verify that the procedures and practices developed under the standard are adequate and are being followed." In our company, corporate level auditors fulfill this requirement by conducting an assessment every 3 years.

However, our analysis and plant experience has shown that auditing MOC every 3 years, although it complies with all legal requirements, is not sufficient to assure that an MOC system is operating at maximum effectiveness. Therefore we concluded that more frequent audits were necessary, either by corporate auditors or by plant personnel.

We concluded that additional auditing by plant personnel was a better solution for several reasons:

- this would provide 2 different levels of audits
- plant personnel have more in-depth understanding of the manufacturing processes
- plant personnel are ultimately responsible for the successful execution of MOC and
- it makes effective use of resources because additional corporate/site auditors are not required.

Although we were convinced that MOC audits by plant personnel were desirable, this did present certain challenges. It was necessary to give sufficient guidance to personnel that, although they had some training, were not full time professional auditors or MOC experts. We had to develop an auditing process and protocol for plant personnel to use which were effective but also practical. In addition, we needed a data collection system and measures which were easy to use and provided relevant practical information on how well MOC was being executed. This includes defining what is level of performance is acceptable and considered to be "good".

Our larger manufacturing plants are organized into divisions and which in turn are divided into departments. Each manufacturing division and department has someone designated as the MOC Coordinator. At smaller sites there may be only one division. Division MOC Coordinators serve on a corporate MOC Team, which has overall responsibility for the MOC process. This is the team that developed the audit. In addition, each plant has a central safety group (at small sites this function may be included in a group with other functions such as environmental).

Audit Process

There are 2 basic questions to be answered by the audit:

(1) Is MOC initiated when required?
(2) After initiation is the MOC process being followed correctly?

Attachment 1 is our MOC Audit Guideline. It contains audit protocols devoted to each of these questions. The protocols contain all the essential information, and are simple and easy to use by plant personnel.

The first protocol is for a management system audit that is normally conducted by the division MOC Coordinator as the lead auditor. This is an annual check step to ensure all critical MOC management systems are working as defined. It includes, among other items, a check of work orders to determine if they were correctly classified as changes or not changes. This audit requires about 4 hours with a group of 3 or 4 people, plus the lead auditor will spend some additional time for preparation and follow up.

The second protocol is for a "3rd party" audit. It is normally conducted by an MOC Coordinator from a different department than the one being audited and an MOC process expert from the site safety group. Generally one department is audited at a time and each department is audited annually, therefore there will be several audits per year for a division. In some divisions, departments are further divided into sections and each section is audited annually. The idea is to have units of reasonable size for a short audit. In the "3rd party" audit at least 5 MOCs are checked in each audit for accuracy, completeness, and following correct practices. As stated, this audit also requires about 4 hours with a group of 3 or 4 people, plus the lead auditor will spend some additional time for preparation and follow up.

Attachment 2 describes the logistics for a "3rd Party" MOC audit – how to get it done.

Important outcomes from audits include identifying bottlenecks, weaknesses in training, and any non-compliance with Division and Corporate policies. Issues are noted as (NC) Non Conformity or (OFI) Opportunity for Improvement.

Data Collection and Measures

Currently we are still collecting data in spreadsheets; however we have a project in progress to develop a database in order to facilitate more in-depth analysis capability. As shown in Attachment 1, there are 8 requirements in the management system audit and 11 requirements in the "3rd party" audit of individual MOCs. We track trends for these 19 requirements to identify systemic problems. We can look at this on a company-wide basis or drill down to individual divisions and departments. Measures include:

1. Total number of management system audits
2. Total number of "3rd party" audits
3. Percent of MOC management system elements functioning properly
4. Percent of MOC procedure requirements properly executed

Measures are reviewed on a quarterly basis. Figure 1 is an illustrative example of the 4th measure, percent of MOC procedure requirements properly executed. It shows how we track trends.

Figure 1 – Illustrative Example of MOC Audit Measure

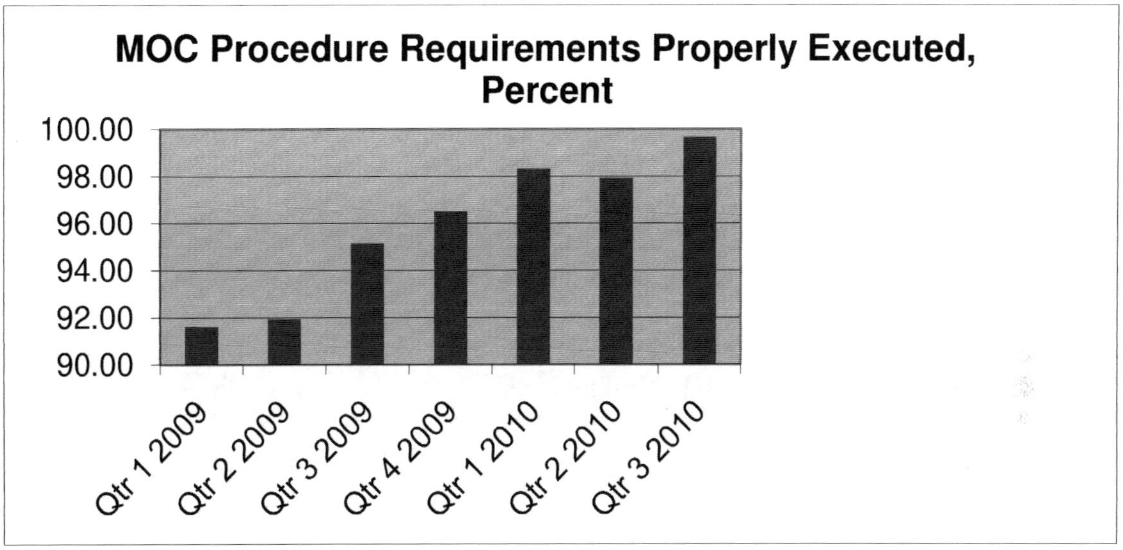

Conclusions

Plant personnel now have been doing these audits for about 2 years (some divisions for a lesser time because they started later). Even though it is often difficult to allocate the time and resources to do the audits, plant personnel have embraced them because they see value - problems have been identified and significant improvements made to MOC. This is really the only way that management has to gauge the effectiveness of MOC. Managers are now reviewing the MOC measures regularly and view the MOC audits as a critical part of our overall safety program. Therefore we feel that the audits have been highly successful.

Attachment 1 - MOC Audit Guideline

I. Assess MOC Management Systems
- Review MOC management systems
- Corrective action resolution plan
- Resolution plan reporting

This assessment will be an annual check-step to ensure all critical MOC management systems are working as defined. The Division MOC Coordinator will normally be the lead auditor. The assessment will require the local MOC experts (possibly the MOC Coordinator, Safety Coordinator and Process Safety Information Coordinator or PSIC) for approximately 4 hours to verify existence of critical MOC management systems. In some cases the same person may fill more than one role. Also any additional people required should be invited. The audit should be a sampling of the entire MOC process including all types of MOCs. The lead auditor should be responsible for: (1) scheduling the audit, (2) documentation of audit findings, and (3) follow-up of findings. A report of findings/observations should be issued to the Department Manager and the leadership team being evaluated. The Leadership Team will prepare a corrective action plan. The corrective action column in the protocol can be used, or the area can use its own system if desired.

MOC Management Systems Audit Protocol

DATE: ______/______/______

DEPARTMENT/AREA: ______________________________

AUDIT LEADER: ______________________________

AUDIT TEAM: ______________________________
____
____

The following items should be checked through originator interviews and/or document reviews. These elements are primarily from company policy. For processes covered by Process Safety Regulations, most of these elements can also be referenced from those regulations. Verify that the following exist:

Elements	Verifi-cation	Comments Corrective Action
1. A system to (a) document and obtain appropriate management approval for all bypasses of safety systems/devices, (b) communicate this information to all affected shifts, and (c) promptly reverse bypasses when no longer needed.		
2. A system to ensure that work orders are correctly classified and MOCs are written for all changes (see definition above). Also to ensure that MOCs are written as appropriate for SORs, SCPs and experiments. There should be an identified person(s) screening work orders on a frequent basis to (a) transfer MOC work orders placed in operations backlog to the appropriate location (to Engineering or other locations, depending upon site procedures) and (b) evaluate completed work orders to determine if changes were made during implementation of corrective action.		
3. MOC process training as required per site/division procedures. All key MOC participants (originators, approvers, processors and MOC Coordinators) should have received training.		
4. A system to determine when a change triggers a revision to the EPA Risk Management Plan (49 CFR 68.190), OSHA PSM program, Seveso II, or other risk management requirements.		
5. A system to determine when a change triggers a revision or review of the Mechanical Integrity, Evidence-Based Inspection, or other preventive maintenance programs for specific types of equipment.		
6. A system to determine when a change affects the reliability of critical safety systems and to ensure review by an appropriate committee.		
7. Originator individual MOC management responsibilities		
8. A system by which the MOC Coordinator manages the process which may include identification of bottlenecks, flagging late temporary MOCs, reporting regularly to management, conducting regular review meetings with key participants, and updating roles as needed.		

II. Assess MOC Process Compliance – 3rd Party

- Evaluate 5 recently closed MOCs – verify closure
- Evaluate 2 MOC WOs – verify closure
- Corrective action resolution plan
- Resolution plan reporting

This assessment will be an annual check-step to ensure MOC process compliance. The auditors will generally be an MOC Coordinator from another department and an MOC process expert from the site safety group. They will evaluate at least 5 recently closed MOCs in each area, and verify closure exists on all required, mandatory fields. Also, evaluate at least 2 MOC work orders and verify closure. The 3rd Party MOC Audit Protocol attached below should be used. A report of findings/observations will be issued to the Division Manager and the leadership team being evaluated. The Leadership Team will prepare a corrective action plan and track to closure. The corrective action column in the protocol can be used, or the area can use its own system if desired.

3rd Party MOC Audit Protocol

DATE: ______/______/______

DEPARTMENT/AREA: __________________________________

AUDIT LEADER: ____

AUDIT TEAM: ____

____

____

Five recently completed MOCs (within the previous year) should be selected at random. Consider permanent and temporary changes, small capital projects, and temporary deviations to shutdowns, interlocks and controls. MOC documentation should be checked and participants interviewed. In addition, a listing of recent work orders should be reviewed to determine if MOCs have been written when appropriate. The area Leadership Team (Department Team or equivalent) is responsible for developing and completing a resolution plan for any findings.

For each MOC the following items should be checked. These requirements are from company policy. For processes covered by the Process Safety regulations, most of these requirements can also be referenced from those regulations (for example, 29 CFR 1910.119 in the United States).

Requirements	How to Verify	Verifi-cation	Comments Corrective Action
1. Experiments and Temporary Changes are not expired. Active Permanent MOCs are not more than 2 years old without a legitimate reason (note: legitimate reasons are rare)	Check Actual Start Date on MOC or Late Reports to see if the MOC is past its expiration date.		
2. MOC Form filled out correctly and verified correct by MOC Coordinator.	Review items on eMOC Form to verify that they are consistent. Refer to site MOC procedure or Safe Practices Manual.		
3. Site and division required checklists are used.	Check MOC form and workflow versus site/division procedure.		
4. Site and division required approvals are done.	Check MOC form and workflow versus site/division procedure.		
5. Safety and environmental checklists were filled out properly with no obvious errors or omissions. Safety and environmental concerns identified on Safety/Environmental checklists and were appropriately resolved and the resolutions documented. Any required safety and environmental approvals were obtained per regulations and site/division procedures. Any required reviews checked on the MOC form are properly documented (for example, Formal Safety Review, Formal Environmental Review, PPE Hazard Assessment, etc.). This can be done in Comments, with an attachment, or in another documentation system (as long as it is cross referenced to the MOC number). This includes PreStartup Safety Reviews (PSSRs) as appropriate - see AP 700-03 Element 3.5.4 and any site/division procedures.	Check Workflow Comments, identify an attachment that may satisfy this requirement or find documentation in another system that is cross referenced to the MOC number. Check Workflow Details Report workflow step date and compare to actual implementation date. Check Form and other external records to verify actual startup date. Verify that Formal PreStartup Safety Review is completed and documented if required.		
6. If the change results in change(s) to plant procedures or practices, verification that the plant procedures or practices required for startup were updated **prior to**	Check Workflow Details Report workflow step date and compare to actual implementation date. Check Form and other external		

Requirements	How to Verify	Verifi-cation	Comments Corrective Action
startup. This requirement would include affected safety, maintenance, and emergency procedures. Approved red-line markups are acceptable for startup but there must be a system in place to assure that the final version is issued in a timely manner and before the MOC is closed. Also verification that any plant procedures not required for startup were updated before the MOC was closed.	records to verify actual startup date. Check documentation of procedure updates through MOC attachments and external filing systems.		
7. For all Facility type changes, verify that all PreStartup tasks (incl. Field Validation for facility changes) were completed **prior to startup** by qualified personnel. Credentials for qualified personnel shall be established by the PSRS/C or equivalent. Modifications to existing facilities shall have field inspections to ensure compliance with design intent, as well as compliance with building and plant permit requirements. Findings from the inspections shall be tracked to completion, with final resolution completed before startup.	Check Workflow Details Report workflow step date and compare to actual implementation date. Check Form and other external records to verify actual startup date.		
8. Personnel affected by a change were notified of and (if needed) trained in the change **prior to startup** of the process per site/division procedure and this notification is properly documented. Documentation of training that corresponds to the Training workflow step (when required) in the departmental training record system.	Check Workflow Details Report workflow step date and compare to actual implementation date. Check Form and other external records to verify actual startup date. If the Training workflow step was completed, check MOC Work package attachments and/or departmental training records for supporting documentation.		
9. All other workflow steps should be completed in the correct order. Proper order should be (a) MOC approvals, (b) design & required reviews, (c) construction, (d) PreStartup tasks and (e) Startup.	Check actual start date by verifying eMOC Form date, workflow step dates, and other document dates (such as Work Orders, SCPs, etc.) to ensure they all correspond to		

Requirements	How to Verify	Verifi-cation	Comments Corrective Action
	requirement. Check Workflow Details Report workflow step dates and compare to actual implementation date. Check Form and other external records to verify actual startup date. Check for possible problems such as: (a) Field validated before W.O. started/opened. (b) Construction not before engineering design. Startup date correct. (c) Field validation not before startup date. (d) MOC Started after construction started or after Effective startup date (as may be evidenced by comments associated with workflow steps or Work Order record). (e) Notifications completed after Effective Startup Date (workflow step completed after real or documented effective startup date).		
10. All PSI updates marked as needed on the checklist are complete. Operations and Engineering PSI Checklist(s) are filled out and attached to MOC work package as required. These should be attached for all PSM and RMP covered processes and other processes as required by site/division procedure.	Check PSI files to verify that all required process safety information updates are complete. Open MOC Work package from MOC Library and check Attachments to see if PSI Checklist(s) have been used and attached.		
11. Site and division requirements for verification of MOC completion were followed.	Check MOC form and workflow versus site/division procedure.		

Attachment 2 –
MOC "3rd Party" Audit Logistics Guideline (How to Get It Done)

Lead Responsibilities:
Preparation

- Reserve a conference room for a half day. Set up on calendar and include other auditor attendees.

Day before

- Run any available reports for MOCs including Completed MOCs, Past Due MOCs and Current Steps of MOCs.
- Print last year's MOC audit.

ALL Responsibilities:
MOC Audit Purpose:

- Identify bottlenecks.
- Identify weakness in training.
- Identify where not in compliance with Division and Corporate policies.

Participants:

- An MOC Coordinator should be the lead auditor.
- Another MOC Coordinator should be the assistant to the lead auditor.
- A Safety Coordinator, MOC Coordinator, PSM Coordinator or Trainer should be given primary consideration for a third participant. In some cases it may be helpful if they are from the area being audited, provided they did not write or validate the MOC being audited.

Frequency:

- Each MOC Coordinator should be the lead auditor for at least five MOCs annually.
- Each MOC Coordinator should be the assistant auditor for at least five MOCs annually.

How will we do an audit?

- Division MOC Team drives the process.
- Division MOC Team will determine schedule for the year.
- Two or more auditors will be required.
- Lead should be an MOC Coordinator.
- We will not audit our own area.
- Each MOC Coordinator will take a turn to lead the MOC audits. The LEAD person will be responsible:
    - Scheduling the audit with appropriate auditor participants and conference rooms.
    - Scheduling a planning meeting (should probably be scheduled about a week prior to the actual audit) to determine specific audit criteria

(includes: specific MOCs that will be reviewed, identifying need for MOC summary reports, work orders, or other documentation)
    - Performing the audit.
    - Document issues and follow up.
- Small group will gather in a conference room and pull up several MOCs on the computer and look at them together. Cover what you can and then you may have to go out into the field, but not usually. MOCs in each cost center should be audited. Each MOC Coordinator in the lead role should audit a minimum of 2 Active and 3 Completed (recently – within previous year) MOCs annually. If the cost center areas assigned in the MOC Audit Schedule don't have any or enough MOCs, audit the ones available. If none are available, then no audit for that cost center.
- Address issues to the Division, maybe even Corporate Team.

Suggested Things to look for and think about when auditing:

- Does the flow make sense?
- Were they trained and the training documented?
- Does change take place before effective date? Shouldn't.
- Proof of Notification? If not attached to MOC, provide other proof of notification.
- Nothing older than 2 years (Corporate Guidelines)

Day of-

- One person usually an assistant, will need to run the computer and research the MOCs and reports.
- One Person usually the leader, will be on a Laptop documenting the findings on the NOTES for MOC Audit Form or if no one has a laptop, make notes by hand and enter them later.
- At the end of the Audit, the NOTES page must be simplified and names removed to send to the MOC Coordinator for that cost center in a Memorandum format.
- Both files should be saved
- The NOTES should be named by date (year, month, day) Audit Notes for Cost Centers (numbers) and Department. Example: ***100614 Audit Notes for 1200 ABC Department***
- The MEMO should be named by date (year, month, day) Audit Memo for Cost Centers (numbers) and Department. Example: ***100614 Audit Memo for 1200 ABC Department***

Lead Responsibilities:

Report

- Send full report, being both Notes and Memo to MOC Coordinators.
- Send the Memo (summary) to Production Manager and Department Head.
- Action Items should be noted as (NC) Non Conformity or (OFI) Opportunity for Improvement
- Action Items should be assigned to the MOC Coordinator of the department audited.
    - Then he or she can follow up or reassign appropriately. The action items should be closed with the action addressing the root problem, not just a temporary fix.

A Conceptual Approach to the Evaluation of Layers of Protection for Inherently Safer System Design

J. Wayne Chastain
Eastman Chemical Company
Kingsport TN
chastain@eastman.com

Robert Kline
Eastman Chemical Company

Keywords: Inherently Safer Design, Layer of Protection Analysis

Abstract

First order inherently safer technology (IST) involves the elimination of a hazard through one of four traditional approaches. While these methods are recognized as preferred means of addressing chemical facility hazards, all hazards cannot be removed through the use of these methods due to technical, financial or end-use constraints. When a hazard still exists after the application of first order inherently safer approaches, second order layers of protection can be applied to appropriately reduce the risks of a process to a tolerable level. For any given application, judgment can be applied to order the possible layers of protection in terms of their quality as second order inherently safer solutions.

In this paper, two unit operations, distillation and chemical storage, have been analyzed using Layers of Protection Analysis (LOPA) to determine the appropriate reliability of risk reduction measures. Once the required reliability was established, additional qualitative and quantitative analysis has been conducted to evaluate, as inherently safer solutions, different layers of protection for efficacy, cost effectiveness, and quality. In each case, estimates have been made for capital cost as well as ongoing operating costs for maintaining the layers of protection to determine an appropriate life cycle cost for each implementation. While the most cost effective approach for a given implementation will vary based on the specific situation to which the method is applied, this paper presents a conceptual approach to analysis that can be applied to most unit operations to indicate the cost and benefit of using layers of protection with a higher degree of second order inherent safety which meet the reliability criteria established by a LOPA.

1. Introduction

The initial concepts of inherently safer design for the chemical processing industry were introduced by Trevor Kletz in the paper "What You Don't Have, Can't Leak" in the aftermath of the Flixborough Disaster *(1)*. The concepts introduced in this paper were developed over time and in 1996 were presented in a more complete form in the CCPS book **Inherently Safer Chemical Processes: A Life Cycle Approach**. The primary concepts of inherent safety presented in this text had not changed significantly from the original concepts introduced by Kletz. The four governing concepts associated with inherent safety presented in the CCPS text are:

- Minimization
- Substitution
- Moderation
- Simplification

These four approaches have been discussed thoroughly in the 1996 CCPS text and in many other papers and texts *(2)*. Detailed discussion of these four methods of inherent safety is outside the scope of this paper.

In 2009, CCPS revised the text, producing a second edition. An important concept introduced in the second edition of the text is that of second order inherent safety. The four standard approaches introduced in the original text were defined as first order inherent safety methods. In contrast to first order methods, which eliminate a hazard, second order methods of inherent safety reduce the risk of an event by minimizing the likelihood or reducing the intensity of the hazard *(3)*.

Second order inherently safer design practices apply the concepts of inherently safer design to the other categories of risk management strategies. Risk management strategies are typically classified into four categories; Inherent, Passive, Active, and Procedural *(4)*. These categories are viewed as decreasing in reliability in the order stated.

2. Cost Effective Safe Design

From a safe design standpoint, not considering cost or end-use constraints, first order inherently safer design practices would always be implemented and the hazards of operation eliminated from designs. However, implementation of these design practices may result in processes which are either not economical, or fail to produce the desired outcome for the customer (end-user). In many cases inflexible application of this principle would result in the either inability to operate some processes or produce certain materials, or in shifting the risk from one operation to another. For this reason, life cycle costs for various options which reduce the overall risk of a process to a tolerable level, without inappropriate risk-shifting, are appropriate for evaluation of design.

First order inherently safer designs must typically be implemented early in the design process. Substitution for a highly hazardous chemical, whether it is a feedstock, solvent or isolated intermediate, can generally only be done economically during the flow sheet development process. Likewise, from a moderation standpoint, the choice of heating media and the selection of reaction paths that operate at lower temperatures and pressures, impact the design of equipment, which then requires these parameters to be selected early in the design process as well. The same thought process also applies to the application of minimization and simplification in the process design.

Conversely, second order methods of inherently safer design, can generally be implemented at a later date within the process with little or no loss of economic efficiency. Safety instrumented systems, relief systems, and other active or procedural controls can added to a design without significantly impacting the overall flowsheet or equipment design. Because these items are often not added until later in a project when the basic design is set, the cost impact of not implementing first order inherently safer design practices may not become clear until a point in a project when reversing those decisions are costly.

Early inclusion of the cost impact of excluding inherently safer design decisions can improve the decision making process. Estimates of the life cycle costs of such decisions greatly improve overall decision making, since the decision to implement a design with greater inherent safety can result in higher capital cost, but a potentially lower operating cost and a lower risk for the process. Cost estimation routines, such as Aspen Technologies Kbase®, which account for the increased cost of relief systems when low, non-inherently safe pressure ratings are chosen for equipment, greatly improve the evaluation of the impact of design decisions.

3. Overpressure Protection

Significant effort, time, and capital are spent in a typical chemical processing facility ensuring that adequate overpressure protection is in place to prevent equipment from failing, releasing its contents, and resulting in potential knock on effects due to overpressure. Sources of overpressure vary depending on the type of equipment being evaluated. A common reference for the sources of overpressure to be evaluated for chemical processing equipment is ISO 23251: "Petroleum, petrochemical and natural gas industries – Pressure-relieving and depressuring systems" *(5)*. This document is an update to the American Petroleum Institute standard API 521. The ISO standard defines 16 cases to be considered for sources of overpressure:

1. Closed Outlets on Vessels
2. Cooling-water failure to condenser
3. Top-tower reflux failure
4. Sidestream reflux failure
5. Lean-oil failure to absorber
6. Accumulation of noncondensables
7. Entrance of highly volatile material
8. Overfilling storage or surge vessel
9. Failure of automatic controls
10. Abnormal heat or vapor input
11. Split exchanger tube
12. Internal explosions
13. Chemical reactions
14. Hydraulic expansion
15. External fire
16. Power failure

In addressing overpressure concerns for equipment there are both first and second order approaches which can be applied to relieve cases or reduce the risk to a tolerable level. Two distillation systems and several storage tanks will be evaluated to demonstrate a heuristic for application of first and second order inherent safety design concepts. This process will evaluate the options for overpressure protection on the basis of degree of inherent safety as well as cost effectiveness and efficacy.

The first order approach which is applicable to overpressure protection, when the chemicals cannot be changed, is to restrict the pressure which can be reached through inherent methods or design the equipment to be adequately robust to handle the pressures which can be generated by upset conditions. The pressures which can result from upset conditions can be reduced by limiting the discharge pressures of pumps by appropriate equipment selection or limiting the supply pressure of steam or the temperature of other heating media. Robust design of equipment entails selection of a pressure rating which is greater than or equal to the highest upset pressure which can be experienced due to an upset *(3)*. This method can typically be applied to all cases with the exception of the external fire relief case and potentially the chemical reaction case when it is applicable.

The traditional means of addressing overpressure scenarios for process equipment is to relieve those cases through a properly sized relief device. In many cases, such relief systems require additional effluent handling which may be as simple as a tailpipe but could include vapor – liquid separation equipment, scrubbers or flaring systems, depending on the hazards presented by a release of the material being relieved. While relief systems are still the most common means of addressing overpressure, the use of safety instrumented systems to address overpressure concerns has been becoming more prevalent. The 2010 edition of the ASME Boiler and Pressure Vessel Code Section VIII paragraph UG-140 effectively cleared the way for the mitigation of relief cases using safety instrumented functions or other means *(6)*.

The 2010 edition of the Code moved what had previously been Code Case 2211 into the main body of the document and clarified the use of Overpressure Protection by System Design. Part A of UG-140 addresses cases where the pressure source is self limiting. Part B addresses cases where the pressure source is not self limiting and is instead mitigated through the application of other controls.

Mitigation of relief cases through application of UG-140 can be implemented by various means, but in most cases where the source of pressure is not self limiting, this mitigation is through the application of safety instrumented functions (SIFs). In cases where first order inherently safer design methods cannot eliminate a relief scenario, or where such methods are cost prohibitive, a comparison can be made between properly relieving the case through an adequately designed relief system or mitigating the case through the application of safety instrumented functions. As mentioned above, if the material being relieved is highly flammable or toxic the relief effluent handling equipment may be needed and must function properly to prevent another hazardous scenario from occurring due to the release through the relief system. Contrasting this with the action of most safety instrumented functions which are used to limit the overpressure from an upset scenario, such SIFs terminate the overpressure scenario without allowing any material to escape from the primary containment. In this case no effluent handling is required for the release.

The use of SIFs to address overpressure contingencies does pose its own risks, however. Relief valves and rupture disks, the primary types of equipment used for mechanical overpressure protection on process equipment, are relatively simple devices. Rupture disks depend on the failure characteristics of the metal or other material from which they are manufactured, which are well defined. Pressure relief valves depend on the functioning of a spring in order to properly relieve systems in the event of overpressure. Safety instrumented systems, however, typically use electromechanical sensors, a logic solver, and the action of one or more final elements. These types of systems are not as simple as the mechanical protection provided by traditional relief systems. The reliability of the SIFs, however, is calculated and documented as the safety integrity level for a SIF. The SIL and associated reliability are primary factors in the design of a SIF.

4. Life Cycle Costs

When evaluating the cost effectiveness of various solutions for overpressure scenarios, the calculation should account not only for the initial capital cost, but the overall life cycle cost for the solution being applied. If the hazard can be eliminated, the life cycle cost is based on the changes to the process required to eliminate the hazard. For cases where the system pressure is designed to withstand overpressure scenarios, there is no life cycle cost over the initial increase in capital cost. For pressure relief systems the life cycle cost includes the maintenance of the relief system. This cost increases with the size of the relief system. Typical maintenance includes the removal of the relief device, testing the device in a shop, executing any repairs associated with the device, and replacing the device in the field. This process may be conducted annually or every several years depending on complex down schedules and the implementation of risk based inspection methods.

Safety instrumented systems also require regular maintenance and testing. Because of the complexity of the SIFs the testing and maintenance costs are higher than those associated with typical relief systems. The testing frequency is used in determining the reliability of the SIF and can vary depending on the design of the system. Increased testing frequency will reduce initial capital expenditures on hardware but will increase the maintenance cost of the SIF.

Before the end of the useful life of a relief system or a component of a SIF, replacement or refurbishment and a return to an as new condition is required. The replacement or refurbishment of the equipment also results in an increased life cycle cost for these active controls used to mitigate or address hazards. However, such actions are required unless the reliability of the system will be allowed to significantly degrade to the point where the required risk reduction is not supplied by the control.

The recurring cost of labor for maintenance and testing and the cost of refurbishment or replacement can be accounted for through a net present value economic analysis during the design of a process. Any organization implementing this methodology would need to determine the appropriate parameters for any such analysis. For the analyses conducted herein, the following assumptions were made to support the NPV analysis:

1. 15 % Cash Flow Rate of Return (CFRR)
2. 3% inflation rate
3. 30 year plant life
4. 38% corporate tax rate (representative of a multinational company)

With these assumptions as the basis for the calculation, the NPV of any annual recurring labor charge associated with testing or inspection is represented by the following equation:

$$NPV = 5.13 * AE \qquad \text{[Eq. 1]}$$

Where:	NPV	=	Net Present Value
	AE	=	Annual Expense

To determine the cost of any required refurbishment or replacement, the calculation for the NPV would be:

$$NPV = 0.74 \times CC \times \left[\frac{1+IR}{1+CFRR}\right]^{n} \qquad \text{[Eq. 2]}$$

Where:	NPV	=	Net Present Value
	CC	=	Capital Cost
	IR	=	Inflation Rate (3%)
	CFRR	=	Cash Flow Rate of Return (15%)
	n	=	number of years

These relationships will be used in the example cases introduced below to help in the comparison of costs for independent protection layers being considered for inclusion to address the risk of the evaluated process.

Simple assumptions of the cost of relief valves and safety instrumented functions have been made to support the economic analysis when comparing active independent protection layers. Any actual application of this methodology should use cost values which are determined from actual purchase costs of equipment and instrumentation. A simple Lang factor of 4 has been applied to all of the purchase costs for converting purchase costs to initial installation costs in the economic analysis. A Lang factor is a simple method of converting between the purchased and installed cost of equipment *(7)*. No Lang factor is applied to replacement or refurbishment cost, since the installation has already occurred. Labor estimates used in this analysis are rough rounded values and would need to be selected appropriate with the values for a given site. A fully allocated cost of $100 per labor hour has been used in the analysis. For situations where cranes are required for removal and replacement of devices, the crane cost has been estimated at $1000 per hour and the number of hours estimated for the operation is given below.

The following table shows the values for purchase cost and the amount of labor and crane time required for removal, testing, and reinstallation of relief devices. The values are based on the body size of the valve assuming standard API body sizes.

Table 1: Relief Valve Sizes and Costs for Economic Analysis

Relief Valve Size	Purchase Cost	Labor Hours	Crane Hours	Maintenance Cost
1" x 2"	$500	3	0	$300
2" x 3"	$1000	4	2	$2400
3" x 4"	$2000	6	2	$2600
4" x 6"	$4000	6	2	$2600
6" x 8"	$8000	9	3	$3900
8" x 10"	$16000	12	4	$5200

For cases where liquid relief will occur through the relief path, a vapor-liquid separator will be included in the cost of the system. This will be added to the base cost calculated by Kbase®. The cost estimate will be based on a purchase cost of $10,000 for the vessel and a Lang factor of 4 will be applied to this cost to determine the installed cost of the separator.

For distillation columns which are not rated for full vacuum, a vacuum protection system will be required since loss of boil up in the column can result in a vacuum condition very quickly as the condenser removes heat and causes collapse of the vapor in the column. Typically a two-tiered approach is taken in the case of a column which is not rated for full vacuum. An inert gas source is normally connected to the column's condenser through a pressure regulator. If this fails, a conservation vent can pull air into the condenser as a last line of defense. These systems prevent the column or associated equipment from collapsing in the event of vacuum. For systems requiring vacuum protection, an equipment cost of $5000 for the nitrogen system will be assumed, giving an installed cost of $20,000 for the nitrogen regulator. In addition $5000 will be included for a vacuum vent, giving a total cost increase of $25000 if the system is not designed for full vacuum.

For safety instrumented functions used to mitigate relief cases per UG-140(b) of the ASME code, this analysis will assume the following for initial cost, annual testing, and refurbishment schedule and cost. For simplicity, the same values will be assumed regardless of the type of safety instrumented function (e.g. pressure, level) and will scale only with the required reliability of the SIF as indicated by the SIL.

Table 2: Safety Instrumented Function Costs for Economic Analysis

SIL	Purchase Cost	Testing Labor	Refurb. Schedule	Refurb. Cost
1	$6000	2	10	$10000
2	$12000	6	10	$20000

All of the values given are for illustration and are not based on any analysis of real costs associated with these systems. The purchase costs indicated for relief devices and safety instrumented functions do need to have a Lang factor applied as indicated above. Any application of this methodology should use best estimate values for the site, instrumentation, and labor costs applicable to the installation. No value has been provided for SIL 3 applications since these typically require significantly greater engineering and are much more situation-specific in design and application.

5. Layers of Protection Analysis

When a scenario, such as a potential overpressure event, has been identified, comparison of the risk of that event against some criteria is a useful method for determining the type and required reliability of controls that should be in place to prevent or mitigate the event. One risk analysis method which has become very popular recently is the Layer of Protection Analysis (8). This risk analysis method estimates a frequency of occurrence for an event consisting of a single cause – consequence pair, by evaluating the initiating event frequency, the reliability of any IPLs, and, in some cases, the impact of conditional modifiers on the frequency of the ultimate consequence of the event.

To use Layer of Protection Analysis in any way other than as a comparative tool, a risk criterion is needed to compare with the frequency for an event calculated by the LOPA. A discussion of risk criteria for use in LOPA can be found in the original LOPA text as well as other sources. The value used in these analyses is a risk criterion of 1e-5 (once in 100,000 years) for an event which could cause localized severe injuries if personnel are present. This criterion is calibrated for the use of conditional modifiers within the LOPA technique.

LOPA is often restricted to order of magnitude reductions of risk. However, the technique can be applied using more accurate values when available. The following examples of the application of LOPA use a precision of a half order of magnitude of frequency or probability of failure on demand.

6. Distillation Examples

The first two cases that are investigated are distillation systems. An example schematic for a distillation system is shown in Figure 1. In each case, the base assumption is that the chemicals involved in the process and the distillation processes are required and cannot be eliminated through the application of first order inherently safer design approaches. In this case, the potential overpressure cases must be addressed through other first order methods such as moderation of steam pressure or robust design of equipment to withstand the potential overpressure scenarios or the application of second order independent protection layers. Second order methods considered are overpressure protection by relief and mitigation of specific relief cases through the application of appropriately designed SIFs that meet the required reliability, as indicated by LOPA.

6.1 Crude Acetic Acid / Acetic Anhydride Splitter

The first case study is a crude acetic acid (HOAc) / acetic anhydride (Ac2O) column taking a feed of 60% acetic acid (on a weight basis) and separating the components. The top specification for the column is 10% Ac2O while the bottoms specification if 10% HOAc. This system was modeled using Aspen Plus 2006.5 using the Wilson-Hayden O'Connell vapor-liquid equilibrium model. All costing was done using the standard Aspen Kbase® utilizing built in cost assumptions. The column was designed to operate under vacuum with a top pressure of 250 torr (absolute) with 14 ideal stages. In all cases the heating media used in the model was steam with a maximum pressure of 125 psig.

Three different cases for this column system were evaluated:

1. Base Case – 20,000 lb/hr of feed with material of construction being 316 SST
2. Low Rate Case – 5000 lb/hr of feed with material of construction being 316 SST
3. High Metallurgy – 20,000 lb/hr of feed with material of construction being carbon steel clad with a corrosion-resistant high nickel alloy.

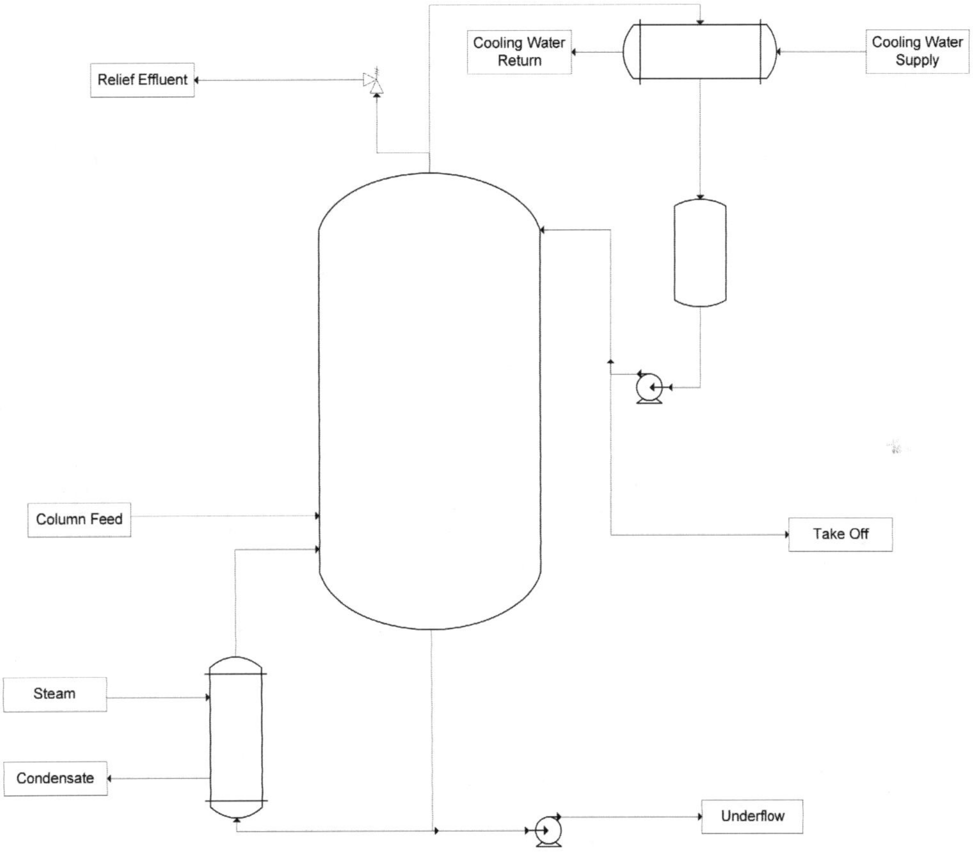

Figure 1: Typical Distillation Column

For each of these cases, three different options were evaluated; designing the systems with a robust pressure rating to eliminate all relief cases but the fire case, designing the systems for the coincident pressure rating and relieving all applicable overpressure cases, and designing the systems for the coincident pressure and mitigating the cases by using SIFs. This distillation process operates at vacuum. Therefore the distillation column and all associated components will be designed for full vacuum. The full vacuum rating for the equipment will give a coincident minimum pressure rating.

For this type of distillation system, the relief cases which would be considered are cases 1, 2, 3, 6, 9, 10, 11, 13, and 15 from the list of relief cases introduced above. Often simplifying assumptions are made in the evaluation of overpressure protection which make cases 2, 3, 6, 9, and 10 collapse to a single case involving full steam flow to the reboiler with no cooling in the condenser or reflux in the column to remove heat. This results in the following cases which require evaluation:

1. Liquid Overfill (Case 1)
2. Loss of Cooling with Excessive Steam Flow (Cases 2,3,6,9,10)
3. Reboiler Tube Rupture (Case 11)
4. Chemical Reaction (Case 13)
5. External Fire (Case 15)

6.1.1 Base Case

The base case for the evaluation of this distillation system is based on 20,000 lb/hr of feed to the column. The sizes of equipment for this case predicted by the Aspen Plus model using Kbase® are shown in Table 3.

Table 3: Base HOAc / Ac2O Equipment Sizes

Equipment	Height (ft)	Diameter (ft)
Column	71	12
Reflux Drum	12	6

Exchangers	Duty (MMBTU/hr)	Area (ft2)
Reboiler	51	12914
Condenser	53	12347

6.1.1.1 HOAc / Ac2O Column Base Case Inherently Safer Design

Assuming that the process in question must use these chemicals and that they must be separated from one another by distillation, the next appropriate inherently safer design is rating the system to withstand possible upset pressures. This approach is typically referred to as inherently robust process equipment and is considered a first order method of inherent safety under the umbrella of simplification. For the case of the HOAc/Ac2O column, the vapor pressures of the components in question are lower than the pressure of the steam in the reboiler. In this case, the highest pressure that could result from loss of cooling with excessive steam flow would be the vapor pressure of acetic acid at the saturation temperature of 125 psig steam, or approximately 57 psig.

In the event of a tube rupture in the reboiler, discounting chemical reaction, the column could be exposed to 125 psig, the supply pressure of the steam.

When all components in the system are rated for the maximum pressure of 125 psig, all relief cases are eliminated from consideration except for Case 15 (External Fire) and, in this case, Case 13 (Chemical Reaction.) The Kbase® predicted installed cost for this system is $2.41 million. This estimate includes the predicted cost for the relief system designed to protect against external fire. SuperChems® by ioMosaic© Corporation was used to predict the relief valve size for external fire for this column. A K-orifice relief valve is needed to protect the column for this case, which would be a 3" x 4" valve.

For the flow rates being considered and the chemicals processed, centrifugal pumps would be the normal means of providing flow to the column. In the case where the design pressure of the column is 125 psig, centrifugal pumps should be able to be selected which can provide the required flow rate but whose maximum pressure for delivery of fluid is below the design pressure of the column. Selecting pumps with this design would improve the inherent safety of the installation. If pumps must be used which can supply higher pressure than the pressure rating of the column, then the relief systems can be designed to address this potential overpressure source or the case may be mitigated through the use of safety instrumented systems.

The acetic acid / acetic anhydride system poses a concern from a chemical reaction standpoint when steam is used as a heating media and cooling water is used as a cooling media. At high temperatures acetic anhydride reacts quickly with water to make acetic acid. However, in the distillation column, the water will be the most volatile component while the acetic anhydride will be primarily present in the liquid phase. Because of the high temperatures in the column and the reaction rate of anhydride with water in the presence of significant amounts of acetic acid, the reaction of any water that ends up in the liquid will be fast. This fast reaction rate should prevent the accumulation of significant amounts of water in the column in the liquid phase. Water entering the distillation column from a tube break in the condenser while the column is operating would stay stratified in the top of the column due to its high volatility compared to acetic anhydride. Chemical reaction may not impact the relief design for the column except in situations where reactants could be pooled. The hazard of chemical reaction should be further considered for a column of this type, but will not be addressed further here.

Based on these assumptions, the maintenance cost for a 3" x 4" relief device would have a net present value of $13,338. This would increase the NPV for the column system accounting for the relief system to $2.42 million.

6.1.1.2 HOAc /Ac2O Column Base Case with Overpressure Protection by Relief System

Kbase® predicts coincident pressures for the various pieces of equipment in the distillation system. The pressures for Case 1 are shown below:

Table 4: HOAc / Ac2O Column System Design Pressures

Equipment	Pressure	Vacuum
Column	80 psig	FV
Reboiler	125 psig	FV
Condenser	60 psig	FV
Reflux Drum	80 psig	FV

The limiting pressure rating in this case is 60 psig. Since the vapor pressure of acetic acid at the temperature of 125 psig steam is less than 60 psig, the overpressure cases related to excessive steam flow with loss of condensing are eliminated. The only cases to be considered are Case 1 (Liquid Overfill), Case 11 (Reboiler Tube Rupture), Case 13 (Chemical Reaction) and Case 15 (External Fire). LOPAs can be conducted to estimate the required risk reduction to meet the criteria used in the analysis. LOPAs or some other risk analysis methodology should be applied even when adequately sized relief systems are going to be provided to ensure that the risk reduction provided by the relief systems does reduce the risk to the criteria. Cases may arise when the predicted risk reduction requirement is higher than the risk reduction provided by a properly sized relief valve. In those cases, adequately sized relief and SIFs may be required to reduce the risk.

For Case 1, when the system pressure is limited to 60 psig, the centrifugal pumps will likely be able to overpressure the column system. Determining the required reliability for active controls to prevent or mitigate this case can be done using the LOPA methodology.

Table 5: LOPA for Liquid Overfill of Column System

Element	Value	Description
Initiating Event	-1	BPCS Failure resulting overfilling of column
Occupancy	-1	Occupancy of 10% or less
Ignition	-0.5	50% chance of ignition of combustibles released to general processing area
Injury	-0.5	50% chance of injury due to fire
Frequency	-3	Event occurrence ~1/1000 years
Target Frequency	-5	1 / 100,000 years (Single Severe Injury)
Differential	-2	Risk Reduction Factor of 100

Since a risk reduction factor of 100 is indicated by the LOPA, a relief device, properly maintained, in clean service does provide adequate protection. Relieving the liquid overfill case was evaluated using SuperChems and the conservative assumption that the peak flow on the supply pump was 50% greater than the normal flow even with the column at relieving conditions. In this case an F-orifice relief valve is required to relieve the liquid overfill case. This meets the required risk reduction. However, protecting the vessel from liquid overfill using a relief system results in a relief effluent case of a significant amount of liquid flow to the surroundings. A liquid overfill case would require appropriate effluent handling directing the flow to an appropriate secondary containment. Providing this effluent handling will increase the cost of the relief option by $40,000.

The next case to be considered is reboiler tube rupture. A LOPA on this case yields the following results:

Table 5: LOPA for Reboiler Tube Rupture in Column System

Element	Value	Description
Initiating Event	-1.5	Reboiler Tube Rupture (Once in 30 years)
Occupancy	-1	Occupancy of 10% or less
Ignition	0	100% chance of ignition due to high energy overpressure event
Injury	-0.5	50% chance of injury due to fire
Frequency	-3	Event occurrence ~1/1000 years
Target Frequency	-5	1 / 100,000 years (Single Severe Injury)
Differential	-2	Risk Reduction Factor of 100

A properly sized relief system in clean service can provide the recommended level of risk reduction, supplying a risk reduction factor of 100. SuperChems® can be used to size a relief valve to address this overpressure scenario. ¾" tubes have been assumed in the reboiler and a total tube rupture is assumed. The flow of steam from the tube break is conservatively estimated as 3850 lb/hr of steam flow. To relieve this flow a J-orifice relief valve is indicated.

The discussion of Case 13 (Chemical Reaction) is the same for the lower pressure rated column as it is for the column with the pressure rating designed to be inherently safer. No changes to this discussion or its impact on relief system sizing are included.

Finally, Case 15 (External Fire) needs to be evaluated for the lower pressure rated column. Again SuperChems® can be used for this evaluation. In this case, an M-orifice valve is recommended. For all of the relief cases for the low pressure column for this scenario, the largest required orifice was the M orifice relief valve recommended for fire case relief. The M orifice is for a 4" x 6" valve.

The base cost from Kbase® for the low pressure column system in 316 stainless steel was $2.27 million. The NPV of the maintenance on a 4" x 6" relief valve, based on the values provided above, is the same as for a 3" x 4", or $13,338. Including this and the cost of the vapor-liquid separator for effluent handling with the Kbase® cost of the installed column, the net present cost is $2.33 million.

6.1.1.3 HOAc / Ac2O Column Base Case with Mitigation of Relief by SIF

In this option, the lower pressure ratings for the column are used, but the relief scenarios presented by Case 1 (Liquid Overfill) and the Case 11 (Reboiler Tube Rupture) are mitigated through the application of SIFs. The LOPAs presented above indicate that an additional risk reduction factor of 100 is required to protect the distillation column in each case. This risk reduction corresponds to a SIL 2 rated SIF. It should be noted that since the governing case for relief design was external fire, any cases mitigating through the use of SIFs will not reduce the cost of the relief device itself. However, restricting the cases handled by the relief system to only external fire may result in a lower cost effluent handling system. The net present value for the two identified SIFs can be found by accounting for their capital cost, cost for maintenance and testing, and cost for refurbishment or replacement of components over the assumed 30 year life of the column. Using the values provided in Table 2 and Equations 1 and 2, the life cycle cost of the implementation for the two SIL 2 SIFs is estimated to be $115,000.

Accounting for this, the cost of the column systems using SIFs to mitigate the Case 1 and Case 9 would be $2.39 million. The table below shows the cost comparison of the three options:

Table 6: Cost Comparison for HOAC / Ac2O Base Case Design Options

Case	Description	Cost	Delta
1.A.1	Inherently Safer Design	$2.42MM	
1.A.2	Low Pressure Design with Relief	$2.33MM	-3.7%
1.A.3	Low Pressure Design using SIFs	$2.39MM	-1.2%

Based on this analysis, the lowest cost option is to use a low pressure column and design the relief system to handle the cases. With the small differential between the inherently safer design option and the slightly lower cost low pressure design options, a case could be made for using the higher pressure rated column.

6.1.2 Low Rate Case

The low rate case for HOAc / Ac2O Column is for a feed rate of 5000 lb/hr. The purpose of this case is to evaluate the impact of smaller equipment on the economic analysis.

Table 7: Low Rate HOAc / Ac2O Equipment Sizes

Equipment	Height (ft)	Diameter (ft)
Column	71	6
Reflux Drum	8	4

Exchangers	Duty (MMBTU/hr)	Area (ft2)
Reboiler	12.75	3229
Condenser	13.25	3087

6.1.2.1 HOAc / Ac2O Column Low Rate Case Inherently Safer Design

The Kbase® cost estimation program indicates that the pressure rating of the low rate case column will be greater than 125 psig in order to obtain a full vacuum rating. The reboiler and reflux drum are also rated at 125 psig with a full vacuum rating. Only the condenser has to be re-rated to a higher pressure to design the system with a robust pressure rating that will prevent all relief cases other than external fire and chemical reaction. The cost of the low rate case column system estimated by Kbase® was $1.14 million. Since the system pressure is already indicated to be inherently safer from a robust pressure rating standpoint, no additional evaluation was done for second order inherently safer independent protection layers for this system.

6.1.3 High Metallurgy Case

In this case the column system for the separation of 20,000 lb/hr of the acetic acid / acetic anhydride feed was constructed from carbon steel clad with a high nickel alloy. The acetic acid / acetic anhydride system can cause corrosion of stainless steel under the conditions found in the distillation system. This option allows the impact of clad equipment on cost of inherently safer design options to be investigated. For this case, the equipment is sized identically to that investigated in the base case for this system.

6.1.3.1 HOAc / Ac2O Column High Metallurgy Case Inherently Safer Design

For this system, a result very similar to the situation for the low rate case was found to exist. Since all equipment was designed to withstand full vacuum, the coincident pressure ratings for all of the equipment except the condenser were greater than 125 psig. The Kbase® installed cost estimate for the system with a 60 psig rated condenser was $15.5 million. Increasing the pressure rating to 125 psig for the condenser caused the installed cost estimate to drop to $15.4million. In this case, the robust pressure design option should be selected with the only relief cases needing additional consideration being external fire and chemical reaction.

6.2 Methanol / Water Column

The next case investigated was a methanol / water separation column. In this case the column was assumed to receive a feed of 6000 lb/hr of a mixture that is 40% methanol and 60% water. The distillate for the system has a specification of no more than 0.5% water. The underflow specification is no greater than 100 ppm of methanol. The system was simulated using Aspen Plus 2006.5 with the built-in Wilson parameters being used to predict the vapor liquid equilibrium. The column is designed to operate at atmospheric pressure fed on the 16th stage with 26 ideal stages.

Six different options were evaluated for this distillation system. Two different steam pressures were evaluated as possible heat sources for the reboiler; 15 psig and 100 psig. Three different design pressures were evaluated; 15 psig, full vacuum and the coincident pressure associated with that vacuum rating, and an inherently safer pressure selected for robust vessel design. Table 8 below shows the Kbase® predicted design pressures for each case.

Table 8: MeOH / Water Column System Pressures

Steam Supply Pressure	Low Pressure Column	Full Vacuum	Robust Pressure
15 psig	**15**	**30**	**85**
100 psig	**15**	**60**	**300**

Kbase® also predicted costs for these systems.

Table 8: MeOH / Water Column System Costs from Kbase®

Steam Supply Pressure	Low Pressure Column	Full Vacuum	Robust Pressure
15 psig	**$1.93MM**	**$2.05MM**	**$2.09MM**
100 psig	**$1.19MM**	**$1.74MM**	**$2.24MM**

In this case the Kbase® estimates include relief systems, but only include the impact of the pressure rating on the relief size estimated for the external fire case. SuperChems® was used to estimate these relief sizes for external fire.

Table 9: MeOH / Water Column System Fire Case Relief Sizes

Steam Supply Pressure	Low Pressure Column	Full Vacuum	Robust Pressure
15 psig	**6"x 8"**	**4"x 6"**	**3"x 4"**
100 psig	**6"x 8"**	**4"x 6"**	**2"x 3"**

SuperChems® was also used to estimate the relief requirements for the abnormal heating / loss of condensing relief case. If this case is not mitigated then it will result in the largest size for the relief system for the column, as is shown in the following table.

Table 10: MeOH / Water Column System Heat Imbalance Relief Sizes

Steam Supply Pressure	Low Pressure Column	Full Vacuum	Robust Pressure
15 psig	(5) 8"x 10"	(2) 8"x 10" + 4"x 6"	X
100 psig	(5) 8"x 10" + 6"x 8"	8"x 10" + 6"x 8"	X

In determining the actual NPV cost estimate for the safety systems for these design options, the purchase price and maintenance cost differential should be taken into account. For the case of liquid overfill, the effluent handling cost will be added to all cases with a design pressure of 60 psig or less. The pumps used to supply feed to the column must be checked to ensure that they cannot generate more pressure than this or other design cases could be impacted by the need for effluent handling of liquid on the relief system. In addition, for the column systems not rated for full vacuum, the cost of vacuum protection systems should also be taken into account. When these costs are properly accounted for, the following values are predicted for the cost of the column system:

Table 11: MeOH / Water Column System Costs Including Relief Systems

Steam Supply Pressure	Low Pressure Column	Full Vacuum	Robust Pressure
15 psig	$2.40MM	$2.27MM	$2.09MM
100 psig	$1.71MM	$1.89MM	$2.24MM

These values include significant purchase and maintenance costs for relief systems. In addition, the relief systems shown in Table 11 for the non robust pressure design column systems could pose a significant hazard if they were to actuate. Large amounts of methanol could be vented in the event of a loss of cooling or abnormal heat input scenario.

The use of a SIF on high pressure to stop steam flow to the column reboiler would address the relief cases in the column caused by heat imbalance. LOPAs conducted on the failure modes which can lead to a heat imbalance to determine the required reliability of a SIF to mitigate those relief scenarios. A loss of condensing LOPA might be represented by Table 12. Likewise, a LOPA on excessive steam flow due to a Basic Process Control System (BPCS) failure could be performed. This LOPA could be represented by Table 13.

Table 12: LOPA for Loss of Condensing in Column System

Element	Value	Description
Initiating Event	-1.5	Cooling Water Loss (Once in 30 years)
Occupancy	-1	Occupancy of 10% or less
Ignition	0	100% chance of ignition due to high energy overpressure event
Injury	-0.5	50% chance of injury due to fire
Frequency	-3	Event occurrence ~1/1000 years
Target Frequency	-5	1 / 100,000 years (Single Severe Injury)
Differential	-2	Risk Reduction Factor of 100

Table 13: LOPA for Excessive Steam Flow in Column System

Element	Value	Description
Initiating Event	-1	Excessive Steam Flow (Once in 10 years)
Occupancy	-1	Occupancy of 10% or less
Ignition	0	100% chance of ignition due to high energy overpressure event
Injury	-0.5	50% chance of injury due to fire
Frequency	-2.5	Event occurrence ~1/300 years
Target Frequency	-5	1 / 100,000 years (Single Severe Injury)
Differential	-2.5	Risk Reduction Factor of 316

This LOPA indicates that if excessive steam flow can result in overpressure of the column system, a properly sized relief system will not provide adequate protection if the relief system supplies a risk reduction factor of 100. A SIL 1 SIF or some other means of risk reduction would be required to address the risk. However, a SIL 2 SIF could be designed to meet this required risk reduction. The following table shows the cost estimates for the methanol / water column system using a SIL 2 SIF to address the heat imbalance scenarios for the column. The cost of a vapor-liquid separator for effluent handling and vacuum protection for columns not rated for full vacuum are included in the cost estimates.

The lowest cost column system for this service, taking the necessary relief systems and other protective measures into account is the low pressure column supplied with 100 psig steam to the reboiler. The increases in cost over that baseline are shown in Table 15.

Table 14: MeOH / Water Column System Costs Including SIFs

Steam Supply Pressure	Low Pressure Column	Full Vacuum	Robust Pressure
15 psig	$2.05MM	$2.15MM	$2.09MM
100 psig	$1.31MM	$1.84MM	$2.24MM

Table 15: MeOH / Water Column System Cost Differentials

Steam Supply Pressure	Low Pressure Column	Full Vacuum	Robust Pressure
15 psig	156%	164%	159%
100 psig	100%	140%	171%

In this case, the robust pressure inherently safer options for design are significantly more expensive than solutions which result in adequately controlled risk based on reasonable risk criteria. However, in comparing the second order inherently safer options, the use of SIF to mitigate the relief cases associated with heat imbalance in the column not only improved the safety of the system, since it eliminated the potential for large releases of flammables through the relief system, but also resulted in a significant reduction in the cost of the system.

6.3 Chemical Storage

The last example considered is that of chemical storage. Three different size tanks are considered; 99,000 gal, 250,000 gal, and 1,000,000 gal. Storage tanks are typically designed to one of two standards. API 650 is the standard for design of tanks with a pressure rating up to 2.5 psig. API 620 is used to design tanks up to 15 psig. Anything with a higher pressure rating is a pressure vessel normally governed by the ASME code. The material of construction used for the tanks in this example was 316 stainless steel.

One significant difference between the API 650 and API 620 codes is the robustness of the tanks if overfilled. An API 650 tank is not rated for liquid above the top weld line. Often, this weld line is designed to be a weak point in the design of the tank, to allow overpressure relief by intentional weld failure. Although a relief event would then necessitate repair of the tank, the majority of the tank contents are maintained in the tank and not released to the surroundings.

API 620 tanks, however, can withstand hydraulic pressure caused by liquid above the top weld line and can relieve the liquid through a properly designed, top mounted relief device to a safe location. A release from a failure in the top weld seam of an API 650 tank, however, could result in a release anywhere along the circumference of the weld. A LOPA can be used to determine if an API 650 tank requires additional protection for the liquid overfill case than the potential relief path through a failure of the top weld seam.

In this case a risk criterion of -5 will be used in the analysis. The material in the tank will be assumed to be flammable or a combustible above its flash point such that a flash fire could occur if a significant amount of material is released from the tank. The tanks in question will be assumed to be in a remote tank farm which will impact the occupancy and the probability of ignition used in the risk analysis.

Table 16: LOPA for Overfilling of Chemical Storage Tank

Element	Value	Description
Initiating Event	-1	BPCS Level Control Failure Resulting in Overfill of Tank
Occupancy	-1.5	Occupancy of 3% or less
Ignition	-1	10% chance of ignition in remote tank farm
Injury	-0.5	50% chance of injury due to fire
Frequency	-4	Event occurrence ~1/10,000 years
Target Frequency	-5	1 / 100,000 years (Single Severe Injury)
Differential	-1	Risk Reduction Factor of 10

With these assumptions, a SIL 1 SIF is recommended to address the liquid overfill case on an API 650 tank. This would not be recommended on an API 620 tank which properly relieved any fluid overfilling the tank to a safe location such as secondary containment or a process sewer.

The NPV of a SIL 1 SIF to prevent overfilling of a tank, based on the cost estimates provided above, would be $28,000. Kbase® was once again used as the means of determining a cost for the tanks. The cost of the SIL 1 SIF can be added to the costs predicted by Kbase® for the various cases to determine the cost effective solution that meets the risk criteria. The results from that analysis are shown in Table 17.

Table 17: Storage Tank Costs Including SIFs for Overfill Protection

	99k gal	250k gal	1MM gal
2.5 psig	$332k	$489k	$1331k
15 psig	$305k	$567k	$1776k

From this analysis, smaller tanks are actually more cost effective if designed to the more rigorous API 620 standard. API 650 tanks are definitely favored for large tanks. Tanks in the 250k gal range are slightly more cost effective as API 650 tanks with a SIL 1 SIF present to prevent overfilling.

7. Conclusions

A methodology has been presented for a review of proposed designs to determine the most cost effective option taking into account the concepts of inherently safer design. The proposed approach first evaluates whether a hazard can be eliminated from a design through application of first order inherently safer design practices. If that cannot be accomplished, then costs are compared between options, taking into account life cycle costs for active controls. Small differentials in cost between inherently safer solutions and the use of active controls should be weighted to favor the solutions which eliminate the hazard.

Overpressure protection of equipment was specifically focused on in three example cases. Two distillation examples as well as chemical storage in general were evaluated. LOPA was presented as a means of comparing alternatives for active controls to address risk differentials caused by the selection of passive, active and procedural design options. Even among active control options, the concepts of second order inherently safer design can be applied instead of relying solely on life cycle cost as the method of determining the best design.

Application of a process like this during the design process will improve the cost effectiveness of facility designs while ensuring that risk decisions regarding inherently safer design are integral to the design process.

8. References

[1] Kletz, T.A. (1978), "What You Don't Have, Can't Leak," *Chemistry and Industry*, September 1978, pp. 287–292

[2] Center for Chemical Process Safety (1996), **Inherently Safer Chemical Processes: A Life Cycle Approach 1st Edition**, New York: American Institute of Chemical Engineers

[3] Center for Chemical Process Safety (2009), **Inherently Safer Chemical Processes: A Life Cycle Approach 2nd Edition**, New York: American Institute of Chemical Engineers

[4] Center for Chemical Process Safety (1993), **Guidelines for Engineering Design for Process Safety**, New York: American Institute of Chemical Engineers

[5] International Organization for Standardization (2006), **ISO 23251: Petroleum, petrochemical and natural gas industries – Pressure-relieving and depressuring systems**, Geneva, Switzerland

[6] ASME Boiler and Pressure Vessel Committee on Pressure Vessels (2010), **2010 ASME Boiler and Pressure Vessel Code, Section VIII, Division 1, Rules for Construction of Pressure Vessels**, New York: American Society of Mechanical Engineers

[7] Lang, H.J. (1947), "Engineering Approach to Preliminary Cost Estimates", *Chemical Engineering*, September 1947, pp. 130-133

[8] Center for Chemical Process Safety (2001), **Layer of Protection Analysis: Simplified Process Risk Assessment**, New York: American Institute of Chemical Engineers

Considerations beyond Basic LOPA

Harold W Thomas
Air Products & Chemicals, Inc.
Allentown, PA 18195
thomashw@airproducts.com

Joseph V. Saverino
Air Products & Chemicals, Inc.
Allentown, PA 18195
saverijv@airproducts.com

Keywords: LOPA, low demand, high demand, continuous demand, time at risk, duration controlled risk, frequency controlled risk

Abstract

When performing a LOPA, practitioners need to consider both the type of demand that will occur on the protection system(s) and when the potential risk is actually present. Most LOPA practitioners make the assumption that the initiating event represents a low frequency demand on subsequent layers of protection. This paper addresses some situations where this is not a good assumption and provides guidance with regards to such scenarios. The paper also provides a thought process that can be used to address time at risk, with some guidance as to when it is appropriate to include in the LOPA as well as when it is not.

Introduction

The type of demand that protection systems witness is an important consideration when performing a LOPA. These initiating events can be categorized as low demand, high demand or continuous demand. A low demand event is one where protection system operates in standby mode, with the potential for unrevealed failures until it is either proof tested or an actual demand is placed upon it. As a rule of thumb, demand rates of less than or equal to once per year can be considered as low demand. A high demand initiating event represents a frequency of demand that exceeds once per year. In this case the protection system still has the potential for unrevealed failures, but the simplified mathematics employed by the LOPA technique will typically over-estimate the risk by computing a hazard rate that exceeds the failure rate of the protective system. A continuous demand initiating event is one where upon failure of the protective system, the hazard will occur. As a high demand rate increases in frequency, the mathematics used to model it fairly rapidly converges with the mathematics of continuous demand mode of operation.

Time at risk can be categorized as continuous, duration controlled or frequency controlled. A continuous risk is one where an initiating failure (event) can trigger a demand that is hazardous at any time if available safeguards do not prevent or mitigate its propagation. The system does not need to be in any particular mode of operation for the hazard to occur. A duration controlled risk is a non-continuous risk where the hazard can only occur *during* the fraction of time a particular operating mode or a particular task is being performed, making the risk proportional to the duration. Duration-Controlled risks are sometimes called *Short-Term Risks* because the duration of the operating mode or task is usually (though not always) short relative to a person's total working time. A frequency controlled risk is a non-continuous risk where the hazard can only occur as a *result of* a particular operation or task, making the risk proportional to the frequency of that operation or task.

Risks are often controlled by more than one type of protective measure. These different types of protections can be thought of as layers which build upon each other. The layers can be divided into three broad categories of Control, Protection, and Mitigation. Control measures include inherent system process and mechanical integrity as well as basic process control. The Protection layer typically consists of alarms, instrumented protection (interlocks), and relief systems. Finally, the Mitigation layer may include such things as fire protection, barriers, deluge systems, emergency response programs, etc.

LOPA is most often used to assess the Control and Protection layers as these include the protections that are most effective in avoiding the risk of personnel injury or equipment damage. The Control and Protection layers are also the most conducive to quantification. The Mitigation layer is most effective against incident escalation and is less easily quantified.

Comparing Low and Continuous Demands

To better understand the difference between low and continuous demands, consider the application of a variable frequency drive (VFD) pump. Assume that in the initial design, a motor was selected with a maximum horsepower rating that matched closely to the needs of the application. Based upon the pump curves, the piping system was designed so that the deadhead pressure represented an inherently safe solution to overpressure. As such, in this example no overpressure protection is provided. If the motor had been provided with excessive horsepower, the designers may have used the same pump curve to justify not including any overpressure protection, as the VFD controller is configured to clamp the speed at some upper limit. In this case, there is some potential for the VFD controller to fail in such a way that there is an over speed. If this were to occur, the pressure could exceed the pressure rating of the equipment if the motor horsepower were sufficiently excessive. In this scenario, the VFD controller is the only protection available, meaning that the protection system is operating in the continuous demand mode. As soon as it fails dangerously, the hazard would manifest itself.

If the design team recognized this potential when the motor was selected, then they could easily add protection via a safety instrumented function and / or a relief device. If this were done, then the protection system would be operating in a low demand mode as the dangerous failure rate of the VFD controller would be significantly less than once per year.

Comparing High, Continuous and Low Demand Modes

To illustrate an example of a high demand, figure 1 shows a compressor feeding a gaseous pipeline.

Figure 1

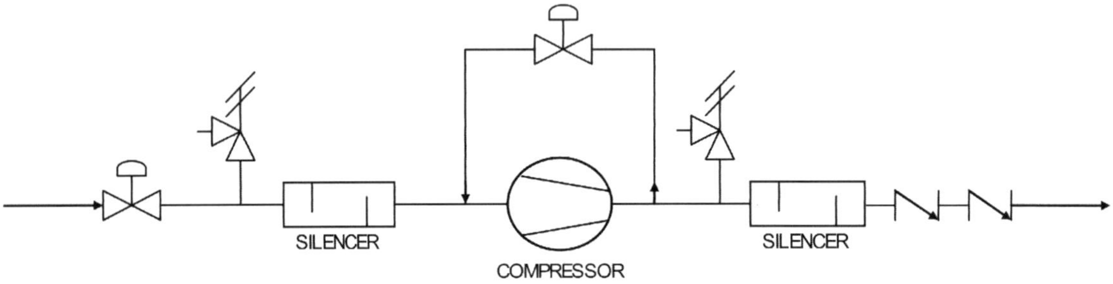

In this example, the compressor is part of a back-up system to ensure the pipeline pressure is maintained during periods of peak usage that occurs for a period of time each day for about an hour. When not needed, the compressor is shut down. When this occurs, the check valves are designed to prevent the suction of the compressor from seeing an overpressure. The initiating event rate on this layer of protection is 365 times per year. Further assume the probability of failing to close is calculated to be 0.01 based upon a proof test frequency where the check valve is also internally inspected to ensure no incipient conditions exist and it is maintained to ensure it continues to operate within its useful life, where failures can be expected to be random. Using the simple LOPA math, the hazard rate would be calculated to be:

HR = (Initiating Event Rate)(PFD) [Eq. 1]
HR = (365)(0.01)
HR = 3.65 demands per year on the relief valve

The number of demands on the suction relief valve is clearly representative of a high demand mode that cannot be accommodated by the simple LOPA math.

The other way to look at this problem is to think of the check valves as a protective system operating in a continuous demand mode, that is, when they fail, the hazard will occur. It is unlikely a hazardous event will immediately occur at the exact point in time that both check valves fail; however, once they are both in a failed state, as soon as the compressor shuts down, the hazardous overpressure demand will occur. When looking at the problem this way, the failure rate of the check valves can be used as the initiating event. Assuming that the failure rate of both check valves failing to close is 0.01 per year, it then represents a low demand initiating event on the suction relief valve layer of protection. Further assuming that the relief valve probability of failing to open is 0.01, then;

$$HR = (\text{Initiating Event Rate})(\text{PFD}) \qquad [\text{Eq. 2}]$$
$$HR = (0.01/\text{year})(0.01)$$
$$HR = 1\text{e-}4/\text{year}$$

Accounting for Duration at Risk

In many LOPA scenarios meant to analyze the potential for personnel harm, if the event were to occur, there is only the potential for harm if the operator is present in the vicinity. In this case, the practitioner might take credit for time or duration at risk.

For example, assume there is the potential for a carbon monoxide leak into a partially enclosed operating area that could potentially overwhelm an operator if they were present. In a significant release, there would also be the potential for steam and hydrocarbons to be released. The overall cloud could impair evacuation attempts. Gas monitor alarms are installed in the operating area to alert personnel not to enter in the event of a leak. Due to the potential hazard, entry into the area is restricted to a total of two hours maximum per week. It can further be assumed that the release frequency occurs randomly at a rate of 1e-4 per year.

Because the area is restricted to two hours per week and diagnostics exist to provide an operator alert so that repairs can be made safely, the duration at risk can be given credit within the LOPA. The time at risk can be computed by dividing 2 hours by the number of hours in a week, assuming the plant is operated 24 hours a day, 7 days a week. The hazard rate is thus:

$$HR = (1\text{e-}4 \text{ major leaks per year})(2 \text{ hr/week})(1/168 \text{ hours/week}) \qquad [\text{Eq. 3}]$$

$$HR = 1.2 \text{ e-}6/\text{year}$$

Frequency Based Risk

In other situations, a risk can only be present if a specific action is taken or procedure is performed. Consider a hazardous gas that is manually filled into a cylinder ten times per year. Each fill operation takes one hour to complete. Operator error can result in overfilling of the cylinder and release of the hazardous product. The probability of operator error has already been determined to be 0.003 per cylinder fill. Therefore the likelihood of overfill is calculated to be 0.03 overfills per year. Since this is a manual operation, the probability of operator exposure is unity.

The cylinder overfill hazard can only occur during the cylinder filling operation. The hazardous demand is very closely associated with the filling operation and is proportional to the *number* of fills and not the *duration* of the filling operation as it is the cylinder nearing the full condition that potentially exposes the operator to a hazardous event. The operator is essentially the protective system who must stop the filling operation before it becomes a hazard. In this case the hazard rate is modeled as:

HR = (10 fills/year)(0.003 probability of operator error) [Eq. 4]

HR = 3.0e-2/ year

In this case, to achieve further risk reduction, a safety layer independent of the operator would need to be implemented. If additional safeguards were added, the 0.03 / year frequency value would be used as the initiating event rate and the protective system would be in low demand mode.

Continuous, Non-continuous, or Frequency Based?

It is not always obvious whether a risk is time related or frequency based. Figure 2 shows a pressure vessel which operates at low pressure with feed coming in through valve F and process gas going out through valve P. The vessel is fitted with an automatic vent valve, V; a relief valve, R; and a high pressure alarm, PAH.

Figure 2

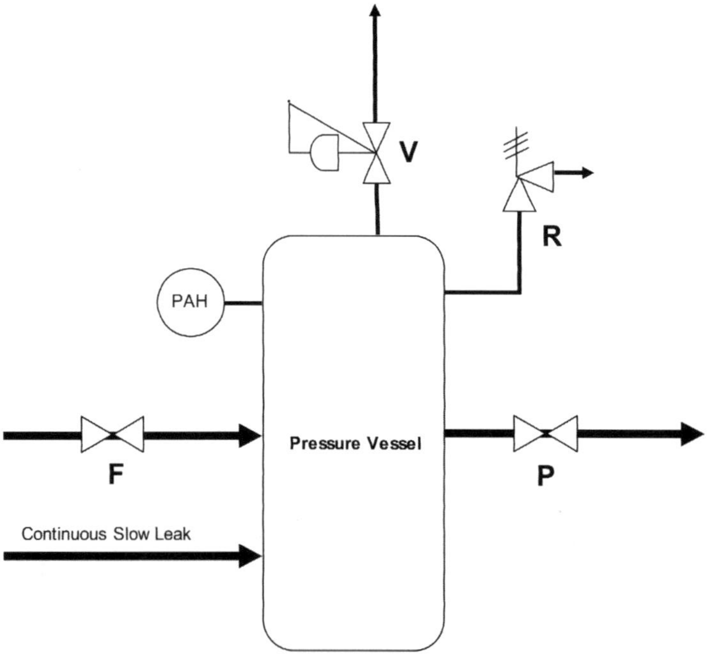

When the plant trips, both valves P and F are designed to close; however a slow leak of gas from a high pressure source is presumed to exist and continues to enter the vessel. This slow leak is not a problem except during the time that valves P and F are closed, which is only during the time that the plant is tripped. Assumptions for this example are listed below:

Assumptions:

- Assume that the slow leak is capable of pressuring-up the vessel to a value well above the relief valve set point (an overpressure condition) within four hours.
- Assume that the vent valve and relief valve are independently capable of relieving the slow leak flow rate.
- Assume that if the vent valve fails, it will be unrevealed, that is no one would know until it is tested per its maintenance schedule.
- Assume that if the relief valve fails it will be unrevealed, that is no one would know until it is proof tested as part of the mechanical integrity program.
- Assume that the PAH will alarm at a pressure such that the operator will have sufficient time to prevent tank overpressure.
- Assume that the vent valve set point is lower than the PAH set point which is lower than the relief valve set point.
- Assume that the plant trips four times per year and remains in that tripped state for six hours each time.

In this case, the hazard can only exist during the time the plant is shutdown so the risk is non-continuous. However, when the scenario is evaluated for what type of non-continuous risk exists, it becomes obvious that there are both elements of frequency and duration based risk. Clearly, the more often the plant trips, the higher the likelihood of vessel overpressure. However, when the plant shuts down, it is at risk for a significant period of time and dependent upon the leak rate in a way that resembles duration based risk.

One resolution to the dilemma would be to assume that the leak rate is always sufficient to overpressure the vessel within the duration of the plant trip. This eliminates duration as a factor and makes this a frequency based scenario.

In this case the frequency can be viewed as a high demand initiating event. As in a previous example, the way to compute the hazard rate for a high demand case is to consider the failure of the first layer of protection operating in continuous demand mode. When it fails, the initiating demand event is placed on the next layer of protection. In this case, if we assume the following failure rates and failure probabilities we can calculate a reasonable approximation of a hazard rate.

Assumptions

- Vent control fails = 0.1/year
- Alarm layer of protection failure PFD = 0.1
- Relief valve fails to function on demand = 0.01

$$HR = (0.1/yr)(0.1)(0.01) \qquad [Eq. 5]$$

$$HR = 1.0e\text{-}4 \text{ per year}$$

Conclusion

It has been shown that the type of demand that protection systems experience is an important consideration when performing a LOPA. Treating high demand initiating events as low demand events using the simple LOPA math can lead to excessively conservative results. However, it is also important to recognize when the potential risk is always present or whether it is only present during specific operations, making it a non-continuous risk.

Understanding whether a risk is a function of time or is frequency controlled is important as it helps define the mathematical approach. If the risk is time-related, understanding whether the initiating event rate is a low demand, high demand, or continuous demand also has a direct impact on the appropriateness of the various mathematical models that can be employed to estimate the hazard rate.

References:

[1] ISA—TR84.00.04—2005 Part 1: Guidelines for the Implementation of ANSI/ISA-84.00.01-2004 (IEC 61511 Mod)

Incorporating and Quantifying Human Activities and Actions in Layer of Protection Analysis

Philip M. Myers
Advantage Risk Solutions, Inc.
4251 N. County Line Rd.
Sunbury, OH 43074
pmyers@ARiskSolution.com

Keywords: Layer of Protection Analysis, LOPA, Human Factors, Human Error Probability, Human Reliability Analysis, Independent Protection Layer, Initiating Event, Risk Analysis, Safety Integrity Level, Process Hazard Analysis, PHA, Buncefield, PSM

Abstract

Layer of Protection Analysis (LOPA) is clearly a tool of choice within the process industries to address risk-based issues and decisions in a simplified manner, while adding a greater degree of understanding and confidence in decisions made. LOPA is effectively used to bring objectivity and a more consistent approach to addressing layers of protection and assessment of risk beyond that afforded in traditional qualitative Process Hazard Analysis (PHA) reviews. LOPA can be used to address a wide range of risk issues and decision making needs – and has become the preferred tool for selecting appropriate Safety Integrity Levels (SILs) for Instrumented Protective Functions and for Safety Instrumented Systems (SISs) specifically.

The human role as a potential initiating event or as part of a protective layer is important in the process industries generally, and plays an even more significant role for batch processing facilities and in non-routine operations. There is a need to both include and quantify human performance in LOPA. Human activities as potential initiating events and within human independent protective layers (IPLs) are reviewed and methods for quantification outlined – including an extension beyond LOPA into Human Reliability Analysis (HRA) and methods that can be used to develop Human Error Probabilities (HEPs) specific and suitable to the operations and process safety culture at a given plant site.

1. Introduction

Layer of Protection Analysis (LOPA) is a highly valued, semi-quantitative risk methodology embraced by the process industries and in widespread use. LOPA uses a relatively simple, scenario-based approach that can effectively address many risk related issues, providing a timely and cost-effective tool to conduct engineering analyses as an aid to decision making. LOPA is typically used to determine if existing layers of protection are sufficient, and to develop risk reduction measures for specific scenarios of concern. A LOPA scenario consists of a single, unique initiating event-consequence pair. Generally used for high consequence or high risk scenarios, LOPA generates additional support and a greater degree of confidence in decisions made as compared to those relying on the use of purely qualitative tools such as Process Hazard Analysis (PHA).

Over the years since the introduction of LOPA to the process industries [1-2], and with the requirements of industry standards for functional safety [3], it has been used extensively, with a wealth of application experience gained. There are now many variations of LOPA in practice – some are highly simplified, order-of- magnitude approaches with simple calculations, while others are more detailed and complex with extensions to quantitative techniques such as Human Reliability Analysis (HRA), Event Tree Analysis (ETA), Fault Tree Analysis (FTA), and Quantitative Risk Analysis (QRA). LOPA has been stretched in many respects, with new developments in and applications for the methodology, and also limitations and problems encountered in practical use of LOPA [4-6].

CCPS' *Layer of Protection Analysis – Simplified Process Risk Assessment* [7] provides a sound starting point for the conduct of LOPAs. Resources continue to be developed expanding lists of initiating events and IPLs for LOPA, and providing additional guidance for use. This effort includes individual company efforts [8], as well as those of industry trade groups, and CCPS specifically [9]. There are also many company-specific LOPA guidance documents and procedures now in use for standardization – addressing topics ranging from the overall program, strategy, and criteria to the basics in conduct, methods, data, calculations, documentation, to handling of special situations that may arise [7, 10]. Additional guidance and materials generally are needed to further improve LOPA quality and consistency, and this is particularly true when addressing human Independent Protection Layers (IPLs) and human Initiating Events (IEs).

Human activities and actions are important, though sometimes challenging, considerations in LOPA. Initially, a number of companies did not take any credit for human actions and interventions in Independent Protection Layers (IPLs). While this is a conservative approach, many companies found it to be too conservative, potentially resulting in unwarranted expenditures to reduce risks through additional IPLs and Safety Instrumented Systems (SISs). Human activities and actions are an integral part of safe process operations (especially for batch and non-routine operations), and generally now are included in LOPA - in terms of both potential initiating events and as part of human IPLs if they meet the required criteria.

There are many continuing developments to address the human aspects since conduct of the first LOPAs. Some of these advances include:

- incorporation of procedural controls as part of human IPLs
- clarification and definition of the required components to qualify as a human IPL
- efforts and methods to increase confidence in human IPLs through –
    - explicit consideration of all modes of operation
    - cross-checking
    - development of critical task lists
    - use of human equivalent "SIF" (Safety Instrumented Function) or Safety Requirements Specification (SRS) sheets
    - testing and validation of human IPLs

- data and tables for human Initiating Events (IEs) and human IPLs
- application of additional data, HRA and advanced methods to specific LOPA scenarios

LOPA is a highly valued risk tool, with many advances and continuing improvements in the handling of the human role in both independent protection layers and in initiating events.

2. LOPA Scenario Frequencies – Initiating Events and IPLs

An area of significant and continuing improvement within the process industries is in the handling of LOPA scenario frequencies – including initiating event frequencies, and guidance for and numerical specification of independent protection layers (IPLs). Many of these developments are applicable to both human and other initiating events and IPLs.

2.1 Protection Layer Components

Protection layers generally include a *sensor* (or means of detection), *decision making*, and a way to take *action* to deflect the undesired consequence, as shown in Figure 1.

Figure 1. Simple Model of a Protective Layer

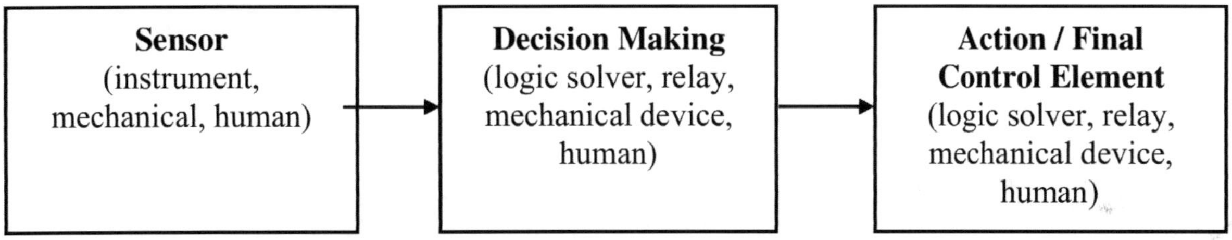

Human IPLs are those that involve people that serve as one or more of the functions depicted above – sensing, deciding, and/or taking the final action.

2.2 Qualification of IPLs

An important aspect of any LOPA evaluation is to determine which safeguards qualify as IPLs, or that with modifications *can be made* into IPLs. Advances have been made in the qualification of IPLs, and industry continues to improve. However, ensuring independence (in particular) has been a struggle, and additional guidance has been and continues to be developed to aid in qualification of IPLs. Human IPLs must meet all of the same criteria. The following simple keywords given in Table 1 can be used to screen candidate IPLs [7, 11].

CCPS (2007) *Guidelines for Safe and Reliable Instrumented Protective Systems* [12] has expanded the list of IPL requirements to a total of seven core attributes: independence, functionality, *integrity*, *reliability*, auditability, *access security*, and *management of change*. These requirements should all be met before concluding a "system" qualifies as an IPL and will be sustained in the planned state. The four additional core attributes of IPLs given in this list of core attributes are *integrity, reliability, access security, and management of change*. *Integrity* is

Table 1. Keywords for Screening of IPLs

"3Ds"	"4 Enoughs"	The "Big I" - Independent
• Detect • Decide • Deflect	• Big enough? • Fast enough? • Strong enough? • Smart enough?	• Of initiating event / cause • Of other IPLs

the expected risk reduction, quantified as the Probability of Failure on Demand (PFD) for the IPL, while the *reliability* (or availability) accounts for the probability that the IPL continues (once activated) to operate when called upon, and for a specified period of time under the stated conditions. As facility changes occur, *management of change* programs direct reviews to identify if / how existing IPLs may be affected. Finally, *access security* for IPLs is also important, to ensure IPLs work as intended when called upon, and to ensure that designed protective systems cannot be used by unscrupulous characters to cause disruptions or harm.

3. Procedural Controls and Human IPLs

Many advances in LOPA are to address procedural controls and human IPLs. It can be difficult for human IPLs to meet all of the qualifying rules – in that they are able to detect, decide and deflect and that they are big enough, fast enough, strong enough and smart enough, as well as independent of the initiating event and other IPLs. There are the problems associated with integrity and reliability – especially given the potential stress and limited time for operations personnel to respond. There is also the issue of auditability and keeping track of all of the necessary aspects of a human IPL. Add in security concerns, and variability when dealing with people, and the changes that occur through modifications to the process and or changes in the workforce or methods of operation. It can be an initially daunting task to take a credible approach and realistically take credit for human IPLs in LOPA. However, advances are being made in addressing human IPLs that include:

- new thinking and guidance for how to identify the necessary components of human IPLs
- consideration of various modes of operation
- means to increase the confidence in and credit that can be taken for human IPLs
- development of critical human task lists
- analysis of human error, including site specific factors
- tools for analysis of human error
- testing and validation of human performance
- collection and analysis of plant data
- integration with HRA and other quantitative risk analysis techniques

A number of companies initially did not take credit for human IPLs in LOPA due the difficulties associated with them and the perceived limited risk reduction value, especially under potential conditions of stress and with limited time to take the correct action. While recognizing human actions as safeguards in PHAs, they were often seen as activities that could not meet the requirements of an IPL. So the potential risk reduction benefit was sometimes "left on the table". However, this was an immediate problem for smaller companies that may have less automation in general, and even for larger companies primarily utilizing batch operations – that often rely much more on human activities both in operation of the plant and as "safeguards". For that matter, it could be a problem for any company when considering operational modes other than normal operations. There is a clear need to include human IPLs in LOPA where it can be justified – but the means to achieve sound human IPLs is not as obvious.

3.1 Human Response and Process Safety Time

When considering human IPLs, it is helpful to first consider the general human response to alarms and abnormal conditions [6], as given in Figure 2.

Figure 2. General Human Response to Alarms and Abnormal Conditions

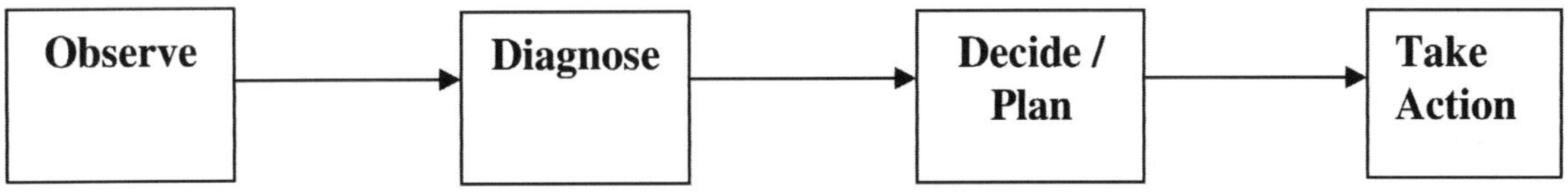

The first in this sequence of steps is to *observe* the condition or alarm. Often this can be quick for control room operators with lighted and audible alarms. On the other hand if it requires an observation in the field, the time can be substantially longer. The next steps are to *diagnose* the situation and then *decide / plan* what to do. The length of time for these steps will depend upon many factors, including the available inputs, familiarity, complexity, the written procedure(s), training, perceived severity or danger, and others. (Note that for a human IPLs to be effective, it is important that the *diagnose* step not involve calculations or complex diagnostics.) The final step is then to physically *take action.*

The time for human response in abnormal, potentially dangerous or escalating situations is an important consideration when evaluating human IPLs versus engineered solutions - and in determining realistic PFDs for human IPLs given the expected range of potential conditions. The "process safety time" is a useful concept for these evaluations. The process safety time is the "time period between a failure occurring in the process or control system and the occurrence of the hazardous event [12]." So the process safety time includes the time required for a person to go through all of the above steps (observe, diagnose, decide/plan, and take action) relative to the time available before the process or situation reaches the "point of no return" – when the action can no longer be taken to prevent the undesired consequence. Figure 3 presents the process safety time for a LOPA scenario.

In considering human IPLs, the person expected to take action must have sufficient time to do so. Reduced available time leads to higher PFDs for human IPLs (i.e. they are more likely to fail under increasing time pressures). Some sources have suggested that the human response time should be less than half of the process safety time to take credit for a human IPL. This criterion essentially builds conservatism into the evaluation. Companies can choose to use this approach,

Figure 3. Process Safety Time – Available time for Human Response

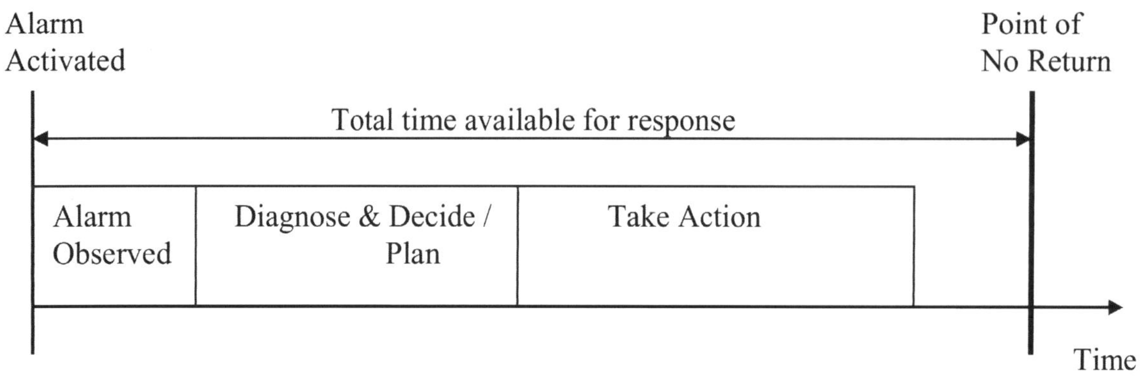

or if comfortable in determining the amount of time for human response to a given scenario, relax the requirement such that the human response time simply must be less than the process safety time.

3.2 Components of Human IPLs

Often cited safeguards in PHAs are procedures and training – both involving people. Yet, neither written procedures nor training are by and of themselves IPLs. They don't make it past initial screening, as they do not detect, or decide, or deflect. Still, is there a way to take credit within LOPA for the positive benefit of human actions and interventions in the process industries? In short, yes, but it requires incorporation of human activities within a larger picture to qualify as a human IPL. Procedural controls or human IPLs include a combination of a field sensor, human / operator, and final control element (e.g. valve, switch, relay) all within a written procedure. Table 2 presents a comparison of active protection layers involving human IPLs [13].

Table 2. Active Protection Layers Involving Human IPLs

Type	**Sensor**	**Decision / Logic Solver**	**Action / Final Control Element**
Human IPL – Control Room	Field sensor for pressure, temperature, etc.	Human action based on BPCS or SIS alarm	Remote activation of control valves, motors, etc
Human IPL – Based in Plant/Field	Field sensor for pressure, temperature, etc.	Human action based on local sensor or alarm, or field observation or sensor	Remote activation as above, or manual operation in the field (e.g. manual valves)

A human IPL or procedural control includes the following components [7, 13]:

- written procedure specifying the required action
- clear communication that the task/action must be performed (and the consequence if not)

- means to detect a problem – inputs, available, clear indication even in emergency situations, and simple to understand
- any graphical or decision aids that may be helpful
- the physical means of interaction with the process (e.g. manual valve) under all reasonably expected conditions, to prevent or alter the outcome (undesirable consequence), and defining the task(s) or what is to be done given the inputs
- training on how to perform the task – regular, documented, drills/tests, all operators capable
- provision of needed materials or tools for the task
- appropriate personal protective equipment (PPE)
- sufficient time - to observe the condition or alarm, to diagnose problems and analyze what should be done, and to correctly perform the task
- successful performance benchmarks
- ability to verify the action/task was performed (and done correctly)

A checklist can be used for evaluation of the quality of procedural based safeguards or human IPLs [13]. Human IPLs, however, must meet the same requirements as for other types of IPLs, and consideration should be given to all seven characteristics: independence, functionality, integrity, reliability, auditability, access security and management of change.

3.3 Increasing Confidence in Human Performance and IPLs

There are a number of developments and ongoing efforts to increase the confidence in human IPLs, including: explicit consideration of all modes of operation, incorporation of independent cross-checking, development of critical task lists, development of "SIF" specifications for human IPLs, testing and validation of human IPLs, and collection and analysis of plant data.

3.3.1 Explicit Consideration of All Modes of Operation

One consideration in dealing with human response and human IPLs is to address the concern regarding the ability of operators to carry out the intended action in all relevant circumstances. While PHAs are intended to include all phases or modes of operation, in practice often they focus mainly on normal operations. LOPA studies following PHAs may "fall into the same trap" of focusing on normal operations and neglecting other modes. One means to increase confidence in human IPLs is to explicitly consider all relevant modes of operation. Use of a matrix similar to that presented in Table 3 can be helpful as a "prompt" to PHA teams and to the LOPA "team" or analyst [6]. While some scenarios identified by the PHA team may be specific to only one mode of operation, it is worthwhile for the LOPA team or analysts to clarify the LOPA scenario and start by considering whether other modes may be of importance. If after reviewing all relevant modes of operation the LOPA team is convinced that the human IPL can be carried out as intended, it adds confidence that an appropriate amount of credit can be taken within LOPA.

Table 3. LOPA Human Response Considerations "Prompt" for Modes of Operation

Mode of Operation	Relevant to LOPA Scenario (Yes/No)	Human Response / Potential IPL Considerations
Start Up – initial and normal		
Normal Operation		
Normal Shutdown		
Emergency Shutdown		
Restart after Turnaround or Emergency Shutdown		
Turnaround / Maintenance		
Abnormal Modes – temporary, reduced production, etc.		

3.3.2 Cross-Checking

Another approach to improving human IPLs is through cross-checking of human actions by independent work groups. The general philosophy is that errors of the person initially carrying out a task can be identified and corrected by a second person with specific cross checking duties. Human errors are generally of omission (not done), commission (performed incorrectly – too much, too little, wrong, action, action out of sequence, etc.), action not in time (too early, too late), extraneous acts (act where there is no task demand), and error recovery failure (failing to recover from a recoverable error is itself a failure). Targeted cross checking of human actions has the potential to reduce errors. Table 4 presents various levels of cross checking [6].

Table 4. Levels of Cross-Checking Effectiveness

Confidence	Dependency	Level of Cross Checking
None	Complete	No justifiable reason why the checker should identify the failure when the person carrying out the original action has not.
Low	High	The checker can determine the correct course of action independent of the first person. However, the checker either has a common link with the first person or there is good reason to believe that the checker will make the same error as the first person.
Improved	Moderate	Checker has a weak link to the first person or there is moderate likelihood the checker will make the same error as the first person.
Highest	Low	Checker has sufficient independence from the first person and action, and the check is designed to highlight possible errors.

In LOPA, it becomes necessary to translate the qualitative assessment of the effectiveness of cross-checking into the calculation of scenario frequencies – or to decide that no credit can be taken. There are a number of sources for the probability of human error associated with carrying out steps in procedures for example.

3.3.3 Development of Critical Task Lists

Critical human tasks lists can be developed to focus attention on tasks with the greatest impact on process safety and to aid in proper management of those tasks to sustain human IPLs. Additional resources can also be directed to training and ensuring that management of change issues do not compromise these critical activities. Once a list is developed, critical tasks can be further reviewed to identify error potential and factors affecting success or failure of the activity. An example critical task list format [6] is given in Table 5.

Table 5. Critical Task List and Initial Evaluation

Critical Activity or Task	**Undesired Consequence**	**Potential Error Leading to Undesired Consequence**	**Factors Affecting the Critical Task and Human Error Potential**
Opening manual routing valve between the transfer pump discharge and a designated receiving tank	Overflow of a storage tank	Opening the wrong valve and thereby transfer materials to the wrong tank	-poor labeling of valves -all communication is by a single channel radio from the control room -significant proportion of new process operators with little on-site experience

3.3.4 Development of Human SIF Specifications or Safety Requirements Specifications (SRS)

Another means to develop confidence in human IPLs is to treat them the same as other IPLs based on engineered systems (e.g. SIS) and develop a simple SIF specification - especially for highly critical human IPLs. . Table 6 provides an example simple template that can be used. An example operator initiated SIF specification is given in Summers [14].

Table 6. Template for Operator SIF Specification

	Description
Process Unit	Identifier
Area / PHA Node	More detailed identifier & tie to PHA
SIF#	Number for tracking
Process Hazard	Detailed description of the complete scenario and undesirable consequence
Functionality	Process parameters that should be monitored, actions to be taken, identified trip/action points, capability for testing, minimum acceptable test intervals, and environmental considerations.
Input	Specific equipment/sensor/alarm – tag # and description
Output	Expected action together with reference to tag numbers and descriptions
Alarm Communication	Specific details of how the operator will be aware of or have an indication of a problem, including equipment details and how it is communicated visually, audibly, etc.
Critical Set Points	Set point for the input(s) – specific tag # and set point
Operator Response	Details what the operator is to do – including references to tag #s
Final Control Elements	Description of equipment to manipulated / used, switch, push button, manual valve, etc.
Time Available for Response (Process Safety Time)	Time available to take the required action.
Independence Considerations	Discussion of any potential concerns, or items for further investigation – include potential common cause failures
Safety Integrity Level (SIL)	Safety integrity requirements

3.3.5 Testing and Validation of Human IPLs

While it is within the mandate of the LOPA process to ensure that credit for human IPLs is realistic, it is the job of the site or site management to ensure that human IPLs are tested and validated. Even if there are multiple references that can be used to justify the Probability of Failure on Demand (PFD) for a given human IPL, if site performance does not concur, it is invalid. Just as other IPLs need to be tested and validated, the same is true with human IPLs. Operators with specific requirements should be trained and re-trained as appropriate, and the human IPLs should be audited by the site to ensure the stated performance can be achieved. If the stated performance cannot be achieved in simulated situations and tests, then credit for human IPLs should be reconsidered and either alternative solutions be employed, or steps should be taken to strengthen the human IPLs. If testing and auditing of human IPLs indicate that the human IPLs are working as intended, additional confidence in them is gained. Sites with a low level of operational discipline will find that the credit that can be taken for human IPLs is especially limited, while those with a high level of operational discipline will be able to take more credit. Bridges [15] outlines a program to collect experimental data related to human response IPLs including the test setup, test plans and statistical sampling, equations for determination of an appropriate sample size, and acceptance criteria. Bridges suggests that while companies may believe collecting plant data on human responses may be difficult, "the actual effort to collect such data is low and the benefits are great, as demonstrated by several chemical plants and refineries…."

4. Human Error Probabilities – Initiating Events and Human IPLs

Human error probabilities and performance shaping factors that affect them form the basis for human performance in terms of initiating events and human IPLs included in LOPA studies. The two basic sources of human error are typically in following procedures and in human response to an alarm or call for action. Of course, human activities have been considered as potential initiating causes or events in LOPA from the beginning of its use. This often takes the form of human errors in following procedures, such as in unit startup for example. However, in the continuing development of LOPA, while a number of companies initially did not include human IPLs (in basic LOPA), most now do incorporate human IPLs when considering if protective layers are sufficient. Human reliability analysis (HRA) and advanced quantitative techniques also can be used to determine the credit that can be taken and to build confidence in human IPLs. Still, there is a wide range of data, practices, and techniques in use when it comes to addressing human activities and the potential to initiate events, to aid in preventing incidents, or to break the sequence that can lead to undesirable consequences as part of a protective layer.

4.1 Human Error as an Initiating Event or Cause

When considering human activities as an initiating cause in LOPA, it may be helpful to evaluate procedures used and calculate the probability of failure. The *Human Reliability Handbook* [16] can provide useful human error probabilities (HEPs), such as those reproduced here in Table 7.

Table 7. Estimated Probabilities of Errors of Omission per Item of Instruction When Use of Written Procedures is Specified

Omission of Item	**HEP**	**EF**
When procedures with check-off provisions are correctly used:		
Short list, < 10 items	.001	3
Long list, >10 items	.003	3
When procedures without check-off provisions are used, or when check-off provisions are incorrectly used:		
Short list, <10 items	.003	3
Long list, > 10 items	.01	3
When written procedures are available and should be used, but are not used	.05	3

The data given in the above table (Table 7) are for a highly idealized, optimized human factors environment atypical of the process industries. Therefore, caution is advised in directly using the HEP value suggested in the table. The Error Factors (EFs) given in the table are used to represent uncertainty bounds that are symmetrical around the mean value. Both lower and upper bounds are considered, with the lower bound intended to represent the 5th percentile and the upper bound the 95th percentile. It should be noted that the uncertainty bounds given are based on judgment and should not be confused with statistical bounds based upon data analysis.

As the uncertainty in this table is symmetrical around the mean HEP value (say .003 for example), the lower and upper bounds can be calculated as follows:

- the lower bound can be obtained by dividing the mean value by the error factor (EF) of 3 giving a lower value of 0.001
- and the upper bound is given by multiplying the mean value by the error factor of 3 to get approximately 0.01 (rounded off).

Swain and Guttman [16] suggest use of a nominal HEP value of 0.003 per step or instruction for errors of omission and also for errors of commission - for use as a first estimate when no other information is available. The following equation can then be used to calculate the probability of failure to correctly complete a procedure.

$$P_{failure} = 1 - (1\text{-HEP})^n$$

where,

$P_{failure}$ = the probability of failure to carry out the procedure as intended

HEP = the human error probability per step

n = the number of steps

4.2 Human Error Probabilities in Human IPLs

An important consideration in LOPA is the benefit of IPLs incorporating human actions (i.e. human IPLs) and accounting for human error probabilities (HEPs) associated with them. The reduction of event frequency or risk is in part limited by human errors associated with these human IPLs. There is a wide range of guidance available for the handling of human error probabilities (HEPs) for human IPLs, ranging from quite simple methods to complex calculations and adjustments that should be attempted only by trained, experienced quantitative risk analysts or human factors specialists. Reviewed here are approaches that can be used to determine PFDs for human IPLs within LOPA.

4.2.1 Simple Approaches and Tables

One very simple approach to human error probabilities - and really the credit that can be taken for operator response - is presented in an International Society of Automation (ISA) text on Safety Integrity Level (SIL) selection [17]. Only three categories – *normal operator response*, *drilled response*, and *response unlikely* – are used to represent the range of possibilities, as reproduced here in Table 8. This represents an incremental step forward to include human activities in a simple manner within a fairly basic LOPA study.

Table 8. Simplified Technique for Estimating Operator Response

IPL Category	Description	PDF
1	**Normal Operator Response** – In order for an operator to respond normally to a dangerous situation, the following criteria should be true: • Ample indications exist that there is a condition requiring a shutdown • Operator has been trained in proper response • Operator has ample time (>20 minutes) to perform the shutdown • Operator is ALWAYS monitoring the process (relieved for breaks)	0.1
2	**Drilled Response** – All of the conditions for a normal operator intervention are satisfied, and a "drilled response" program is in place at the facility. Drilled response exists when written procedures, which are exactly followed, are drilled or repeatedly trained by the operations staff. The drilled set of shutdowns forms a small fraction of all alarms where response is so highly practiced that its implementation is automatic. This condition is rarely achieved in most process plants.	0.01
3	**Response Unlikely** – All of the conditions for a normal response intervention probability have not been satisfied.	1.0

Additional tables have been developed to address human actions and the credit that can be taken or the associated human error probability (HEP) - either as a PFD or as a risk reduction factor. All data presented in the following tables are in the form of a PFD, even if it the data may have been given as order of magnitude risk reduction factors or as the number of IPL credits in the reference document. CCPS [12] provides guidance on human actions as IPLs, as reproduced here in Table 9.

Table 9. Examples of Operator or Supervisory Activity PFDs

IPL	Condition	PFD [a]
Process Related Rounds and Inspections	Frequency of operator rounds must be sufficient to detect and prevent the hazardous event. Operator is trained to recognize and respond to unacceptable out-of-range values. If a specific process variable is being monitored, the operator should record the specific value displayed by equipment independent of the initiating cause.	10^{-1}
Observational	Frequency of operator rounds must be sufficient to detect and respond to the hazardous event. The need to take response must be obvious to the operator through normal visual or hearing range, e.g. loud noise, high vibration, serious leaking, etc.	10^{-1}
Review	Independent inspection / verification and sign-off that a required operator action was performed as intended (e.g. valve line-up is confirmed as correct).	10^{-1}
Action	An operator action that uses a different operator, relying on independent observation.	10^{-1}
Corrective Action	An operator action taken based on a scenario where the propagation is so slow that the operator has sufficient time to gather further information (e.g. laboratory tests, product quality, and material balance) as necessary to recognize the error and to correct it.	10^{-1}

[a] Note that to claim the tabulated PFDs - The operator should be trained and tested on the procedure, which should list the process condition(s) that clearly indicate the need to take action. The procedure should provide a list of the action(s) required by the operator when the process condition(s) are unacceptable, the time available for the operator to take such action, and the consequences if action is not taken.

Additional data is available from a number of references for the PFDs of various human activities and in response to alarms, given stated time constraints [7, 12, 13, 16-18]. NUREG CR-1278 [16, 18] provides data for the probability of failure of diagnosis as a function of time after a compelling signal of an abnormal situation for a control room operator. The figure shows that the probability of failure is fairly level for the first 10 minutes, drops off significantly in the 40 to 60 minute range, and then again levels off, as can be seen in Figure 4. Table 10 then presents a summary of PFDs for human IPLs given in more recent texts that address LOPA.

Figure 4. Probability of Failure by Control Room Personnel for Correct Diagnosis After an Abnormal Situation – Probability of Failure Versus Time in Minutes

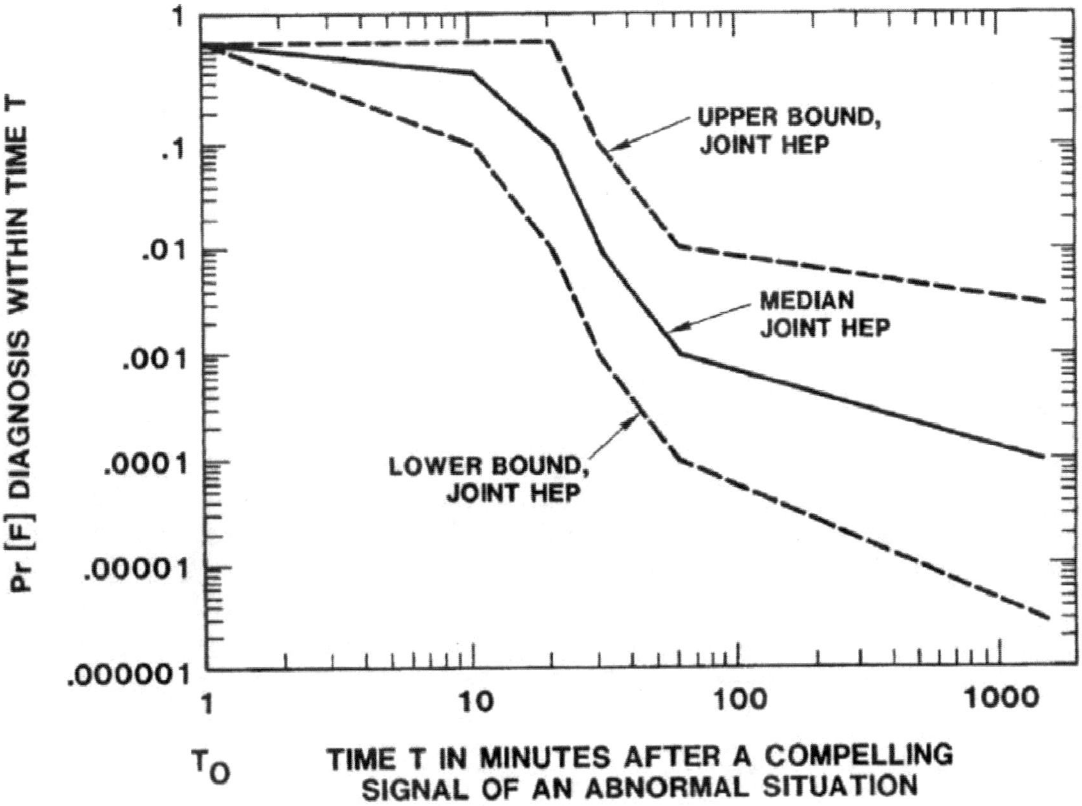

Table 10. Summary – Probabilities of Failure on Demand for Human Actions or Response (in Human IPLs)

Time Available After Alarm [b] *(or initial observation)*	**Conditions and/or Description**[c] *(Assuming adequate documentation, training and testing procedures)*	**PFD**
CCPS Layer of Protection Analysis [7]		
10 minutes	Human action with simple, well-documented action with clear and reliable indications that the action is required.	10^{-1} (data range 10^{-1} to 1.0)
40 minutes	Human response to BPCS indication or alarm. Simple, well-documented action with clear and reliable indications that the action is required. [The PDF is limited by ISA-84.00.01-2004 (IEC 61511 Mod]	10^{-1} (> 10^{-1} allowed by IEC / ISA)
40 minutes	Human action with simple, well-documented action with clear and reliable indications that the action is required.	10^{-1} (data range 10^{-1} to 10^{-2})
CCPS Guidelines for Safe and Reliable Instrumented Protective Systems [12]		
any response time	Operator action is complicated, e.g. large number of alarms generated by initiating cause or the required response is not documented in a written procedure or the operator is not trained on the written procedure.	1.0
< 10 minutes	Operator must troubleshoot to determine what the appropriate response is.	1.0
2 – 10 minutes	Drilled and practiced response, also known as a "never exceed, never deviate" response. If the alarm is received, the operator must execute the safe state action without delay. Alarm is independent of the BPCS.	10^{-1}
≥ 10 minutes	Operator response does not require troubleshooting or investigation prior to action. Alarm may be implemented in the BPCS or independent of the BPCS [d].	10^{-1}
≥ 40 minutes	Operator response requires minor troubleshooting or investigation prior to action. Alarm may be implemented in the BPCS or independent of the BPCS [d].	10^{-1}
24 hours	Multiple operators can take action. Alarm should be automatically repeated at an interval necessary to ensure that each shift is notified of the process condition. Minor troubleshooting may be performed prior to action. Alarm is independent of the BPCS.	10^{-2}

[b] The operator response time should consider the time it takes to recognize the alarm, to diagnose the problem, to complete the required action and for the process to reach the designated state. This is compared to the allocated process safety time, which considers how rapidly the process moves from the alarm condition to the hazardous event.

[c] The required action must be clearly indicated by the alarm, the response covered by a written procedure, and the operator trained and tested on the procedure.

[d] As long as independence from the initiating cause and other IPLs is demonstrated, allocation is influenced by the operator interface design and the importance of the operator response. In all cases, the operator should receive the information in a clear, unambiguous, and prioritized manner.

Additional data is given in Table 11 for probabilities of failure on demand (PFDs) for human or procedural controls [13]. The probabilities in this table that include cross-checking functions are also supported by calculations in the referenced document based upon data from Swain [16]. Also provided in Swain are tabulated and graphed probabilities of failure for control room personnel to diagnose an abnormal event(s) based upon the available time, as previously given in Figure 4.

Table 11. **Procedural Controls (or Human IPLs) and PFDs**

Time Available After Alarm *(or initial observation)*	**Conditions and/or Description** *(Assuming adequate documentation, training and testing procedures)*	**PFD**
< 10 minutes	Single operator with less than 10 minutes to diagnose and take action on a clearly annunciated alarm event.	1.0
10 - 40 minutes	Single operator with 10 minutes to 40 minutes to diagnose and take action on a clearly annunciated alarm event.	10^{-1}
40 minutes – 8 hours	Single operator with 40 minutes to 8 hours to diagnose and take action on a clearly annunciated alarm event activated by a SIS.	10^{-2}
10 – 40 minutes	Single operator completing a simple routine task with between 10 minutes and 40 minutes to complete the task.	10^{-1}
10 – 40 minutes	Two operators acting independently completing a short (10 items or less) written checklist procedure (done-by / checked-by) with between 10 minutes and 40 minutes to complete the task.	10^{-2}
> 10 minutes	Two independent operations groups completing a task with one group independently checking the work of the other (operations checking maintenance) with adequate time (greater than 10 minutes) to detect any issues.	10^{-2}

4.2.2 Advanced Methods and Human Reliability Analysis (HRA)

One means to address human error and factors that may affect it is to begin with actions or critical task lists and make adjustments for specific conditions at the site. Figure 5 show the steps in going from a critical task list (or other means used to consider the key human tasks or activities for a given LOPA scenario) to assessment of the human error probability. This type of detailed analysis is highly dependent upon comprehensive knowledge of the data used and the application of factors to more appropriately account for conditions expected in the process industries – and should be carried out only by experienced quantitative risk analysts and human factors specialists, or using tools developed by them.

Figure 5. Process for Assessing Human Error Probability – Including Performance Shaping Factors or Error Producing Conditions

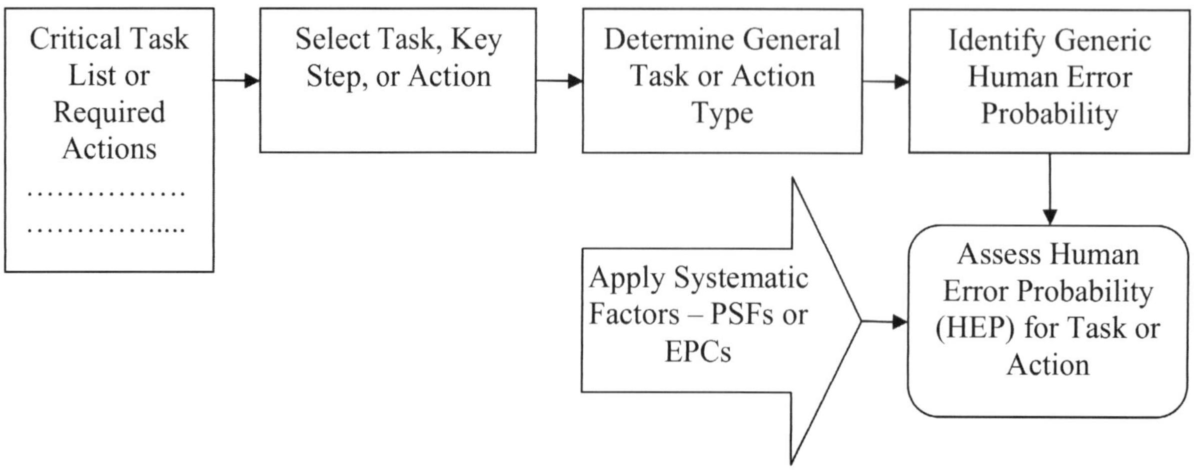

The general process for assessing human error probabilities (HEPs) is to begin with a list of human activities or tasks of interest for LOPA scenarios. An activity or task can be selected, and the general task type and corresponding basic human error probability can be identified from human performance data such as given in Gertman [19], Hunns and Daniels [20], Williams [21], Swain [16], and others. This basic human error data is typically for highly idealized human factors conditions, so there is a need to adjust them to conditions expected in a process plant. This is accomplished through use of Performance Shaping Factors (PSFs) for Error Producing Conditions (EPCs) that serve as multipliers to the base human error values. PSFs or EPCs may be associated with the human-machine interface, individual human factors, the work environment, task demands, task characteristics, instructions and procedures, stresses, sociotechnical factors, and others. The Human Error Assessment and Reduction Technique (HEART) developed by Williams [21-25], provides PSFs for various EPCs expected to be encountered in a process plant. The HEART technique has been found to be useful due to its basis on sound human factors science and its simplicity in use. Other techniques have been developed that take a similar approach in using EPCs or PSFs as multiplying factors for base human error rates such as in SPAR-H [19]. This type of analysis can also be included in simple tools to support LOPA studies [26]. Major chemical and other companies in the process industries have utilized outside risk analysts and human factors experts as well as Subject Matter Experts (SMEs) to develop additional tools for the estimation of human error probabilities. These tools can include a given a set of conditions and parameters, and the factors that affect HEPs for a plant site using techniques such as HEART or SPAR-H. These tools typically are applied by SMEs or risk analysts for a limited number of scenarios that are critical or may require larger investments. In these cases, application of these tools can aid in reducing conservatism that is otherwise included "by design" in the practice of basic LOPA.

Please note that the base error rates from the sources mentioned above are typically from highly optimized human factors environments and programs and <u>should never</u> be used directly, without modification for human error rates in the process industries. They are idealized values with the expectation that significant factors will be applied to develop a value appropriate for use at a process plant.

4.3 Human Error Probabilities – Norms and Practical Lower Limits in LOPA

In the determination of PFDs for human IPLs, the "norm" is to use a value of 10^{-1} if the operator response can be performed within the process safety time (PST) for the scenario, and the PST is 20 minutes or more to allow ample time to accomplish the sequence detect/observe-diagnose-plan-act. A value lower than this is typically not allowed in basic LOPA [15]. According to Bridges [27], with optimized human factors at a facility (currently atypical for process plants), the lower limits of human error tend to be approximately 1 mistake in 100 steps for most procedure-based tasks (as the procedures are often longer than 10 steps), and a 1 in 10 chance (or a little better) for diagnosis and response to a critical alarm.

When it comes to assignment of safety integrity levels for safety instrumented systems involving human IPLs, the typical assignment is SIL 1 when it can be justified. It is common for companies to identify a large number of human responses in LOPA studies. Some companies believe that if they have a written procedure and training they can use a PFD of 10^{-1} for a human IPL. However, in order for a human IPL to qualify as an IPL it must meet all of the same characteristics and conditions as for other active components of IPLs. As a result, to use a PFD of 10^{-1} for a human IPL, it must also be tested and validated to ensure that the credit is justified and sustainable over time. The European Process Safety Centre Process Safety Leadership Group (PSLG) that followed up the Buncefield Disaster (2005) in the United Kingdom and the subsequent reviews of application of LOPA to tank overflow scenarios [6] – concurred with the Engineering Equipment and Materials Users' Association (in EEMUA 191) [28] recommendation that LOPA should not take credit for SIL 2 or higher integrity levels for IPLs involving human action. Due to the complexities involved, base human error rates, likely performance shaping factors or error producing conditions at a process plant site, as well as the difficulties in testing, SIL 2 and SIL 3 ratings are difficult to justify and validate. Lower values for human IPL PFDs are possible, especially given longer available response times, but should be the result of human reliability analysis (HRA). HRA is beyond the capabilities of many personnel conducting basic LOPA studies, and expert resources should be used in this type of effort. Further, a SIL 4 IPL realistically is not possible today in the process industries given the human aspects and potential errors in in-situ testing.

5. Limitations and Problems in Practice – Incorporating the Human Role

LOPA clearly is a beneficial tool of choice with many successful applications in the process industries – however, it does have limitations and, in practice, problems arise in its use [4, 29-32]. Briefly reviewed here, are problems related to incorporation of the human role in initiating events and independent protection layers.

5.1 Limitations

There are a number of limitations in the conduct and application of LOPA ranging from its design as an engineering analysis tool, the focus on single initiating event – consequence pairs, to its simple approach and the balance between accuracy and science, apparent limitations in use of LOPA rules, to a limited base of experts once studies go beyond basic LOPA – this is particularly true when addressing human factors and operating errors. There is a strong case for using a skilled quantitative risk analyst (with experience in human factors and human reliability) in the conduct of LOPAs - or at a minimum in providing supporting tools and quality assurance for them. This need is apparent both from LOPA experience and common problems encountered in the United States [4], and from the United Kingdom (UK) Health and Safety Executive (HSE) Buncefield study - considering a review of multiple LOPAs conducted and the associated identified problems with them [5-6].

5.2 Problems in Practice – US Experience – Human IPLs and IEs

A significant amount of experience has been gained in the conduct of LOPAs throughout the process industries over a period of years. From this experience, some common problems in the conduct of LOPA or in its application are apparent [4]. These are problems that occur in the practice of LOPA…"where the rubber meets the road." Generally, it has been more difficult to properly address human actions and activities within LOPA. Several key, specific problems (and cautions) related to the human role in IPLs and IEs within LOPA follow.

Taking excessive credit for human actions and human IPLs – While, initially, many companies did not take any credit for human actions and interventions as part of IPLs, now when human actions or human IPLs are included in LOPA, often too much credit is taken. Companies tend to take credit for a factor of 10 reduction (i.e. PFD of 10^{-1}) if they have a written procedure and a general training program at the site. However, human IPLs must meet the same criteria as other active IPLs. As a result they must be complete, maintained and tested. Assuming a universal PFD of 10^{-1} for human actions or IPLs without regard to determining if all necessary operators can capably carry out the action and without specific training and testing, leads to highly optimistic results.

Not considering the culture (and operating discipline) – Another related problem in LOPA is not considering the culture (and operating discipline) of the company, plant site, and possibly even the specific unit. This can be a mistake. Consider a situation where a particular scenario includes a human IPL and consideration is given to adding a SIS to address the risk gap. If the process safety culture at the site is poor (or the operational discipline is low), with the addition of a SIS, the operator may no longer (at least reliably) carry out the action in the human IPL, instead assuming that the SIS will take care of it. As a result, the company may invest in the SIS with no actual benefit in terms of risk reduction. If there is a risk gap that cannot be met without a SIS, and the safety culture is poor, it may be prudent to overdesign the SIS to a higher SIL, so that even if the operators do not reliably carry out the required human action, the process is still adequately protected as intended.

Over specification of SISs leads to "alarm overload" – At least partially as a result of the expanding use of LOPA there has generally been a trend of over specification of safety instrumented systems that has also lead to a ballooning in the number of alarms. This has created the need for entire standards [28, 33] to deal with alarm handling and alarm management systems. "Alarm overload" has become a very real concern in the process industries and in other settings [34]. The possibility for "alarm overload" should be considered within the context of human IPLs.

Stretching LOPA for complicated scenarios and complex calculations – LOPA is being used as the tool of choice for many assessments of risk. However, its use is being stretched to scenarios with dependencies and potentially complex sequences of events. In some cases it is simply better to use fully quantitative techniques such as human reliability analysis (HRA), fault tree analysis (FTA), event tree analysis (ETA), and quantitative risk analysis (QRA) directly.

Failure to ensure IPLS are independent – One of the most challenging aspects of LOPA is to ensure that protective layers are, in fact, independent. This rule is all too often violated. Special care should be taken to ensure IPLs meet all of the required characteristics – especially independence – prior to taking credit for them in LOPA. Common independence problems involve using the same operator or operating group more than once, or using the same operator involved in the initiating cause in a human IPL [11].

Use of Data Without Understanding its Basis or Applicability – "Blind" use of data (probabilities of failure for initiating events or IPLs) from handbooks or standards, assuming they apply to your situation, can quickly lead to invalid results in LOPA. Use of the data must be defensible for its applicability to the company, plant site, and process.

Failure to maintain – Often data is selected for initiating events and IPLs based upon the design and initial conditions. An important component in the lifecycle of the process is to maintain human performance and IPLs in a condition that assures the desired protection. In considering the human aspect, reductions in the workforce, departure or transfer of senior operators, reorganizations, hiring of new personnel, use of temporary staff and other circumstances may impact human IPLs. Care must be taken and a plan should be in place to maintain the IPLs in the needed or desired state – a component that is sometimes missing.

Failure to validate, test, fully document, and audit – A problem that occurs in LOPA is that data is selected for initiating events or IPLs and used in LOPA without regard to validation for the specific plant. It doesn't matter what value is selected for use in a LOPA if the plant data is clearly inconsistent. The electronic or "paper trail" for IPLs and initiating frequencies should include all relevant documentation so that the entire IPL can be audited.

5.3 Problems in Practice – UK HSE Buncefield LOPA Review – Human IPLs and IEs

Following the major incident on December 11, 2005 involving explosions and fires at the Buncefield Oil Storage Depot in the UK, LOPAs were conducted at fuel storage sites throughout the UK. The UK HSE report *A Review of Layers of Protection Analysis (LOPA) Analyses of Overfill of Fuel Storage Tanks* [5] presents detailed information from review of seven LOPA studies, while tabulated data are provided for a sample of 15 LOPAs. Problems in the conduct of LOPA were identified and communicated to industry. A summary of the types and range of problems encountered in the LOPA review is given [5, 6, 35].

A specific concern raised in the executive summary of the UK HSE report [4] - related to human IPLs and IEs - is stated as "Human factors appear to dominate a number of initiating event (IE) frequencies and conditional modifier (CM) error probabilities in all the LOPA studies assessed in this work." When taken together with the other main findings, including concerns regarding the quality of data and data sources, the wide variation in the degree of rigor applied to the LOPA studies, inconsistencies in how dependencies between initiating events and protection layers are handled, invalid logical arguments, and omission of supporting information, there were obvious concerns. Relating to human activities, the report also indicated: overly optimistic human error probabilities, a failure to show independence in terms of operator activities and responses to alarms, use of generic data without consideration of whether it was appropriate for the site, no justification for HEPs for operator responses to alarms, and confusion over whether the claimed probability of failure on demand for operator response protective layers also included reliability data for mechanical failures of equipment operated (i.e. final elements such as pumps and valves).

6. Summary

LOPA is clearly a tool of choice within the process industries to address risk-based issues and decisions in a simplified manner, while adding a greater degree of understanding and confidence in decisions made. It can be used to address a wide range of risk issues and decision making needs. LOPA is effectively used to bring objectivity and a more consistent approach to addressing layers of protection and assessment of risk beyond that afforded in qualitative PHA reviews. LOPA provides a timely and cost-effective means to analyze many high consequence and high risk scenarios to aid the decision making process.

A great deal of work has been done and progress made to advance methods for addressing human activities within LOPA – both as a potential initiating events and as part of human independent protection layers. Progress includes specific consideration of various modes of operation, development of critical task lists, additional guidance for the necessary components of human IPLs, incorporation of cross-checking, development of human "SIF" specifications, analysis and tools for calculation of human error probabilities, incorporation of site specific factors that affect human performance, testing and validation of human performance and IPLs, beginning efforts for collection and analysis of data for the process industries, and integration with other quantitative risk analysis techniques. These methods, techniques and data can be utilized to fully address the human role in independent protection layers and in initiating events within LOPA. However, as the complexity of LOPA increases, so does the possibility for errors in use. The experience of quantitative risk analysts and human factors specialists may need to be fully integrated into LOPA studies to avoid common problems and pitfalls, or at a minimum be utilized to provide expert guidance, supporting tools and aids, and for LOPA quality assurance. Provision of additional guidance, aids, and training to LOPA practitioners is also justified.

7. References

1. Center for Chemical Process Safety, Guidelines for Safe Automation of Chemical Processes, New York: American Institute of Chemical Engineers, 1993.
2. Arthur M. Dowell III, "Layer of Protection Analysis: A New PHA Tool After HAZOP, Before Fault Tree Analysis," 12th Center for Chemical Process Safety International Conference and Workshop on Risk Analysis in Process Safety, Atlanta, GA, 1997.
3. The Instrumentation, Systems, and Automation Society (ISA), ANSI/ISA-84.00.01-2004 (IEC 61511 Mod), Functional Safety: Safety Instrumented Systems for the Process Industry Sector (Parts 1-3), Research Triangle Park: ISA, 2004.
4. Philip M. Myers, "Layer of Protection Analysis – Developments, Applications and Limitations," 2010 Mary Kay O'Connor Process Safety Center International Symposium, College Station, TX, 2010.
5. Health and Safety Executive, Health and Safety Laboratory, A Review of Layers of Protection (LOPA) Analyses of Overfill of Fuel Storage Tanks, Research Report RR716, HSE Books, UK, 2009.
6. Health and Safety Executive, Safety and Environmental Standards for Fuel Storage Sites, Process Safety Leadership Group Final Report, HSE Books, UK, 2009.
7. Center for Chemical Process Safety, Layer of Protection Analysis: Simplified Process Risk Assessment, New York: American Institute of Chemical Engineers, 2001.
8. Glenn G. Young and Glenn S. Crowe, "Modifying LOPA for Improved Performance," American Society of Safety Engineers, Seattle, WA, 2006.
9. Center for Chemical Process Safety, Guidelines for Independent Protection Layers and Initiating Events, Hoboken: John Wiley & Sons, Inc., 2010.
10. W. Kent Goddard, "Use of Layers of Protection Analysis (LOPA) To Determine Protective System Requirements," 8th Process Plant Safety Symposium and the 2nd Global Congress on Process Safety, Orlando, FL, 2006.
11. Arthur M. Dowell III, "Is it Really an Independent Protection Layer?," 12th Process Plant Safety Symposium and 6th Global Congress on Process Safety, San Antonio, TX, 2010.
12. Center for Chemical Process Safety, Guidelines for Safe and Reliable Instrumented Protective Systems, Hoboken: John Wiley & Sons, Inc., 2007.
13. Raymond Freeman, "Use of Procedural Based Controls in Layer of Protection Analysis," 23rd Center for Chemical Process Safety International Conference and the 4th Global Congress on Process Safety, New Orleans, LA, 2008.
14. Scott Sandler and Angela Summers, "Operator Initiated Action as an Independent Protection Layer," 7th Process Plant Safety Symposium and the 1st Global Congress on Process Safety, Atlanta, GA, 2005.
15. William Bridges, "LOPA and Human Reliability – Human Errors and Human IPLs," 12th Process Plant Safety Symposium and 6th Global Congress on Process Safety, San Antonio, TX, 2010.
16. A.D. Swain and H.E. Guttman, Handbook of Human Reliability Analysis with Emphasis on Nuclear Power Plant Applications, Final Report, NUREG CR-1278, 1983.
17. Ed Marszal and Eric Scharpf, Safety Integrity Level Selection: Systematic Methods Including Layer of Protection Analysis, Research Triangle Park: The Instrumentation, Systems, and Automation Society, 2001.

18. A.J. Oswald et al., Generic Data Base for Data and Models Chapter of the National Reliability Evaluation Program (NREP) Guide, EGG-EA-5887, Informal Report, 1982.
19. D.I. Gertman et al., The Spar-H Human Reliability Analysis Method," NUREG CR-6883, 2005.
20. D.M. Hunns and B.K. Daniels, The Method of Paired Comparisons, 6th Symposium on Advances in Reliability Technology, Report NCSR R23 and R24, UK Atomic Energy Authority.
21. J.C. Williams, "A Data-Based Method for Assessing and Reducing Human Error to Improve Operational Performance," IEEE Conference on Human Factors in Power Plants, Monterey, CA, 1988.
22. J.C. Williams, "HEART – A Proposed Method for Achieving High Reliability in Process Operations by Means of Human Factors Engineering Technology," Symposium on the Achievement of Reliability in Operating Plant, Safety and Reliability Society, Southport, UK, 1985.
23. J.C. Williams, "A Human Factors Data-Base to Influence Safety and Reliability," Safety and Reliability Symposium '88, Elsevier Applied Science, pp 223-240, 1988.
24. J.C. Williams, "Human Reliability Data – The State of the Art and the Possibilities," Reliability '89, Vol. 1, pp.315/5/1 – 3B/5/16, 1989.
25. J.C. Williams, "Toward an Improved Evaluation Analysis Tool for Users of HEART," 7th Center for Chemical Process Safety International Conference on Hazard Identification and Risk Analysis, Human Factors, and Human Reliability in Process Safety, Orlando, FL, 1992.
26. Robert J. Stack and Paul Delanoy, "Evaluating Human Response to an Alarm for LOPA or Safety Studies," 25th Center for Chemical Process Safety International Conference and 6th Global Congress on Process Safety, San Antonio, TX, 2010.
27. William Bridges, "Human Factors Elements Missing from Process Safety Management (PSM)," 25th Center for Chemical Process Safety International Conference and 6th Global Congress on Process Safety, San Antonio, TX, 2010.
28. Engineering Equipment Materials Users' Association, Alarm Systems: A Guide to Design, Management and Procurements, EEMUA 191 (Second Edition), EEMUA, 2007.
29. Karen A. Study and John W. Champion, "LOPA Misapplied: Common Errors Can Lead to Incorrect Conclusions," 10th Process Plant Safety Symposium and 4th Global Congress on Process Safety, New Orleans, LA, 2008.
30. Arthur M. Dowell III, "Layer of Protection Analysis: Lessons Learned," Instrumentation, Systems, and Automation Society, ISA 2002.
31. William Bridges, "Key Issues with Implementing LOPA (Layer of Protection Analysis) – Perspective from One of the Originators of LOPA," 11th Process Plant Safety Symposium and 5th Global Congress on Process Safety, Tampa, FL, 2009.
32. J. Wayne Chastain, "Use and Misuse of Enabling Conditions and Conditional Modifiers in Layers of Protection Analysis (LOPA)," 12th Process Plant Safety Symposium and 6th Global Congress on Process Safety, San Antonio, TX, 2010.
33. International Society of Automation, ANSI/ISA-18.2-2009 Management of Alarm Systems for the Process Industries, Research Triangle Park: ISA, 2009.
34. Joan Lowy, Associated Press, "Drama in the Cockpit: Qantas Crew Faced 54 Alarms, November 18, 2010.

35. Richard Gowland, "The Buncefield (U.K.) Fire and Explosion: Improving Layer of Protection Analysis Practice to Determine the Required Degree of Protection to Meet Regulator Requirements," 43rd Loss Prevention Symposium and 5th Global Congress on Process Safety, Tampa, FL, 2009.

The Interrelationship of Risk Criteria and Conditional Modifiers

Angela E. Summers, Ph.D. PE
SIS-TECH Solutions, LP
12621 Featherwood Drive, Suite 120
Houston, TX 77034
281-922-8324 (phone), 281-922-4362 (fax)
asummers@sis-tech.com

William H. Hearn, PE
SIS-TECH Solutions, LP
12621 Featherwood Drive, Suite 120
Houston, TX 77034
281-922-8324 (phone), 281-922-4362 (fax)
whearn@sis-tech.com

Keywords: Risk criteria, risk analysis, hazards, hazardous events

Abstract

Risk analysis assesses the likelihood and consequence of events. The acceptability of the identified risk is determined by comparing it to a specified risk tolerance. The criteria applied depend on the analysis boundary, which may be loss of containment or extend to the harm posed by the loss of containment. Risk analyses generally begin with a determination of the likelihood that a hazardous event could result in loss of containment or some other undesirable consequence. These analyses require estimation of the likelihood that the initiating event will occur and the probability that the protection layers will not operate as required. Conditional modifiers are considered when the analysis is evaluating the likelihood that harm may be caused by the loss of containment.

Various methods for performing risk analyses are discussed in several CCPS publications including Chemical Process Quantitative Risk Analysis (1), Hazard Evaluation Procedures (4), and Layers of Protection Analysis (8). However, the link between the selected risk criteria as described in Guidelines for Developing Quantitative Safety Risk Criteria (3) and the factors considered in the analysis is not clearly described in these texts. Recognizing this opportunity, this paper begins with a brief introduction to risk analysis concepts to provide a foundation for a discussion of the typical analysis boundaries and associated risk criteria. Then, it discusses how the analysis boundary and risk criteria affect the consideration of protection layers, enabling conditions, and conditional modifiers.

1. How is risk measured?

Risk is the result of deviations from expected operation and is intimately related to the process design and site safety culture. Loss prevention seeks to identify these deviations and to reduce their frequency of occurrence or impact should they occur. The result of each deviation can be judged by analyzing the design and historical performance. Some process deviations are significant enough that a hazardous event occurs. An effective process safety management program prevents deviations from propagating into hazardous events. The rigor of the management system determines whether the hazardous event risk is reduced as low as reasonably practicable.

The level of a process risk is directly related to the frequency of process deviation and the consequence of exceeding the equipment safe operating limit. Process risk can be minimized using inherently safer principles in the design of the process to reduce the magnitude of the consequence and in the operational and management strategy to reduce the process deviation frequency (7). Varying levels of rigor are applied in estimating the risk from hazardous events (4). Generally, the specific method is chosen based on the following considerations:

- Regulatory requirements
- Company policy
- Lifecycle phase
- Information available
- Process complexity
- Previous experience with process
- Required degree of risk discrimination
- Consequence severity

Risk ranking within the Process Hazards Analysis (PHA) has traditionally been qualitative, supported by a risk matrix that relates frequency and consequence to priority (Figure 1). The risk rank is used to determine the priority and criticality of recommendations made to reduce risk and to address safeguard deficiencies.

The event likelihood (Figure 1) is estimated given the listed causes and identified safeguards that prevent the deviation from exceeding the safe operating limit or reduce the hazardous situation that leads to harm. The frequency estimate is complex, because it requires that team members estimate the likelihood of simultaneous failure of many systems that rely on manual and automated actions. Significant consequence events often have multiple safeguards with varying degrees of independence, diversity and capability. The difficulty for the team in performing the estimate is that hopefully the hazardous event has never happened and given the number of things that must go wrong seems unlikely to ever happen. Process upset, shutdown, near miss, and incident investigation reports can be used to better understand the likelihood of various aspects of the event, such as cause likelihood or safeguard failure (5).

Priority					
CONSEQUENCE SEVERITY					
5	LR	MR	MR	HR	HR
4	TR	LR	MR	MR	HR
3	TR	TR	LR	MR	MR
2	TR	TR	TR	LR	MR
1	TR	TR	TR	TR	LR
	R	U	O	L	V
	EVENT LIKELIHOOD				

Where:

Likelihood:
V=Very
L=Likely
O=Occasional
U=Unlikely
R=Rare

Consequence:
Increases from 1 (first aid) to 5 (community impact)

Priority
HR - High Risk - Take immediate action
MR - Medium Risk - Take action at next opportunity
LR - Low Risk - Verify procedures and safeguards
TR - Tolerable Risk - No action necessary

Figure 1: Example PHA Risk Matrix For Ranking Hazardous Event Risk

The team qualitatively ranks the consequence severity (Figure 1) by considering the hazardous situation posed by the event. The consequence severity ranking must not consider the action of safeguards; it should be ranked based on the harm that results when everything that could go wrong, has gone wrong. The inherent difficulty is that team experience and incident records include the operation of safeguards that can reduce the harm caused by hazardous situation, such as fire and gas systems or emergency response procedures. Since consequence severity is often used to screen events for further analysis, it is important that the documentation reflect the hazardous situation without consideration of any safeguards including proactive, reactive, and emergency response activities. Consequence modeling can be used to better understand the hazardous situation and impact zone (2,10).

The next level of risk analysis rigor estimates the event likelihood using a semi-quantitative technique (8) called Layers of Protection Analysis (LOPA). LOPA allows the risk to be estimated at various points along the incident sequence, so it can provide quantitative estimates of the process risk, hazardous event risk, and harmful event risk. LOPA is generally applied to hazardous events that have a consequence severity involving:

- Community injury or fatality
- Serious worker injury or fatality
- Significant environmental impact
- Significant business interruption or equipment damage

In LOPA, the team examines how causes lead to process deviations (or initiating events) to understand how they propagate into a hazardous or harmful event. The analysis may include enabling conditions that are required for the deviation to propagate, such as environmental conditions, co-incident equipment failures, and process operating conditions. Evaluating enabling conditions is critical to understanding multiple (generally double) jeopardy events, i.e., more than one thing must be wrong for the process deviation to occur. The risk rank is generally performed using a risk matrix or by calculation (Figure 2).

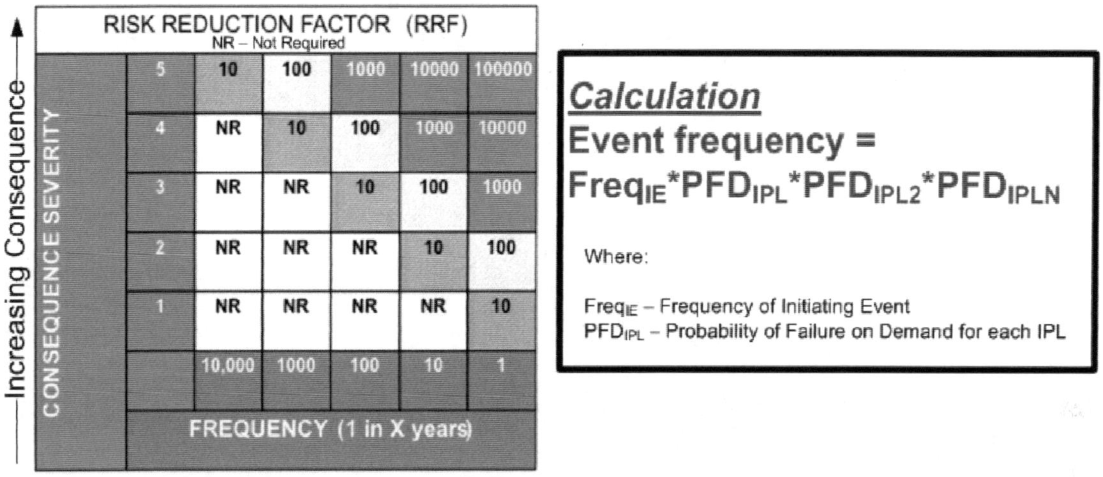

Figure 2: Example risk matrix and risk equation

Some events are too complex for the team to estimate the likelihood using qualitative or semi-quantitative (e.g., LOPA) methods. As examples, runaway reaction events may cause extremely high pressure initiated by multiple independent causes, or flare load mitigation systems may be restricted by system design that limits the number of vessels that can simultaneously relieve. Qualitative risk estimates and semi-quantitative evaluations need clear independence, a small number of causes, and low complexity protection layers in order for the simplified rules and estimation techniques to be effective. Quantitative risk analysis (QRA) techniques, such as fault tree analysis, should be used to determine the frequencies of these more complex events (3). Due to the analysis complexity, QRA requires personnel with special training and expertise, rather than the diverse team used in the case of PHA and LOPA risk ranking.

2. How is risk reduced?

The event risk is compared to the company risk tolerance to determine whether additional risk reduction is required. The selected risk tolerance level should fall within the range of internationally accepted risk measures as discussed below in "What are typical target risk criteria?" If the process risk does not satisfy the chosen risk criteria, independent protection layers (IPLs) are used to close the gap by decreasing the hazardous or harmful event frequency. IPLs are engineered and/or procedural safeguards that are designed and managed to meet seven (7) core attributes: independence, functionality, integrity, reliability, auditability, management of change, and access security (8, 6).

IPLs are intended to stop propagation of the hazardous event to the probable harm caused by the event. An onion-skin diagram (6) is often used to illustrate the typical order of IPL deployment (Figure 3). As the event propagates through the onion-skin of IPLs, the impact on the process operation becomes greater as does the uncertainty of the final outcome.

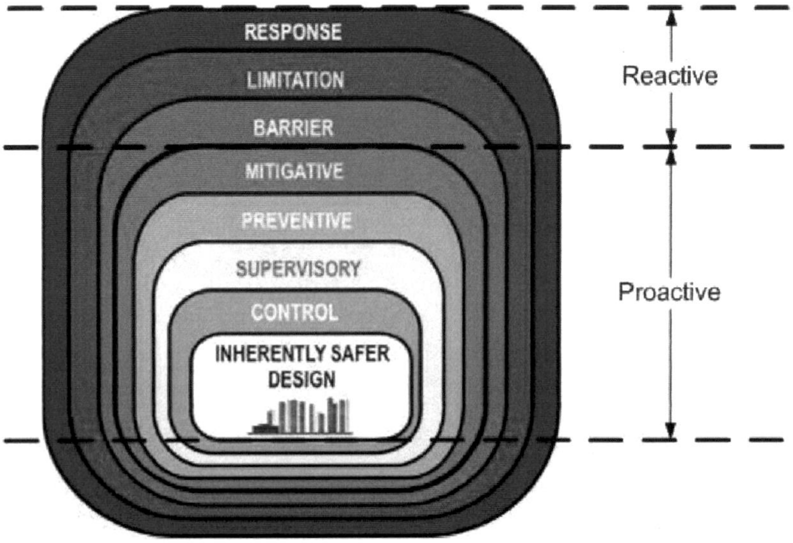

Figure 3: Protection layers represented as onion-skin

Some layers stop a process deviation from exceeding the equipment safe operating limit. The inherently safer design, control, supervisory, preventive and mitigation layers proactively avert loss of containment or equipment damage (Figure 4). A well-designed function acting to prevent the hazardous event can have a high certainty of effectiveness, since the function can be designed specifically for the purpose and the outcome can be predicted using engineering principles.

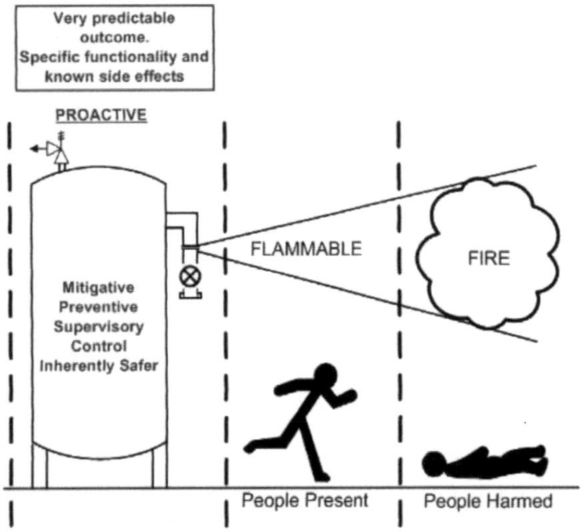

Figure 4. Relationship Between Proactive IPLs and Hazardous Event Risk

Other layers moderate the hazardous situation after a hazardous event (release) occurs (Figure 5). A hazardous situation may expose people, property or the environment, etc. to one or more hazards. Barriers and limitation layers are reactive layers and take action after loss of containment has occurred. Barriers contain the released materials (or energy) and must be designed specifically for the situation to be effective. For example, the design of an explosion barrier must consider the degree of overpressure created by the hazardous event. Limitation functions principally act to reduce the severity of the hazardous situation by monitoring for unacceptable atmospheres and taking action to isolate/de-inventory and/or to evacuate non-essential personnel. Functions acting to moderate the hazardous situation have more uncertainty in their outcome, because their effectiveness is impacted by the specific hazardous situation.

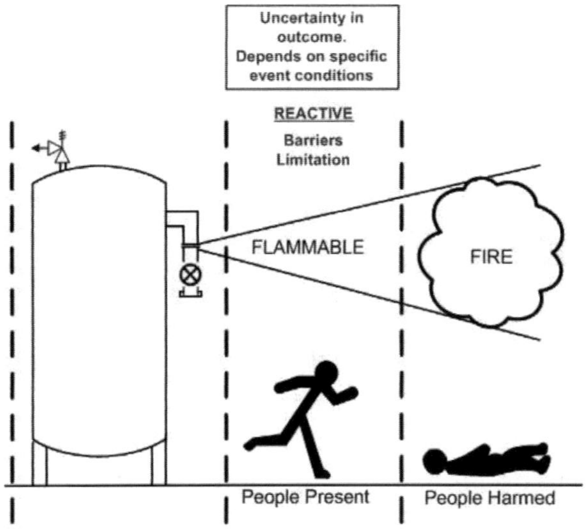

Figure 5. Relationship Between Reactive IPLs and Hazardous Situation Risk

Finally, a properly executed emergency response plan can reduce the harm caused by the hazardous situation by preventing escalation of the situation (Figure 6). For example, putting out a fire stops the exposure to surrounding equipment and structures, preventing further damage. Emergency response plan activities have the highest uncertainty, since they act when the hazardous situation has already started causing harm. Essentially, these activities prevent a bad situation from getting worse. Unfortunately, there are many cases where emergency responders have been injured during response. Effective planning, training, coordination, and communication are extremely important to succeed in ending the incident with minimal loss.

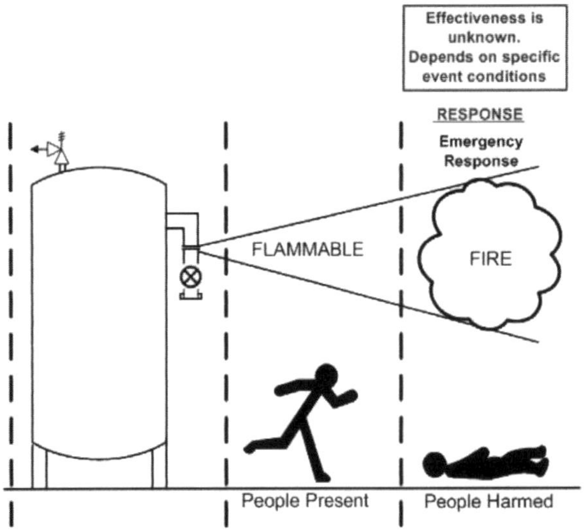

Figure 6. Relationship Between Response IPLs and Harmful Event Risk

3. What are typical target risk criteria?

Guidelines for Quantitative Safety Risk Criteria (CCPS SRC) discussed several criteria used to judge the acceptability of process risk (3). Precise definition of the scenario and proper use of the risk evaluation boundary is critical to the proper application of the risk criteria. The risk criteria depend on the analysis boundary (i.e., the scenario definition). As the boundary moves from measuring hazardous events to measuring harm, more complex analysis is required.

The owner/operator's risk tolerance establishes the minimum required risk reduction and enables prioritization of investments related to IPLs. Internal practices should explain how the criteria are used throughout the lifecycle to ensure consistent application. The risk should be reduced below the tolerable risk level, unless a deviation is justified and formally approved by management.

Many different criteria are applied in risk analysis throughout the world. Some are mandated by government regulation or industry practice. This section presents some order of magnitude risk criteria found in CCPS SRC Section 4.10. Similar data is presented in CCPS Layers of Protection Analysis (8) Appendix E and CCPS Safe and Reliability Instrumented Protective Systems (6) Table 3.1. This data pertains to fatality risk only, but similar criteria can be established for injury, environmental, and economic risks. *The data presented is intended to provide examples of the type of criteria used in various risk analysis and should not be considered a requirement from any publication.*

Typical criteria used in process risk (Table 1) vary by orders of magnitude depending on the analysis boundary. The maximum frequency of each hazardous event focuses on prevention of loss of containment events (i.e., the hazardous event risk). In contrast, maximum individual risk focuses on reduction of harm (i.e., the harmful event risk). The criterion for maximum individual risk assumes that no individual person is exposed to more than 100 hazardous events that could result in a fatality.

Table 1: Example Risk Criteria (3)

Criteria	**Worker**	**Public**
Maximum Individual Risk - All events	10^{-3} fatality/yr	10^{-4} fatality/yr
Maximum Individual Risk - Each event	10^{-5} fatality/yr	10^{-6} fatality/yr
Maximum Hazardous Event Frequency – Each event	10^{-4} event/yr	10^{-5} event/yr

3.1 Hazardous event criteria

Hazardous event criteria focus on preventing initiating events (or deviations) from propagating to loss of containment, thereby reducing the frequency of releases and the resulting hazardous situations (Figure 7). Effectiveness can be tracked using operational records and near miss/incident data (5). Hazardous event criteria may consider enabling conditions that are necessary for the propagation of the initiating event to the hazardous event. As discussed previously, proactive IPLs are used to stop the propagation of the initiating cause to the hazardous event. Typically IPLs are selected to reduce the frequency of the hazardous event to less than 10^{-4}/yr where a worker fatality is deemed possible and 10^{-5}/yr where a community fatality is possible.

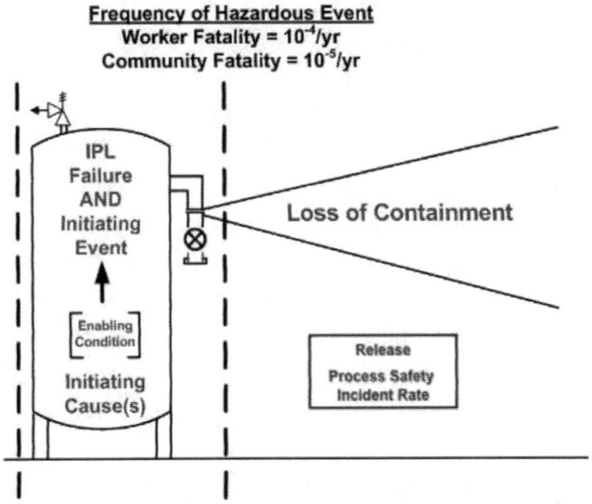

Figure 7. Single Hazardous Event Criteria and Analysis Boundary

3.2 Harmful event criteria

Harmful event criteria are generally based on maximum individual or societal risk and may be apportioned to different external events, such as electrocution, falling, or impact, and process events, such as loss of containment (3). These risk statements focus on fatality rates, which can be verified using injury statistics (5).

The analysis of harmful events extends the evaluation beyond loss of containment to the harm posed by the incident. Consequently, the analysis will often take into account reactive IPLs, in addition to proactive IPLs and enabling conditions that occur prior to or concurrent with the initiating cause. When cumulative risk is evaluated, the fractional time spent in each operating mode may also be considered. Since harmful event analysis is estimating the risk of direct harm, conditional modifiers are often evaluated, such as probability of occupancy, ignition, and fatality.

Misuse of conditional modifiers and enabling conditions can result in underestimation of the risk (9). Conversely, when conditional modifiers and enabling conditions are not included in the analysis of harmful events, the risk estimate will be conservatively overstated, as these modifiers are implicitly assigned a value of unity (≈ 1). The value assumed for any enabling condition or conditional modifier should be justified by analysis and the basis documented to support plant policies and procedures. For example, if occupancy is considered, unit access procedures, impact zone analysis and historical access records should substantiate the occupancy assumptions. Operating modes, conditional modifiers, and reactive and response layers are highly interrelated, so the consideration of these factors in the risk analysis should be performed by a skilled analyst to ensure that the factors are not taken into account multiple times.

Risk is a function of frequency and consequence. This paper is primarily focused on frequency estimation and the different techniques in evaluating its acceptability. However, consequence severity is often used to screen events for more rigorous frequency analysis. The PHA team is generally asked to estimate the consequence severity without the operation of safeguards. Since the team relies on operating experience and incident history to make their determination, they holistically include various factors that influence the severity, including operating modes, reactive and response layers, and conditional modifiers. The analyst should ensure that the estimated consequence severity does not take these factors into account before using them for frequency reduction. For example, if the consequence severity associated with a release of a toxic material considered the presence of gas detection systems that prevent entry into an area, the likelihood estimate should not also consider the system's presence.

The evaluation of harmful event frequency may consider reactive and response layers, i.e., actions to reduce harm, in addition to proactive layers. It is important to ensure the independence of these layers and the conditional modifiers, since they are often interrelated. For example, a fire and gas detection system may be used to initiate evacuation of personnel, thereby reducing occupancy. The risk evaluation should not use both a lower probability of occupancy term and the fire and gas system as an IPL, since the reduced occupancy is the outcome of the successful activation of the fire and gas system. Likewise, the use of classified equipment could be part of the basis for the likelihood of ignition, but then cannot also be considered as a separate protection layer. Due to the complexity of distinguishing the protection layers from the conditional modifiers, procedures must consider how reactive layers and conditional modifiers will be treated in the risk assessments.

Harmful event criteria are applied in 2 ways: single scenario risk or cumulative risk.

- Single scenario risk assesses individual harmful events, e.g., high pressure leads to a release of flammable material exposing an operator to a fire (Figure 8). Typical risk criteria for single harmful events are a maximum individual risk of 10^{-5}/yr and a maximum societal risk of 10^{-6}/yr.

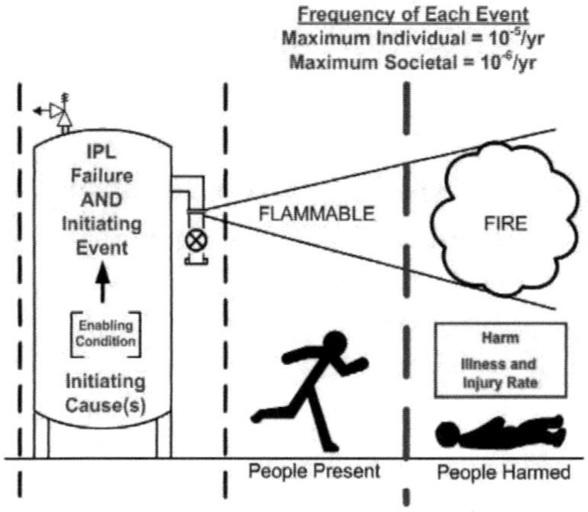

Figure 8. Single Harmful Event Criteria and Analysis Boundary

- Cumulative risk sums all hazardous events that an individual could be exposed to and determines an overall frequency (Figure 9). Since it is adding the risk associated with all of the events, cumulative risk estimates may consider the fraction (%) of time that the process is in certain operating modes. The operating mode may affect the occupancy, so occupancy should be assessed separately for each operating mode rather than using an average occupancy. All operating modes should be considered so that the overall risk of the process operation is determined.

  Cumulative risk analysis is complex, since it requires skills to assess the independence of the events and to account for common cause and systematic failures between events and protection layers properly. These analyses are best performed by specialists rather than by a team of diverse personnel. Typical risk criteria for cumulative risk are a maximum individual risk of 10^{-3}/yr and a maximum societal risk of 10^{-4}/yr.

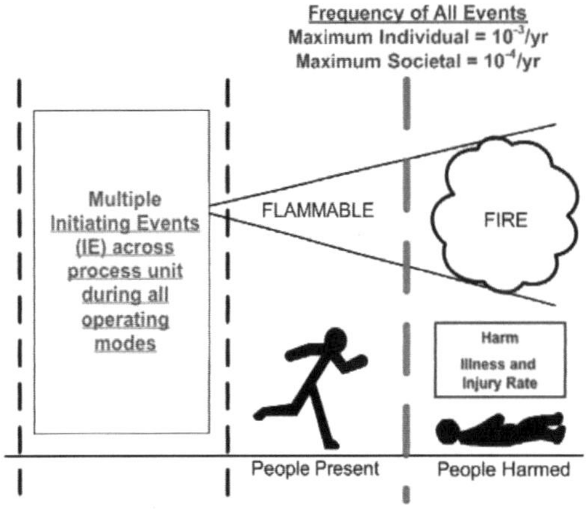

Figure 9. Cumulative Harmful Event Criteria and Analysis Boundary

3.3 Comparison of Risk Criteria

The CCPS Safety Risk Criteria Book provides two examples to illustrate the relationship of maximum hazardous event frequency and individual risk. These examples are provided below:

<u>Maximum Hazardous Event Frequency to Single Event Criteria for Worker</u>

Maximum Hazardous Event Frequency for Worker = 10^{-4} events/yr

- Probability of Ignition = 1.0
- Probability of Occupancy: Operator works 40 hrs/wk = (40/168) = 0.24
- Probability of Fatality: Has 50:50 chance of surviving incident = 0.5

Maximum Individual Risk for Worker: 10^{-4} events/yr * 0.24 * 0.5 = 10^{-5} fatality/yr

<u>Maximum Hazardous Event Frequency to Single Event Criteria for Public</u>

Maximum Hazardous Event Frequency for Public: 10^{-5} events/yr

- Probability of Ignition = 1.0
- Probability of Occupancy: Person at home = 1.0
- Probability of Fatality: Has 90% chance of surviving incident = 0.1

Maximum Individual Risk for Public: 10^{-5} events/yr * 1.0 * 0.1 = 10^{-6} fatality/yr

4. Summary

The risk analysis associated with many PHAs, such as HAZOP and What-if/Checklists, focuses on preventing loss of containment and uses criteria related to maximum hazardous event frequency. Alternatively for significant consequence events, some owner/operators and regulatory authorities require the assessment of direct harm (i.e., injury or fatality rate) and use criteria related to harmful event frequency, such as maximum individual risk or societal risk.

Depending on the analysis boundary, the factors considered during the risk assessment change. For hazardous event criteria, the assessment considers factors that affect the potential for loss of containment and release of hazardous chemicals. These factors may include proactive layers that stop the propagation of the hazardous event and enabling conditions that are necessary for the event to occur. Conditional (or frequency) modifiers are not applicable to hazardous event risk.

When harmful event criteria are used, additional factors that reduce the potential harm may be considered. These factors may include enabling events affecting the hazard presence, reactive layers including barriers and limitation systems that reduce the magnitude of the hazardous situation, and conditional modifiers affecting the likelihood of harm, such as the probability of occupancy, ignition and fatality.

5. References

[1] CCPS/AIChE, *Guidelines for Chemical Process Quantitative Risk Analysis*, Second Edition, New York (2000).

[2] CCPS/AIChE, *Guidelines for Consequence Analysis of Chemical Releases*, Second Edition, New York (1999).

[3] CCPS/AIChE, *Guidelines for Developing Quantitative Safety Risk Criteria*, New York (2009).

[4] CCPS/AIChE, *Guidelines for Hazard Evaluation Procedures,* Third Edition, Wiley-Interscience, New York (2008).

[5] CCPS/AIChE, *Guidelines for Process Safety Metrics*, Wiley-Interscience, New York (2010).

[6] CCPS/AIChE, *Guidelines for Safe and Reliable Instrumented Protective Systems*, Wiley-Interscience, New York (2007).

[7] CCPS/AIChE, *Inherently Safer Chemical Processes: A Lifecycle Approach*, Second Edition, Wiley-Interscience, New York (2009).

[8] CCPS/AIChE, *Layer of Protection Analysis: Simplified Process Risk Assessment*, Concept Series, New York (2001).

[9] Chastain, Wayne, "Use and Misuse of Enabling Conditions and Conditional Modifiers in Layers of Proteciton Analysis (LOPA), American Institute of Chemical Engineers, 2010 Spring Meeting, 6th Global Congress on Process Safety, San Antonio, Texas, March 22-24, 2010.

[10] Summers, Angela, Bill Vogtmann, and Steve Smolen, Consistent Consequence Severity Estimation, American Institute of Chemical Engineers, 2010 Spring Meeting, 6th Global Congress on Process Safety, San Antonio, Texas, March 22-24, 2010.

How Layer of Protection Analysis in Europe is affected by the guidance drawn up after the Buncefield Accident.

Richard Gowland

Technical Director, European Process Safety Centre

Rugby, England
Rtgowland@aol.com

Keywords: Buncefield, Vapour cloud, Overflow, Initiating Events, Enabling Condition, Independent Protection Layer, Conditional Modifier.

Abstract

The Buncefield fire was described as the biggest fire in Western Europe since the second world war. It has had a major effect on the regulator's standards for storage of flammable materials. These standards include significant changes in design requirements for overfill prevention, secondary and tertiary containment, operations, emergency planning, land use planning and risk assessment.

In the case of risk assessment the Health and Safety Executive's preferred methods include Layer of Protection Analysis. The initial reports in 2007 included examples on how this might be applied. In the event, these examples led several operators and supporting consultants to produce what the regulator considers inadequate LOPA studies. The Process Safety Leadership Group requested that a multi interest group be formed under the chairmanship of the European Process Safety Centre to produce suitable detailed guidance. The work of this group formed appendix 2 of the final report published December 11 2009.

The paper describes:

- the process by which the guidance was produced,
- the testing it received and
- the guidance itself

The published guidance goes significantly further than what might be considered 'standard' LOPA practice. It is quite clear that some aspects will conflict with the originators' original intentions. However, it is clear that the current 'state of play' in the U.K. is a development of LOPA which is still operable and valuable. The key advantages of LOPA as a method are retained while at the same time introducing some aspects which the regulator perceives as essential.

How Layer of Protection Analysis in Europe is affected by the guidance drawn up after the Buncefield Accident.

Richard Gowland: Technical Director, European Process Safety Centre, www.epsc.org

Prior to the Buncefield explosion, the most likely consequences from the overflow of an atmospheric storage tank could have been assumed to be a flash fire and/or pool fire potentially leading to a single fatality on site. Following the explosion at Buncefield, the most severe human safety consequence should now be assumed to be an explosion that may cause damage to occupied buildings accompanied by a flash fire and multiple pool fires with resultant more severe effects on people and the environment. In the case of the environment, there were effects on the environment caused by the overflow, fire and by the necessary emergency response measures.

When risk assessment has been carried out for storage and distribution facilities possible failures and their possible consequences are normally used as a starting point for addressing the likelihood of their occurring. Layer of Protection Analysis (LOPA) is one of a number of techniques which can be used for risk assessment.

Key words: Buncefield, Layer of Protection Analysis, Independent Protection Layer, Conditional Modifier,

Scenario Based risk assessment:

LOPA has traditionally been based on Scenario-based safety risk assessment, where the technique's calculation estimates the frequency with which a hazardous scenario consequence could occur. For example - a certain number of fatalities within the total exposed population. Using the scenario basis allows a detailed examination of a particular set of risks without necessarily considering all the risks at an establishment. The scenario risk calculation is not an individual risk calculation which might be required for a Control of Major Accident Hazards (COMAH) Safety Report.

Individual Risk assessment:

Individual Risk assessment, where the calculation is typically performed for a specified individual (often characterised by "the person most at risk" and referenced to a specific job role or a physical location).

Societal Risk assessment:

Where the scenario contributes significantly to the societal risk from the establishment an assessment should be made and included in the Safety Report for the establishment. LOPA is not optimal for calculating societal risk.

Estimating the consequences of a Buncefield-type explosion.

1.1 Effects on people and structures:

Given the limitations on current understanding, it is appropriate to apply the precautionary principle as outlined in 'Reducing Risks, Protecting People'.[1]

Currently there is no widely available methodology for estimating the size, shape and rate of development of the flammable cloud that could be formed from a storage tank overflow. Nor can the behaviour of the explosion in the near-field be reproduced by more commonly used models such as the multi-energy model.

In estimating the spread of the flammable cloud, the simplest assumption is to assume that it spreads in all directions equally. This assumption is conservative and is considered reasonable if there are no topographical factors influencing directionality. At wind speeds of less than 2m/s, it is assumed that the wind direction is too variable and hard to measure reliably to have a significant directional impact. However, the spread of the flammable cloud at Buncefield was influenced by local topography and the cloud did not spread equally in all directions even under very low wind-speed conditions. The following distances (Table 1) are considered to be a conservative approximation of the hazard zones for a Buncefield-type explosion and, in the absence of other information, are recommended as a method by which operators can determine relevant hazard zones. Table 1 also shows how these zones are interpreted

Table 1: Interpretation of Hazardous Zones for a Buncefield-type explosion

Zone Name	Zone Size (measured from the tank wall)	Comment
A	r < 250m	HSE research report RR718 [2] on the Buncefield Explosion Mechanism indicates that overpressures within the flammable cloud may have exceeded 2 bar (200 kPa) up to 250m from the tank that overflowed (see Figure 11 within RR718). Therefore within Zone A the probability of fatality should be taken as 1.0 due to overpressure and thermal effects unless the exposed person is within a protective building specifically designed to withstand this kind of event.
B	250m < r < 400m	Within Zone B there is a low likelihood of fatality as the overpressure is assumed to decay rapidly at the edge of he cloud. The expected overpressures within Zone B are 5-25 kPa (see RR718 for further information on overpressures). Within Zone B occupants of buildings that are not designed for potential overpressures are more vulnerable than those in the open air.
C	r > 400m	Within Zone C the probability of fatality of a typical population can be assumed to be zero. The probability of fatality for members of a sensitive population can be assumed to be low.

Note:

The distances are radii from the tank wall as this is the location of the overflow (see diagram below). Bund layouts can vary significantly, so measuring the distances from the bund wall would not provide a consistent approach.

The zones within Table 1 are provided as a conservative basis. The zones may be adjusted on a case by case basis, due to site specific factors

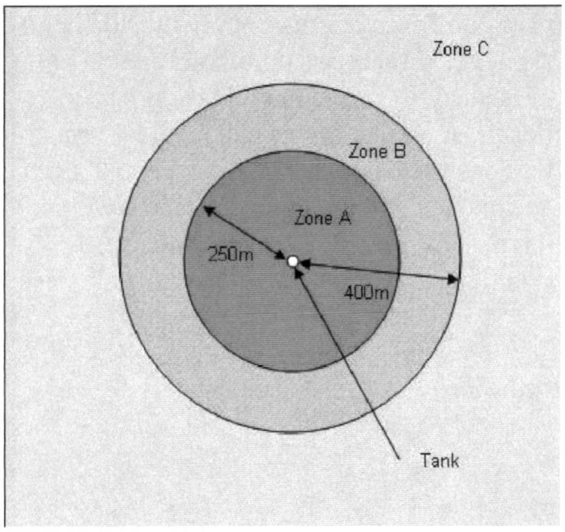

Other factors that should be considered when estimating the consequences to people are:

a. Hazards resulting from blast overpressure can be from direct and indirect sources. For example, indirect sources of fatal harm resulting from an explosion can be missiles, building collapse or severe structural damage. (as occurred at Buncefield).

b. People onsite and offsite within the relevant hazard zones should be considered as being at risk. People within onsite buildings such as control rooms or offices that fall within the hazards zones as described above should be considered at risk unless the buildings are sufficiently blast-rated.

c. The base case should be "normal night time occupancy". Populations just beyond the 250m, e.g. a school or old people's home, should also be considered

1.2 Environmental consequences

This paper also considers the environmental risks associated with a storage tank overflow.

2. RISK TOLERANCE CRITERIA

Risk tolerance criteria can be defined for human risk and for environmental risk on the basis of existing guidance.

a. Scenario-based safety risk assessment, where the realistic potential number of fatalities is compared to the risk matrix detailed in Table 2 or an equivalent

b. Scenario-based environmental risk assessment. This type of analysis may be used to determine whether good practice measures (such as those in the COMAH Containment Policy)[3] are reasonably practicable for existing establishments, or whether the existing operations are As Low As Reasonably Practical (ALARP).

2.1 Scenario-based safety risk assessment

LOPA, like most risk assessment tools, is suitable for this type of risk assessment, using the following approach:

1. Determine the realistic potential consequence due to the hazardous scenario (in this case the number of fatalities due to an explosion following an overflow from a specific tank);

2. Estimate the likelihood of the scenario; and
3. Locate the consequence and likelihood on the following (or similar) risk matrix (Table 2).

Likelihood of 'n' fatalities from a tank explosion per tank per year	**Risk Tolerability**		
10^{-4}/yr - 10^{-5}/yr	Tolerable if ALARP	Tolerable if ALARP	Tolerable if ALARP
10^{-5}/yr - 10^{-6}/yr	Broadly acceptable	Tolerable if ALARP	Tolerable if ALARP
10^{-6}/yr - 10^{-7}/yr	Broadly acceptable	Broadly acceptable	Tolerable if ALARP
10^{-7}/yr - 10^{-8}/yr	Broadly acceptable	Broadly acceptable	Broadly acceptable
Fatalities (n)	**1**	**2-10**	**11-50**

Table 2 Risk matrix for scenario-based safety assessments

Table 2 is based on the Health and Safety Executive Semi-Permanent Circular "Guidance on 'as low as reasonably practicable' (ALARP) decisions in control of major accident hazards (COMAH)", SPC/Permissioning/12 [4]

This assessment should be repeated for each "in-scope" tank in turn. Where there is a large number of "in scope" tanks (for example 10 or more) the aggregate risk from all of the tanks should be considered.

2.2 Scenario-based environmental risk assessment

There are currently no published environmental risk criteria for the UK with the same status as those for safety in "Reducing Risks, Protecting People". However, information on tolerability of environmental risk has also been produced for options assessment in section 3.7 of "IPPC H1: Integrated Pollution Prevention and Control (IPPC) and Environmental Assessment and Appraisal of BAT", Version 6 July 2003[5]. Operating companies seeking to demonstrate compliance with the COMAH Regulations should adopt an approach consistent with the information provided in Table 3 and with that in their COMAH Safety Reports and PPC Permit applications.

	Category	Acceptable if frequency less than	Acceptable if Reduced as Reasonably Practical and frequency between	Unacceptable if frequency above
6	Catastrophic	10^{-6} per year	10^{-4} to 10^{-6} per year	10^{-4} per year
5	Major	10^{-6} per year	10^{-4} to 10^{-6} per year	10^{-4} per year
4	Severe	10^{-6} per year	10^{-2} to 10^{-6} per year	10^{-2} per year
3	Significant	10^{-4} per year	10^{-1} to 10^{-4} per year	10^{-1} per year
2	Noticeable	10^{-2} per year	~ 10^{+1} to 10^{-2} per year	~10^{+1} per year
1	Minor	All shown as acceptable	-	-

Table 3

Categories from Table 3 have been aligned to COMAH terminology as follows,

"Acceptable if frequency less than" equates" to the "Broadly Acceptable region"
"Acceptable if Reduced as Reasonably Practical and frequency between" equates" to the "Tolerable if ALARP region"
"Unacceptable if frequency above" equates to the "Intolerable region"
Note: The issue of Environmental Risk tolerability data remains in discussion. For example, the tolerability criteria are now stated by the Environment Agency to be for the whole establishment.

Table 3a from information in Integrated Pollution Prevention and Control (IPPC) Document

Category	Definitions	
6	Catastrophic	• Major airborne release with serious offsite effects • Site shutdown • Serious contamination of groundwater or watercourse with extensive loss of aquatic life
5	Major	• Evacuation of local populace • Temporary disabling and hospitalisation • Serious toxic effect on beneficial or protected species • Widespread but not persistent damage to land • Significant fish kill over 5 mile range
4	Severe	• Hospital treatment required • Public warning and off-site emergency plan invoked • Hazardous substance releases into water course with ½ mile effect
3	Significant	• Severe and sustained nuisance, e.g. strong offensive odours or noise disturbance • Major breach of Permitted emissions limits with possibility of prosecution • Numerous public complaints
2	Noticeable	• Noticeable nuisance off-site e.g. discernible odours • Minor breach of Permitted emission limits, but no environmental harm • One or two complaints from the public
1	Minor	• Nuisance on site only (no off-site effects) • No outside complaint

Table 3a
Heading and introduction from Section 3.7 in "IPPC H1: Integrated Pollution Prevention and Control (IPPC) and Environmental Assessment and Appraisal of BAT", Version 6 July 2003. [5]
Table 3 Risk matrix for environmental risk

3. Initiating Events

3.1 Identifying Initiating Events

Potential causes of tank overflow should be considered in each of the following categories:

Equipment failures: for example failures of level measurement systems (gauges, radar devices, suspended weights), valves and other components; also failures of site services and infrastructure (e.g. loss of power, utilities, communications);

Human Failures: in particular errors in executing the steps of the filling operation in the proper sequence or omitting steps; and failures to observe or respond appropriately to conditions or other prompts. Possible errors may include but are not limited to:

- Incorrect calculations of the ullage in a tank (leading to an overestimate of how much material can be safely transferred into the tank);
- Incorrect verification of dips or incorrect calibration of level instrumentation;
- Incorrect routing of the transfer (sending material to the wrong tank);
- Incorrect calculation of filling time or incorrect setting of stop gauges;
- Failure to stop the transfer at the correct time (e.g. missing or ignoring the stop gauge and/or succeeding alarms).

External Events: for example:

- Changes in the filling rate due to changing operations on other tanks or due to changes within a wider pipeline network;
- Failure to terminate filling at the source (remote refinery, terminal or ship) on request from the receiving terminal.

3.2 Special Considerations

Failures of the Basic Process Control System (BPCS) as initiating events

The term "Basic Process Control Function" (BPCF) was developed to differentiate between the **functional** requirement for process control (what needs to be done) and the **delivery** of the functional requirement through the Basic Process Control System. The terminology is intentionally analogous to the terms "Safety Instrumented Function" and "Safety Instrumented System". Although the definitions in BS EN 61511[6] are not always explicit in this area, the authors considered that a BPCS can include either a fully automated control system or a system that relies on one or more people to carry out part of the BPCF. The BPCS is considered to comprise all the arrangements required to effect normal control of the working level in the storage tank, including operational controls, alarms through the BPCS and the associated operator response. For the purposes of the LOPA and the type of scenario under consideration, the BPCS would typically include several of the following:

- A level sensor on the tank;
- Field data marshalling and communications systems;
- Input/output cards;

- Central processing units (logic controller, processing cards, power supplies and visual displays);
- Operators and other workers required to perform the normal control function required to control the level of the storage tank;
- Communication arrangements between operators if more than one operator is required to carry out the control function;
- Final elements (which may be a remotely or locally operated valve or pump)

BS EN 61511 sets a limit on the dangerous failure rate of a BPCS (which does not conform to IEC 61511) of no lower than 1E-5/hr. This limit is set to distinguish systems designed and managed in accordance with BS EN 61511 from those that are not.

The performance claimed for the BPCS should be justified, if possible by reference to actual performance data.

- In the first, and most conservative, approach no credit is taken for any component of the BPCS as a protection layer if the initiating event also involves the BPCS. The failures involving the BPCS may be lumped into a single initiating event or may be separately identified. This approach is consistent with simple applications of LOPA. This approach fully meets the requirements of BS EN 61511.

- The second approach is to allow a single layer of protection to be implemented where there is sharing of components between the BPCS as an initiator and the BPCS as a layer of protection. Where credit for such a layer is claimed, the risk reduction factor is limited to 10 and the analysis must demonstrate that there is sufficient independence between the initiating event and the protection layer This approach meets the requirements of BS EN 61511 providing all the associated caveats are applied and adequate demonstrations are made.

3.3 Estimating Initiating Event Frequencies:

- Where the initiating event is caused by the failure of an item of equipment, the failure rate per year (in hours/year) may be derived from the failure-to-danger rate of the equipment item.

- Where the initiating event is caused by the failure of a person to carry out a task correctly and in a timely manner, the initiating event frequency is calculated as the product of the number of times the task is carried out in a year and the Human Error Probability (HEP) for the task. In this case, the time at risk is already included in the number of times the task is carried out in a year and no further factor should be applied.

- Where the initiating event is taken to be the failure of a BPCS control loop (when it does not conform to BS EN 61511), the minimum frequency which can be claimed is 1E-05 dangerous failures per hour.

- It is important that where probabilities or frequencies are assigned numerical values, these values are supported by evidence.

4. Enabling Events / Conditions

Enabling events and conditions are factors which are neither failures nor protection layers but which must be present or active for the initiating event to be able to lead to the consequence. Examples may include:

- The number of tank-filling operations carried out in a year (which may change as commercial circumstances change);
- The proportion of tank fills which are carried out where the batch size is capable of causing the tank to overflow (it may be that the tank under review normally runs at a very low level and would not normally be able to be filled to the point of overflow by typical batch sizes);
- The tank operating mode (if the tank is on a fill-and-draw operating mode so that the level is more or less static);

Crosschecking may qualify for inclusion in the risk assessment:

Tank-filling operations may include a number of cross-checking activities as part of the operation. These may include checks before the transfer starts (eg routing valve line-up, tank dips, available ullage) and periodic checks during the filling operation (eg to confirm the filling rate.

Cross-checks may be represented in the LOPA as factors which modify the initiating event frequency For example, if the initiating event 'Wrong tank line up' where a flow is misdirected is considered, the frequency of misrouting may be adjusted if a suitably rigorous cross-check is carried out. Cross-checks can provide an opportunity to detect and respond to an error condition and may be considered a s part of a prevention layer.
Credit for such activities can only be taken if general rules such as effectiveness, independence, 'testability and auditability' are complied with.

5.0 PROTECTION LAYERS

The LOPA methodology relies on the identification of **Protection Layers**, and in specifying protection layers it is important that all the rules for a protection layer are met. A valid protection layer needs to be:

- Effective in preventing the consequence;
  - A protection layer must be **effective**. This requires that the layer has a minimum functionality that includes at least:
    - A means of detection of the impending hazardous condition,
    - A means of determining what needs to be done, and finally
    - A means of taking effective and timely action which brings the hazardous condition under control. and
- Independent of any other protection layer or initiating event; and

- Capable of being functionally tested:
    - Whole loop testing (sensor, logic solver, final control element)
    - Complete operator action where required to respond to a detected process deviation. This includes the deviation sensing element, the receipt and understanding by the operator and his action to stop the scenario from developing further; and
- Auditable:
    - Sufficient record of reliability for failure frequencies and protection layer testing results.

The Basic Process Control System as a Protection Layer

It may be possible to take credit for the BPCS as a protection layer if sufficient independence can be demonstrated between the required functionality of the BPCS in the protection layer and any other protection layer or the initiating event. Clauses 9.4 and 9.5 of BS EN 61511-1 and BS EN 61511-2 present the requirements on the BPCS when used as a protection layer.

Response to alarms

The Buncefield Standards Task Group (BSTG) Final Report required operators of gasoline storage tanks to review and where necessary revise the settings of the level alarms on their gasoline storage tanks in accordance with a prescribed methodology. Where the alarm settings meet the requirements of the BSTG Final Report[7], it is considered legitimate to consider operator response to a high level alarm under suitable conditions. As with other protection layers, the alarm itself is only part of the protection layer. The full protection layer needs to include the alarm, the operator, the machine-operator interface, any communications systems (if communications between operators is required to deliver the required alarm function) and a final element such as a block valve.

Safety Instrumented Systems

In LOPA studies, the normal convention is that the need for Safety Instrumented Systems (SIS) is determined when all other protection layers have been considered. If an existing SIS complies with BS EN 61511 then a reliability performance consistent with the SIL-rating of the SIS and its design and operation can be claimed.

Other Safety Related protection Systems where relevant may be considered .It is possible to argue that some other protection layers can be considered so long as they meet the requirement for a protection layer set out in the Process Safety Leadership Group Final Report 'Fuel Storage Sites' [8]

6. Mitigation Layers:

Mitigation Layers are Protection Layers representing **intentional** design or operational measures which become effective once primary containment has been lost. Examples include overflow detection, bunding, fire protection and emergency response.

7. CONDITIONAL MODIFIERS (CM).

A Conditional Modifier is a factor which may influence the frequency of a scenario proceeding to the full consequence. For example, it could be assumed that a tank can overflow only when material is transferred into it, either deliberately or in error. If such transfers occur for only part of the time, a Conditional Modifier can be used to allow for this.

The same principles of independence, effectiveness and auditability which apply to protection layers also apply to conditional modifiers. It is important to make sure that the conditional modifier, as defined in the LOPA, is effective in its own right in preventing the consequence without relying on the performance of another conditional modifier or protection layer.

7.1 CM 1 – Probability of calm and stable weather

The basis of this report is that the development of a large vapour cloud with the kind of compositional homogeneity that is believed to have existed at Buncefield required low wind speed and stable atmospheric conditions.

7.2 CM 2 – Probability of ignition of a large flammable cloud

This conditional modifier represents the probability that the ignition of the vapour cloud from a storage tank overflow is delayed until it is sufficiently large to cause a widespread impact. Alternative outcomes are an earlier ignition that causes a localised flash fire, or safe dispersal of the cloud without ignition.

7.3 CM 3 – Probability of explosion after ignition

Factors such as ambient temperature; cloud size, shape, and homogeneity; topography; congestion (including that from vegetation); droplet size; fuel properties; and weather conditions may have a significant effect on the probability of an explosion compared to a fire.

This conditional modifier is intended to represent such factors. Research currently being carried out by the Health and Safety Laboratory, when complete, will add more precision to determining this Conditional Modifier.

7.4 CM 4 – Probability that a person is present within the hazard zone

This conditional modifier can be used to represent the probability that the normal pattern of work or living of a person means that they are only present in the hazardous area at the time of a tank overflow for part of the time. Obviously for a large event like an explosion, this is likely to tend towards 100%.

7.5 CM 5 – Probability of fatality

This conditional modifier may only be used if a single value can be specified for the hazardous scenario.

7.6 CM6 – Probability of the environmental consequence,

This conditional modifier is included to account for any factors additional to those considered elsewhere in the LOPA (for example seasonal factors, if not implicitly included in other factors within the LOPA) that may influence whether the hazardous scenario can cause the defined environmental consequence.

8. COMPLETING THE STUDY OF THE SCENARIO SET (CASE)

The process is repeated for the other studies needed where a given hazardous event can be caused by two or more initiating events. It must be remembered that the resulting predicted frequency of the unmitigated hazardous event is aggregated over all relevant initiating events.

It is important that a **sensitivity analysis** should be carried out to explore the sensitivity of the predicted risk levels to the assumptions made.

9. Concluding remarks:

For the full story please go to appendix 2 in : http://www.hse.gov.uk/comah/buncefield/fuel-storage-sites.pdf

References:

1. 'Reducing Risks, Protecting People'· (U.K. Health and Safety Executive)
http://www.hse.gov.uk/risk/theory/r2p2.pdf

2. HSE Research Report RR718. The Buncefield Explosion Mechanism Phase 1 and 2
http://news.hse.gov.uk/2009/06/25/rr718-buncefield-explosion-mechanism-phase-1-volumes-1-and-2/

3. COMAH Containment Policy U.K. Environment Agency. http://www.environment-agency.gov.uk/static/documents/Business/containmentpolicy_1961223.pdf

4. Health and Safety Executive Semi-Permanent Circular "Guidance on 'as low as reasonably practicable' (ALARP) decisions in control of major accident hazards (COMAH)", SPC/Permissioning
www.hse.gov.uk/comah/circular/perm12.htm

5. Integrated Pollution Prevention and Control (IPPC) and Environmental Assessment and Appraisal of BAT", Version 6 July 2003 (U.K. Environment Agency) www.ni-environment.gov.uk/ippc_h1.pdf

6. BS EN 61511 BS EN 61511-2:2004 (British Standards Institute)
Functional safety. Safety instrumented systems for the process industry sector. Guidelines for the application of IEC 61511-1 shop.bsigroup.com/en/ProductDetail/?pid..

7. BSTG Final Report (Health and Safety Executive)
www.hse.gov.uk/comah/buncefield/bstgfinalreport.pdf

8. Process Safety Leadership Group Final Report 'Fuel Storage Sites'
http://www.hse.gov.uk/comah/buncefield/fuel-storage-sites.pdf

Using LOPA for SIL Assignment: A Tale of Two Plants

Michael S. Schmidt
Bluefield Process Safety, LLC
Mike.Schmidt.BluefieldSafety@gmail.com

Dan Kilpatrick
CF Industries, Inc.
DKilpatrick@CFIndustries.com

Keywords: ammonium nitrate, AN, equipment failure, human error, LOPA, safeguards, SIL assignment, UAN

Abstract

Two plants operated by the same company have "identical" ammonium nitrate (AN) solution pump installations. Yet the teams performing the Layer of Protection Analysis (LOPA) concluded that the safety instrumented functions (SIFs) to protect those pumps needed different safety integrity levels (SILs). Despite the similarity of the installations, the teams discovered legitimate differences that warranted different conclusions about what SIL assignment should be. For those with a specific interest in AN, this paper reviews the consequences of AN pump hazards, their initiating causes, and the types of independent layers of protection (IPL) that can be brought to bear. Of more general interest will be the discussion of the impact of risk tolerance criteria, frequency modifiers, and IPLs on final SIL assignment, and why dictating a fixed SIL assignment to certain types of hazards or installations is inappropriate, especially in the development of industry standards or government regulations.

1. Introduction

Before being acquired by CF Industries, Terra Industries operated several nitrogen products facilities, including Port Neal, Iowa, and Yazoo City, Mississippi. Each of these facilities have units that convert natural gas and air to ammonia and from that ammonia produced upgrade products: nitric acid, urea, ammonium nitrate (AN), and urea ammonium nitrate solution (UAN). Units in each plant oxidize ammonia to nitric acid, and additional units neutralize nitric acid with ammonia to produce AN. AN is used as a fertilizer and as a component in industrial blasting agents. A standard operation at all of these facilities is the transfer of liquid AN by pumping.

AN, an oxidizing agent, can be combined with fuel oil or other organics to form the blasting agent ANFO. AN, which in its pure form is normally very stable, can also be made to deflagrate or explode under certain severe conditions. There are seven critical process parameters that have an effect on the stability of AN: time, temperature, contamination, confinement, concentration, pH, and density. These conditions have the potential to exist when a liquid AN pump is running but not pumping AN, either because the pump is blocked in, is deadheaded, has low suction pressure, has lost feed, or because of other reasons.

Production, engineering, and safety personnel at Terra Industries are acutely aware of this and other hazards associated with AN. As is often the case in loss prevention programs, risk reduction measures were initially applied to those hazards where the consequence severity of an event was perceived to be gravest. Various incidents, including the 1994 disaster at the Port Neal facility, influenced the perception of consequence severity. Typically, the likelihood component of risk was not given as much consideration, if any. Analysis looked at what might happen, rather than how likely it was to happen. Since most risk reduction measures serve to reduce likelihood, not to reduce consequence, every new analysis found the same consequences and the same hazards received attention. Lesser hazards were not identified as being as important, could never be identified as being as important, and finite resources were consistently directed to the hazards with the worse consequences.

Terra Industries' facilities used safety instrumented systems as one of many measures to reduce risk. With the introduction and general adoption of the Safety Instrumented System (SIS) standards for the process industries (IEC 61511 and ISA S84), the personnel at the individual Terra facilities concluded that Safety Instrumented Functions (SIFs) installed in their plants should be assigned SILs and installed in an SIS. Given the similarities between the plants and between specific hazards, it was tempting to simply decide to prescribe Safety Instrumented Functions (SIFs) with predetermined Safety Integrity Levels (SILs). The LOPA methodology was originally perceived as a tool to decide what the predetermined SILs would be. The LOPA teams quickly recognized that this was an inappropriate course of action.

2. Ammonium Nitrate

2.1. Manufacturing Process

A nitrogen products facility consists of two main process areas: ammonia and upgrade products. The ammonia process converts natural gas and air into ammonia and by-product carbon dioxide. In the upgrade processes, ammonia and carbon dioxide are converted to urea, ammonia is used as a feedstock to produce nitric acid, ammonia is reacted with nitric acid to form AN, and urea and AN are combined to make UAN.

AN production processes vary from plant to plant, depending on the process technology used and the desired final product mix. However, all AN plants can be divided conveniently into two sections: a wet section and a dry section. The wet section includes neutralization, evaporators, and where applicable, a dissolved AN liquor system. Plants designed to produce only UAN, such as the Port Neal facility, contain only a wet section. The dry section includes evaporation, prilling, cooling, coating, and screening operations.

Wet Section processing begins with neutralization of aqueous nitric acid with ammonia vapor to form an AN solution. Typical of inorganic acid-base neutralizations, the reaction proceeds rapidly and liberates a large amount of heat, creating steam from the water present.

The ammonia for the neutralizers comes from several sources including

- Ammonia vapor captured by refrigeration units
- Ammonia vapor from anhydrous ammonia storage
- Ammonia liquid from anhydrous ammonia storage
- Off-gas from urea production, which will also contain carbon dioxide and water vapor

In addition to AN produced by neutralization, an n AN solution known as weak AN liquor is produced from several sources, primarily

- Floor wash from spilled prills
- Dissolved fines and oversized particles from the size screening process
- Overflow and blowdown from various scrubbers located in the processing area

The weak AN liquor produced from these sources normally ranges from 20% to 60% AN. The weak liquor system consists of two separate systems, a clean system and a dirty system. The purpose for having two separate systems is to prevent organic contaminants from reaching the solution fed to the evaporators, which operate at elevated temperature. A dirty weak AN liquor system recovers any AN prills and particulates containing organic coating agent and also captures all floor sump collections. Weak AN liquor collected in a dirty system, once filtered, is consumed in the production of liquid urea-ammonium nitrate fertilizer. .

Dry Section processing produces solid AN in the form of prills. AN solution from neutralization feeds to evaporators, where it concentrates to 97.5 to 99.9% melt. The essentially anhydrous AN melt is pumped to prill tower spray headers, where it distributes uniformly across the tower. This spraying action causes the melt to form small spherical droplets, or prills, which are cooled and crystallized as they fall through the tower by air flowing counter-currently upward..

2.2. History of Ammonium Nitrate Disasters

Novices in process safety quickly learn about a half dozen or so significant chemical disasters, among them: Flixborough, Bhopal, Seveso, Pasadena, Piper Alpha, and Texas City. (1) Long before "Texas City" evoked thoughts of the BP refinery explosion in 2005, it referred to the explosion on one day in 1947 of the freighters *Grandcamp* and *High Flyer*, both of which were loaded with AN. The explosions and resulting fires led to the deaths of at least 576 people. (2) Additional disasters involving AN have made AN manufacturers acutely aware of the hazard, and there is a keen interest in applying process safety tools to manage the risk of AN.

Popular literature is full of stories about AN disasters. Perhaps the best example is the summary of disasters in the Wikipedia article, "Ammonium nitrate disasters". (3) It lists 23 disasters, beginning with the 1918 explosion at the Morgan Ammunition Depot in Sayreville, New Jersey and concluding with the 2009 fire at the El Dorado Chemical Company fire in Bryan, Texas.

The Morgan Depot explosions and fire at the T.A. Gillespie Company shell loading plant occurred on October 4, 1918, a little over a month before the armistice that brought World War I to an end was signed. This disaster is perhaps more appropriately attributed to the handling of explosives during wartime than an AN production disaster.

The next two disasters both occurred in Germany in 1921. In Kriewald, two wagon loads of AN had aggregated into solid masses. On July 26, workers used small explosive charges to dislodge the AN and the whole mass exploded, killing 19 people. The practice of using blasting agents to dislodge aggregated masses was not uncommon. It was this same procedure that led to a massive explosion at the BASF plant in Oppau two months later, on September 21, only eight years after the first ammonia plant was built there. The explosion at Oppau resulted in over 500 fatalities.

2.2.1. Transportation and Warehousing

The majority of the catastrophes on the list are transportation-related. Shipment by truck, train, and freighter have all led to AN disasters. Typically, there was an accident that led to a fire. The fire spread and suppression efforts were unsuccessful. Eventually, as firefighters worked to put out the fire and onlookers watched, the AN exploded.

The second largest category of AN disasters on the list involve warehousing. Again, the incidents began with an accident that led to a fire. Many of the warehouse fires, however, did not result in an explosion. This may be because stationary facilities are better prepared for the possibility of an accident, or it may be because transportation accidents are more likely to involve the uncontrolled release of motor fuel and other combustible substances.

2.2.2. Production and Processing

After considering transportation and warehousing catastrophes, there are a little more than a handful of AN disasters on the Wikipedia list that are process related: Oppau, Nixon, Tessenderlo, Papua, Port Neal, and Toulouse.

The Oppau, Germany disaster was triggered by the detonation of an explosive charge in Silo 110 of the BASF plant there. A nominally 50:50 mixture of ammonium sulfate and AN, called mischsaltz, was stored there, although the mixture was not uniform. The mischsaltz had agglomerated into a solid mass, and BASF technicians were using explosive charges to dislodge the agglomeration. The explosion occurred at 7:32 am on Wednesday, September 21, 1921 and resulted in 561 deaths. Damage occurred as far away as Ludwigschafen and Mannheim. Although tests had demonstrated the procedure of using blasting agents to dislodge agglomerated AN mixtures could be used successfully, subsequent investigations showed that mixtures at lower humidity, lower bulk density, and higher AN concentration are more likely to explode. (4)

The Nixon, New Jersey disaster began with an explosion in a building the Ammonite Company leased from the Nixon Nitration Works late on Saturday morning, at 11:30 am on March 1, 1924. Ammonite operated a process to salvage the contents of artillery shells and recycle the AN as fertilizer. The process involved recovering the AN by crystallization. At the time of the explosion, there were fifteen rail cars on site, each containing 90,000 gallons of AN solution, as well as one million gallons of AN solution in storage. The explosion killed 14 employees and 4 others, and set the Nixon Nitration Works, which manufactured nitrocellulose film, ablaze. Speculation afterwards was that trace amounts of trinitrotoluene (TNT) from the artillery shells sensitized the AN, although officials from the Ammonite Company disputed this, arguing that TNT was present at less than 0.2%, so could not have had an effect. (5)

The Tessenderlo, Belgium disaster at the plant operated there by Produits Chimiques de Tessenderlo (now the Tessenderlo Group) occurred at 11:27 am on Wednesday morning, April 29, 1942. This was during the occupation of Belgium by the Nazis, who were very interested in the AN output from the plant. Details are sketchy—investigation of a chemical plant explosion in Europe during World War II would not have been a top priority—but many writers repeat the account that the explosion was the result of another attempt to disaggregate a solid mass of AN with explosive charges. (3) (6) The explosion killed 189 people. (7)

The Papua, New Guinea disaster at the Porgera Gold Mine killed 11 workers when the sensitized AN emulsion plant where they were working exploded. Sensitized AN emulsion is a form of blasting agent used in mining, and so specifically formulated to be explosive. The first explosion occurred at 9:45 am on Tuesday, August 2, 1994. The facility was evacuated so that when a second, larger explosion occurred at 11:02, there were no additional fatalities. (3)

The Port Neal, Iowa disaster occurred at 6:13 am, Tuesday morning, December 13, 1994, the result of an explosion in an AN neutralizer. The U.S.EPA concluded that the explosion was the result of the following process conditions (8):

- Strongly acidic conditions in the neutralizer and rundown tank
- Prolonged application of 200 psig steam to the neutralizer nitric acid spargers
- Creation of bubbles and low density zones in the neutralizer
- Lack of flow in the neutralizer and rundown tank
- Presence of chlorides in the neutralizer and rundown tank

There were four fatalities, all plant employees. Some believe the number of deaths would have been much higher had the explosion occurred a couple of hours later, after the start of the day shift. (3)

The Toulouse, France disaster resulted from an explosion in a warehouse at the Azote de France (AZF) fertilizer plant in Toulouse. Although a terrorist attack was originally rumored to be the cause, the official conclusion following the judicial investigation was that off-spec AN was contaminated with sodium dichloroisocyanurate (SDIC), and that the chlorinated compound initiated decomposition and the subsequent explosion at 10:15 am on Friday, September 21, 2001. The explosion resulted in 31 fatalities, including 21 employees at the plant. (6)

There is little to unify these six AN processing disasters. One involved working with AN that was formulated as a blasting agent (Papua). Two involved subjecting solid AN to shock (Oppau and Tessenderlo). Three involved the presence of very different sensitizing contaminants (Nixon, Port Neal, and Toulouse). No clear pattern emerges. One thing the events do have in common is that the explosions occurred in the morning, between 6 am and noon. Coincidence? Probably. Time of day is certainly not a basis for a safety program.

2.3. Hazards of Ammonium Nitrate

AN is normally stable and can be used quite safely. A century of disasters, however, have made it clear that there are certain hazards associated with AN. While explosion is the hazard with which the public is most familiar, there are other hazards as well. AN presents all three of the major types of process hazards: fire, toxic exposure, and explosion.

2.3.1. Fire

The fire triangle consists of three components necessary for combustion: fuel, oxidizer, and ignition source. AN itself does not burn. However, it is a strong oxidizer and will promote the combustion of other materials, even materials that might otherwise not be expected to burn. Because it is an oxidizer, it will promote combustion, even in conditions where air is excluded or inerting atmospheres have been introduced. (9)

2.3.2. Toxic Exposure

AN melts at 170 C (337 F). Once molten, AN undergoes reversible, endothermic dissociation to ammonia and nitric acid per the following reaction:

$$\mathbf{NH_4NO_3}_{\,(liquid)} \leftrightarrow \mathbf{NH_3}_{\,(gas)} + \mathbf{HNO_3}_{\,(gas)} \quad \text{[Eq. 1]}$$

Several additional irreversible exothermic decomposition reactions can also occur simultaneously. Below 300 °C (572 °F), the predominant exothermic decomposition reaction is:

$$\mathbf{NH_4NO_3}_{\,(liquid)} \rightarrow \mathbf{N_2O}_{\,(gas)} + \mathbf{2H_2O}_{\,(gas)} \quad \text{[Eq. 2]}$$

This reaction is commonly used in the commercial preparation of nitrous oxide. Ammonia, nitric acid vapors, and nitrogen oxides are all toxic. The smoke from decomposing AN is typically tinted yellow or brown by the presence of nitrogen oxides, and poses an unusually toxic inhalation hazard. (10)

2.3.3. Explosion

Above 300 °C (572 °F) several highly energetic irreversible decomposition reactions can become significant:

$$\mathbf{2NH_4NO_3}_{\,(liquid)} \rightarrow \mathbf{N_2}_{\,(gas)} + \mathbf{2NO}_{\,(gas)} + \mathbf{4H_2O}_{\,(gas)} \quad \text{[Eq. 3]}$$

$$\mathbf{2NH_4NO_3}_{\,(liquid)} \rightarrow \mathbf{2N_2}_{\,(gas)} + \mathbf{O_2}_{\,(gas)} + \mathbf{4H_2O}_{\,(gas)} \quad \text{[Eq. 4]}$$

When AN undergoes moderate overheating without confinement or the presence of contaminants, the endothermic dissociation reaction can proceed at a rate that will enable it to absorb the heat liberated by the exothermic decomposition reactions. However, when placed under a condition of confinement, accumulation of the decomposition gases results in increasing pressure, which suppresses the endothermic reversible dissociation [Eq. 1]. Heat liberated by the exothermic decomposition reactions becomes greater than the heat absorbed by the dissociation reaction. This can cause the temperature of the decomposing mixture to increase, further accelerating the decomposition process and potentially leading to explosive behavior (11).

Dangerous contaminants of ammonium nitrate either catalyze the decomposition reactions described above or act as fuels. Catalyzing contaminants include halide salts, particularly chlorides (12), and transition metals, especially chromium (13). Certain combinations of contaminants have been demonstrated to have a synergistic effect where the net destabilizing effect of the combination is greater than the individual contributions. (14) Free acids, from either external sources or generated by the dissociation reaction, also catalyze the decomposition of AN. (15) Therefore, maintaining the pH of AN solution above the neutral point is of critical importance, particularly when handling contaminated AN. Organic contaminants act as a fuel source to the oxidizing properties of AN, and greatly energize the decomposition process. (16) The overall effect of contaminants is to sensitize AN by lowering the 'onset temperature', or the temperature at which self-sustained decomposition (SSD) can occur.

While the fire triangle is quite familiar, the explosion pentagon is a less well known concept. In addition to the three components of the fire triangle—fuel, oxidizer, and ignition source—the explosion pentagon also includes dispersion/suspension and confinement. (17) By itself, in the open, AN will not explode. Contaminated with a fuel, confined within machinery, and exposed to an ignition source, however, and AN becomes a powerful and dangerous explosion hazard.

AN poses a danger of explosion when it is heated in a confined space. This includes drains, piping, vessels, and equipment. (9) Neither friction nor impact during routine handling is believed to cause explosion, but severe shock is reported to. (9) However, shock alone does not routinely trigger an explosion. Greiner reports that in his workshops on AN safety, he routinely impacts high density AN fertilizer with a heavy stainless steel device to no effect. (18) Likewise, the BASF plant Oppau, Germany used blasting agents to dislodge agglomerated AN/ammonium sulphate mixtures over 20,000 times with no ill effects before such a procedure resulted in the destruction of the plant and 561 fatalities. (4)

The AN manufacturing industry generally recognizes seven critical process parameters that have an effect on the stability of AN: time, temperature, contamination, confinement, concentration, pH, and density. (19) The potential of decomposition or explosion increases when any of these seven critical process parameters are exceeded beyond safe boundaries. In summary, organic compounds, such as oils or waxes, low pH, and certain inorganic contaminants are of particular concern. Organics are combustible and can make an AN explosion more energetic. Low pH and certain inorganic compounds such as chlorides, chromium, copper, cobalt, and nickel are reported to sensitize AN to decomposition. Bubbles, as may be introduced with compressed air or steam, or by cavitation, can have the effect of lowering the density of AN solutions and are considered an explosion sensitizer. (20)

2.4. Ammonium Nitrate Pumps

Flow of AN solution or molten AN through a centrifugal pump is a common and routine operation in a nitrogen products facility. In the wet section, AN solution pumps are required for transferring from the neutralizers, for feeding to the evaporators, and throughout the weak liquor system. There are also AN melt pumps in the wet section, which serve to transfer AN melt from the evaporators. In the dry section, AN melt pumps are required to feed the prill towers.

2.4.1. AN Pumps and the Explosion Pentagon

Consider AN pumps in terms of the explosion pentagon: oxidizer, fuel, ignition source, dispersion/suspension, and confinement. The AN represents the oxidizer, and if contaminated, the fuel. At sufficiently high temperatures, the exothermic decomposition of AN is more than adequate as a surrogate for the combustion reaction typically characteristic of explosions. When blocked in, an AN pump and its associated piping is clearly the kind of confinement that causes concern. It is the two additional points of the explosion pentagon – ignition source, dispersion/suspension – where pumps can play a particular role.

The concern with AN solution pumps and AN melt pumps is that they have the potential to be blocked in while they contain liquid AN, yet still be running. The action of the pump transfers mechanical energy to the fluid where some is lost as heat. Under flowing conditions, this is inconsequential. Under non-flowing conditions, this heat accumulates. AN decomposes to gases, and in a confined space, such as in a blocked in pump and its associated piping, pressure can increase. The action of the impeller assures that the AN is well dispersed within the pump casing. The spinning impeller itself can be the source of very high shear and cavitation, causing vapor bubbles to form and rapidly collapse, contributing the sort of shock that can trigger an explosion in highly sensitized AN.

2.4.2. Initiating Causes of AN Pump Explosions

The initiating cause of any undesired consequence, by its very nature, is a failure or abnormality. When things are operating normally (which presumably is the same as correctly), undesired consequences do not occur. Otherwise, undesired consequences become normal. There are many different failures that can lead to an AN pump explosion, but they all have one characteristic in common: flow of the AN melt or AN solution has stopped when it should not have.

Equipment failures are the type of failure most often considered first. While AN solutions remain liquid at much lower temperatures, AN melt freezes below 170 C, plugging lines and stopping flow. Above 210 C, uncontaminated AN can begin SSD. In a contained system like a pipeline, SSD can lead to an explosion. Either extreme—freezing or decomposition—may result from the control failure of thermal protection like heat tracing.

Other equipment failures can also lead to stopped flow, the most obvious being a pump failure. In the instance of AN melt, a pump trip will stop flow. That in itself will not lead to an explosion. However, when AN melt stops moving in a pipe, there is an increased likelihood that it will freeze in spots. Whether in the suction line to or the discharge line from an AN pump, a frozen spot serves to stop flow. When the pump is eventually restarted, the impeller will turn, which allows thermal energy to accumulate in the stagnant AN blocked into the pump casing.

Another equipment failure that can lead to stopped flow is the Basic Process Control System (BPCS) level control loop on the feed tank. When it fails closed, for whatever reason, the pump suction becomes empty, so that the turning pump impeller no longer moves liquid, but instead causes cavitation and heat buildup in the small heel of liquid remaining in the pump casing.

Human errors that can lead to stopped flow were of even graver concern than equipment failures. These errors may have been simple errors made during routine or low stress tasks, or they may have resulted from a failure to follow detailed written procedures.

One category of human errors was valving errors. These errors included failing to open block valves—either suction or discharge—after performing maintenance and before returning the pump to service, and the error of misaligning valves for pumps that serve more than one purpose or in installations that included an installed spare.

Another category of human error involved running equipment when it should not be run. This category of human error included unintentionally running a pump when it should not be running, running the wrong pump, and inadvertently running a feed tank dry. Any of these could result in an otherwise dry pump running while it contains a heel of AN.

A third category of human error involved failing to recognize a change in operating conditions from something normal and safe to something abnormal and unsafe. This included filters becoming fouled, accumulated solids, and lines becoming frozen.

3. Safeguards and Independent Layers of Protection

The teams at both facilities identified several different safeguards with the potential to mitigate the risk of an AN pump explosion, which they universally determined to be caused by a pump running when there was no flow through the pump. These safeguards included procedural, mechanical, and instrumented functions. The utility of each type of safeguard as an independent layer of protection (IPL) depended on the nature of the failure that was the initiating cause.

3.1. Procedural Safeguards

The LOPA teams identified three categories of procedural safeguards that would mitigate risk, given the appropriate circumstances.

3.1.1. Valve Alignment

The initiating cause most commonly noted by the LOPA teams was that of misaligned valves following maintenance and prior to putting a pump back into service. This typically would include either the manual suction valve or the manual discharge valve. While the frequency of the initiating cause depended on the nature of the maintenance procedures and training, there was also a safeguard in the procedures followed by operations in accepting the pump back into service. To be considered an IPL, this operating procedure needed to be completely independent of maintenance, written, and periodically reviewed with any operator expected to perform the procedure.

3.1.2. Attended Loading and Unloading

Another procedural safeguard was that of attended loading and unloading. While of limited applicability, the team agreed that it was appropriate to take credit for this as an IPL in those cases where solution was received or shipped because it was possible for an operator to detect a problem and intervene on a timely basis. Loading stations were remote from the pumps, so an operator could monitor flow without standing in the immediate vicinity of an AN pump, thus becoming a victim in the event of an incident.

3.1.3. Operator Rounds

An important procedural safeguard was that of field operator rounds. Because uncontrolled AN decomposition does not begin immediately upon loss of flow, there is ample opportunity for field operators to detect unsafe conditions with time to act. These conditions included overflowing sumps, overflowing feed tanks, pumps running noisily indicating they are running dry, sump pumps running while their sump is empty, absence of pump discharge pressure, high temperature on local gauges, and high differential pressure across filters or strainers. For certain hazards, one or more of these could be used to indicate loss of flow. Credit for rounds was only taken after considering the procedural basis and frequency of rounds, the type of liquid being pumped (there is more time to detect a problem when AN solution is being pumped than when AN melt is being pumped), and whether the things checked on rounds would actually be effective in detecting loss of flow.

3.2. Mechanical Safeguards

The LOPA teams identified four types of mechanical safeguards that would mitigate risk, given the appropriate circumstances.

3.2.1. Self-Draining Pumps

One safeguard was that of a pump and pump installation designed to be self draining. This safeguard was considered an IPL in cases where the initiating cause was related to hold-up of AN in the pump when it was shut down.

3.2.2. Relief Devices

The second safeguard was of a relief device on a pump that prevented it from deadheading, thus assuring a minimum flow through the pump. This safeguard was considered an appropriate IPL in cases where AN solution (but not AN melt) was being pumped and the initiating cause was a blocked discharge. Although there is a potential for liquid in a close recirculation loop to heat as the pumping energy is absorbed, experience with AN is that a properly designed relief with sufficient pipe length and appropriate discharge point avoids this concern.

3.2.3. Kickback Lines

The third safeguard was that of a minimum flow, or "kickback", line. Again, this safeguard only counted as an IPL when the initiating cause was a blocked discharge, and then only when there were no block valves in the kickback line, or if block valves were unavoidable, when the block valves were included as part of the car seal program at the plant. As with relief devices, there is a potential for liquid in a close recirculation loop to heat as the pumping energy is absorbed, experience with AN is that a properly designed kickback line with sufficient length avoids this concern.

3.2.4. Equipment Configuration

The last safeguard the team identified involved equipment sizing and operating set points. One example in particular involved the minimum level permitted in a feed tank. Below that level, solids carried over and increase the frequency at which a filter fouled, which in turn caused a loss of flow. Changing the configuration of the equipment reduced the likelihood of some initiating events.

3.3. Instrumented Safeguards

Instrumented safeguards fell into three categories:

- Alarms that could prompt effective operator action
- Control loops, process interlocks, or process permissives in the BPCS that could keep or put the process in a safe state
- SIL-rated SIFs that were or could be installed in an SIS

3.3.1. Alarms

Depending on the specific pump installation, there were several alarms that could prompt an operator response while allowing sufficient time to act:

- Low flow alarm (indicating that liquid is not moving through the pump)
- Low tank level alarm (indicating that the pump is about to be deprived of suction feed)
- High filter differential pressure alarm (indicating that the suction filter is fouling and restricting flow)
- Low pump motor amps alarm (indicating that pump is deadheading or starved)
- High tank level alarm (indicating that liquid is not moving)

Not all alarms were available or appropriate in every situation. The team only took credit for alarms when a defined and timely response by an operator was possible. Even when more than one alarm was available and appropriate, the team only took credit for one as an IPL, because the response to separate alarms would be through the same system by the same operator, so would not be independent. Even then, no alarm was considered an IPL if any of the component devices were either part of the initiating cause or part of another IPL.

3.3.2. Control Loops, Process Interlocks, and Process Permissives

Another category of instrumented safeguards consisted of the automated controls found in the BPCS. Not only did they automatically sense process conditions, but they responded immediately without prompting. This was especially valuable when the time to respond was too short to be reliably executed by an operator. These included

- High pump casing temperature trip to stop pump
- Low tank level trip to stop pump
- Low flow trip to stop pump

- ❖ High line temperature trip to shut off steam to jacket

A low flow trip to stop a pump was only rarely used because of complications of linking cause and effect; starting a pump that shuts down on "no flow" requires bypasses which pose their own requirements. Likewise, the interlock that closed the steam valve to jacketed pumps and traced lines was only used in those few instances where pumps were jacketed and lines traced.

The treatment of IPLs in the BPCS was consistent with that found in *Layers of Protection Analysis* (21), §11.2. Specifically, BPCS control loops in a common BPCS were considered independent if they used independent sensors and field elements and did not share input/output cards, or processors. This was limited to two BPCS control loops in a common BPCS. This interpretation also provided that demands initiated by the failure of a BPCS control loop limited credit to only one other BPCS control loop as an IPL, again only if it used independent sensors and field elements and did not share input/output cards, or processors.

3.3.3. SIL-rated SIFs

Typically, any type of process interlock or process permissive can be upgraded to a SIL-rated SIF, once the requirements of the SIS standards are met. This begins with wiring the field devices to a safety logic solver, rather than a BPCS, but includes considerably more. In identifying functions that could serve as SIL-rated SIFs, the LOPA teams were only interested in the SIFs that could be enabled at all times. That is, SIL-rated SIFs were not to be phase-dependent or to depend on momentary bypasses for correct operation. This limited the types of SIFs to two:

- ❖ High pump casing temperature trip to stop pump
- ❖ Low tank level trip to stop pump

However, given the appropriate architecture and proof test intervals, either type of SIF was capable of meeting any SIL-rating required.

4. Layers of Protection Analysis and SIL Assignment

The LOPA methodology has been well described in *Layers of Protection Analysis.* (21) Before beginning analysis, it is important to calibrate the LOPA tool with the Risk Tolerance Criteria (RTC) used by the organization. While LOPA is essentially a tool for analyzing likelihood, that is hazardous event frequency, it is most useful when that frequency can be compared against benchmarks. This allows the LOPA team to establish whether the likelihood of a particular hazardous event is too high to be tolerable. The calibration should also include frequency to be used for various types of initiating causes, and the average probability of failure on demand (PFD_{AVG}) to be used for various IPLs. Initiating frequency and PFD_{AVG} are expressed in orders of magnitude. This avoids the haggling that sometimes plagues group activities like LOPA.

Once the LOPA tool has been calibrated, the team follows a series of steps for each hazard:

- ❖ Determine the consequence severity
- ❖ Identify the initiating cause and determine its frequency

- Identify enabling conditions and frequency modifiers that allow the chain of events to lead from initiating cause to hazardous event, and their probabilities
- Identify IPLs for which credit may be taken
- Compare the resulting risk with the RTC as a ratio.
- Determine how much additional risk reduction is required and suggest risk reduction measures

When the ratio of resulting risk to the risk tolerance criteria is less than one, no additional risk reduction measures are required. When the ratio is greater than one, the ratio is known as the risk reduction factor, and this represents the amount of risk reduction that is required to achieve the tolerable risk.

4.1. Risk Tolerance Criteria

The first reason that two plants of identical design may come to different conclusions about the required risk reduction for a particular hazard is because they are using different RTC. While there are mandated RTC in a few countries, the United States is not one of them. Hence, it is up to every organization to establish its own RTC. Guidance is available (21) (22), but RTC can easily vary as much as an order of magnitude from one organization to another. This alone can result in two identical hazards requiring different amounts of risk reduction.

This was not the case here. Both plants used identical RTC.

Although RTC may address several parameters—personal safety, community impact, environmental harm, asset damage—the RTC established by Terra Industries were directed to the parameter of personnel safety. The RTC were expressed as tolerable frequencies for specific consequences, and benchmarked to a hazard with a probable consequence of 1 fatality per event. Other consequences were then adjusted by an order magnitude for each successive category.

Table 1. Terra Industries Consequence Categories for LOPA

Safety Consequences	Severity Level
≥ 10 fatalities per event	A
≥ 1 fatality per event	B
≥ 1 disabling injury per event	C
≥ 1 reportable injury per event	D
≥ 1 first aid injury per event	E
< 1 first aid injury per event	F

4.2. Consequences

The teams at each plant evaluated the consequences of hazardous events independently. The evaluation took into consideration the location of the AN pumps in relation to where operators were likely to be when in the area, and the mitigating effects that pump pits or retaining walls would have.

At one plant, the LOPA team uniformly determined the consequence to be Severity Level C—one or more disabling injuries per event. That is not to say that fatalities were ruled out as a possible consequence, but that the most probable consequence—the consequence that best corresponded to the determined likelihood—would not be as bad as a fatality.

The LOPA team at the other plant considered a range of consequences as possible, depending on the unit in which the AN pump was located. In some units, where the AN pumps were located rather remotely and shielded, the team determined the consequence to be Severity Level C—one or more disabling injuries per event. For most pumps, the LOPA team determined the consequence to be Severity Level B—one or more fatalities per event. By this, the team was stating its belief that if an event occurred, someone who happened to be in the area would be killed. In some instances, the LOPA team concluded that the probable consequence was Severity Level A—ten or more fatalities per event.

These differences in consequence assessments had the predictable effect on required risk reduction for individual hazards.

4.3. Initiating Causes

Each LOPA scenario addressed a single cause-consequence pair, for which the LOPA team identified the initiating cause. The frequency of that cause then became the starting point for estimating the frequency at which the final hazardous event would occur. An initiating cause was one of two types:

- Ongoing, where the initiating cause can result from a random failure of a component or system. A common example is a control loop failure. The initiating causes described in section 2.4.2 as equipment failures were all treated as ongoing initiating causes.
- Opportunity, where the frequency of the initiating cause is related to the number of opportunities that occur in a year. A common example is failing to properly restore block valves to their operating positions after performing maintenance, where the number of times maintenance is performed each year equates to the frequency of the opportunity. The initiating causes described in section 2.4.2 as human error were all treated as opportunity-based initiating causes. In the case of opportunity-based initiating causes, the challenge to the LOPA team was to accurately estimate the likely number of opportunities per year.

A third type of initiating cause, “Scalable”, was also available to the team, but its use was never warranted. A scalable initiating cause is like an ongoing initiating cause, where there is a failure in an ongoing operation, but the frequency of the initiating event is related to scale. A leak in a pipeline is an example, where the longer the pipeline, the more frequently a leak may occur.

At one plant, the LOPA team characterized the initiating cause as a general ongoing cause, the loss of flow, with an annual frequency of 0.1 occurrences per year.

The LOPA team at the other plant took a more detailed approach, making the effort to identify the specific failure that was the initiating cause. For ongoing causes, the LOPA teams had an extensive menu of potential failures from which to choose, each with a set frequency. As it happened, only a handful of ongoing causes were used in the LOPA scenarios.

Table 2. Ongoing Initiating Cause Frequencies

Initiating Cause	Frequency (1/yr)
Pump trip	1
Unit trip	1
Basic process control loop failure	0.1
Control valve fails in direction of design	0.1
Heat tracing failure	0.1

In the case of opportunity-based initiating causes – human errors – the LOPA team was required to identify the number of opportunities there were for making the error, and to identify the nature of the error. Initiating causes arising from human errors made while performing a non-routine task while under high stress were considered to have a probability of failure of 100%. This is conservative, but as an order-of-magnitude estimate, it is a better estimate than 10%. Human error made while performing routine tasks or low-stress, non-routine tasks were considered to have a probability of failure of 10%.

Table 3. Opportunity-Based Initiating Cause Probabilities

Initiating Cause	Probability
Human error - High-stress, non-routine task	1
Human error - routine or low-stress, non-routine task	0.1
Operator failure to execute routine written procedure	0.01
Failure of procedure that includes independent review	0.001
Lockout/Tagout procedure failure	0.001

Procedures were distinguished from tasks in that they were written, detailed instructions on exactly what needed to be done, and in what order. The probability of failure of an operator to execute a routine written procedure so as to initiate a hazardous event was considered to be 1%.

Finally, the most detailed procedures, which include checklists and are reviewed independently by someone other than the person performing the procedure, were considered to have a probability of failure of 0.1%. A proper Lockout/Tagout procedure would be this kind of procedure.

4.4. Enabling Conditions and Frequency Modifiers

In the next step of the analysis, the LOPA teams considered factors that would reduce the likelihood that an initiating event would lead to the chain of events that would result in the final hazardous consequence. The worksheets explicitly invited consideration of four types of enabling conditions and frequency modifiers:

- Time at risk
- Occupancy factor
- Ignition probability
- Vulnerability

The worksheet also allowed the LOPA teams to consider other specific frequency modifiers.

4.4.1. Time At Risk

The first frequency modifier considered was the time at risk. While a pump might be expected to trip once a year, if it is only running for 876 hours out of a year, the annual frequency is immediately reduced from 1/yr to 0.1/yr. This frequency modifier was appropriate when addressing random failures of an ongoing operation. In most cases, this frequency modifier was not used, because in most cases, the piece of equipment at risk operated year round. Many facilities would acknowledge the reduced time at risk associated with outages or turnarounds; neither of the Terra facilities chose to.

It was not appropriate to consider time at risk when addressing an opportunity-based failure. Regardless of the time the operation was running, the frequency of the initiating cause for opportunity-based failures was set by the number of opportunities at risk, not the time at risk.

4.4.2. Occupancy Factor

The second frequency modifier considered was occupancy factor. Because the LOPA teams were just addressing safety consequences, only the time someone was present to be hurt counted toward occupancy. If someone was present to be hurt during an event for only 1 hour for every 100 hours of operation, then the occupancy factor was 0.01. Had the LOPA teams also been addressing environmental or asset consequences, it would have been inappropriate to apply occupancy factors to those analyses; the environment and the facilities are always present, whether or not personnel are.

The occupancy factors used by the LOPA teams were highly dependent on the units in which the pumps were installed and how those units operated. In some units, personnel are always present, meaning the occupancy factor was 1. In some instances, the teams concluded that a failure would invariably lead to maintenance personnel being present, in which case the occupancy factor was again 1. Other occupancy factors were also used, and are shown in Table 4.

Table 4. Occupancy Factors Used in LOPA Scenarios

Occupancy	Occupancy Factor
Personnel always present	1
Personnel in area 8 hours, 200 days a year	0.18
Personnel in area 5 minutes every hour	0.08
Personnel in area 5 minutes every 2 hours	0.04
Personnel in area 2 minutes every hour	0.03
Personnel in area 1 hour every month	0.0014

4.4.3. Ignition Probability

Ignition probability was not used in any of these LOPA scenarios. The scenarios themselves assumed the pump as the source of ignition, given the initiating cause.

4.4.4. Vulnerability Factor

The LOPA worksheets also allowed for use of a vulnerability factor. Vulnerability factors account for the probability of a fatality or injury in the event of an exposure. Since the consequence of each scenario was already defined in terms of whether or not a fatality or injury would occur, it was not appropriate in these analyses to further consider vulnerability.

4.4.5. Other Enabling Conditions or Frequency Modifiers

In addition to the four standard frequency modifiers, the LOPA team considered other enabling conditions or frequency modifiers. These could have included weather related factors, which would be different from one location to another, or the probability of being beyond a certain operating level in a vessel, which could vary from facility to facility. While there was no need to consider either of these specific enabling conditions, on occasion, one team did consider the probability of a sensitizing contaminant being present in an AN solution being pumped.

4.5. Independent Layers of Protection

Section 3.3 above describes the safeguards which the LOPA teams credited as IPLs. Whether or not credit was taken depended first on whether the safeguard was present. When present, taking credit depended on whether the safeguard was independent of any other IPL for which credit was being taken, on whether the safeguard was actually effective against the hazard described in the scenario, and on whether the performance of the safeguard could be audited.

The credit taken for IPLs was consistent between both plants, and between different units within each plant. Table 5 shows the IPLs used when applicable and the PFD_{AVG} applied when credit was taken.

Table 5. IPLs and Associated PFD_{AVG}

IPLs	PFD_{AVG}
Procedural controls	0.1
BPCS control loop, process interlock, process permissive	0.1
Heat tracing	0.1
Human response to alarm or field condition, with at least 20 min to respond	0.1
Human response to field condition, with at least 40 min to respond	0.01
Kickback (minimum flow) line	0.01
Relief valve	0.01
Self-draining pump	0.1

When SIL-rated SIFs already existed, the PFD_{AVG} credited to them was the PFD_{AVG} calculated for the SIF, given components, the architecture, and the current proof-test intervals. Therefore, it could be any value, not simply an order of magnitude value.

4.6. SIL Assignment

There were many ways in which the AN pumps at the two plant could vary in terms of residual risk and hence, SIL assignment. Although analyses at both plants used the same RTC (which could have been different, but were not), the frequency of the initiating cause varied by orders of magnitude. This was especially true when considering opportunity-based initiating causes and the nature of the error that would lead to such a failure. Other differences hinged on the enabling conditions that applied to each scenario and the number and type of IPLs that happened to be in place for each installation.

As a result, of the 102 AN pumps examined by the LOPA team—AN solution pumps as well as AN melt pumps—almost half required no additional risk reduction. About half as many pumps fell into each successive order of magnitude of risk as fell into the previous, lower, order of magnitude of risk. The number of SIL-rated SIFs, however, did not correspond to the number of pumps in each risk category. This is because it was possible in many cases to apply more than one additional layer of protection to reduce risk. In fact, for the five pumps discovered to require an additional RRF between 1,000 and 10,000, two or three additional layers of protection were applied so that no SIL 3 SIFs were required at all.

Table 6. Distribution of Required Risk Reduction for AN Pumps

Required Risk Reduction Factor (RRF)	Number of Pumps
No additional risk reduction required	50
1 < RRF ≤ 10	25
10 < RRF ≤ 100	15
100 < RRF ≤ 1,000	7
1,000 < RRF ≤ 10,000	5

The risk reduction measures applied depended on the nature of the hazard. Where a blocked discharge was the initiating cause for a pump to run with no flow, a candidate IPL was the installation of a kick back line when there was none. Where low tank level resulting in low pump suction was the initiating cause for a pump to run with no flow, a candidate IPL was the installation of a low level pump shutdown, either as a process interlock in the BPCS or as a SIL-rated SIF in the SIS, depending on the amount of risk reduction required. The most commonly applied risk reduction measure, because of its effectiveness regardless of the nature of the initiating cause, was the installation of a high temperature pump shutdown, either as a process interlock in the BPCS or as a SIL-rated SIF in the SIS, depending on the amount of risk reduction required. The number of times that certain risk reduction measures were applied is shown in Table 7. The total does not add to 52 because in many cases, more than one measure was applied.

Table 7. New Risk Reduction Measures Applied Following LOPA

Risk Reduction Measure (IPL)	Number of Pumps
Non-SIL high temperature pump shutdown	25
SIL 1 high temperature pump shutdown	20
SIL 2 high temperature pump shutdown	4
Non-SIL low tank level pump shutdown	8
SIL 1 low tank level pump shutdown	2
Kick back line	12

5. Conclusion

Although process units may be similar in many regards, may in fact be identical, there are many reasons why a cookie-cutter approach to applying SIFs with predetermined SIL-ratings is misguided and inappropriate. This is especially true in the development of industry standards or government regulations.

Even when there is agreement about the consequences of a hazard and about the tolerable risk—which should not be considered a given—there can be significant differences in the nature and frequency of initiating causes, in the probability that necessary enabling conditions exist, or in the number and quality of safeguards deployed that can be credited as IPLs. These differences can combine to shift residual risk up or down by one or more orders of magnitude, the difference between SIL 2 or SIL 3 on the one hand, and between SIL 1 or no additional IPL required on the other hand.

In the case of AN pumps, there are a number of questions to answer:

- What RTC is used?
- What are the initiating causes for a pump to be running with no flow?
- What is the frequency at which those initiating causes occur?
- What is the consequence of an AN pump explosion?
- What is the probability that conditions enabling the consequence exist?
- Which safeguards are already installed and for which can credit as IPLs be taken?

It is only in the case that the answers to all of these questions are identical that identical solutions are justified. LOPA allows us to do this analysis. It also requires us to do this analysis. In the absence of this analysis, we should not presuppose the answers.

6. References

[1] Crowl, Daniel A. and Joseph F. Louvar. *Chemical Process Safety-Fundamentals with Applications.* 2nd Ed. Upper Saddle River, NJ : Prentice Hall, Inc., 2001. ISBN 0-13-018176-5.

[2] Texas City Disaster. *Handbook of Texas Online.* [Online] [Cited: December 07, 2010.] http://www.tshaonline.org/handbook/online/articles/lyt01.

[3] Wester, Eric. Ammonium Nitrate Disasters. *Wikipedia, the free encyclopedia.* [Online] October 14, 2010. [Cited: December 21, 2010.] http://en.wikipedia.org/wiki/Ammonium_nitrate_disasters.

[4] French Ministry of Environment - DPPR/SEI/BARPI. Explosion in a nitrogenous fertiliser plant - Oppau Germany. *ARIA (Accident Analysis, Research, and Information) database.* [Online] March 2008. [Cited: December 20, 2010.] http://barpipdf.geniecube.info/14373_gb.pdf. No 14373.

[5] 1924 Nixon Nitration Works disaster. *Wikipedia, the free encyclopedia.* [Online] November 25, 2010. [Cited: December 21, 2010.] http://en.wikipedia.org/wiki/1924_Nixon_Nitration_Works_disaster.

[6] Grande Paroisse, A Total Company. The Different Theories. *Grand Paroisse, AZF web page.* [Online] [Cited: December 21, 2010.] http://en.azf.fr/the-azf-trial/the-different-theories-800295.html.

[7] Tessenderlo Group. History. *Tessenderlo Group webpage.* [Online] [Cited: December 21, 2010.] http://www.tessenderlo.com/tessenderlo_group/profile/history/.

[8] Thomas, Mark J., Alan Cummings, and Mariano Gomez. *Terra Industries, Inc. Nitrogen Fertilizer Facility, Port Neal, Iowa.* Region 7 - Emergency Response and Removal Branch. Kansas City, Kansas : United States-Environmental Protection Agency, 1996. p. 108, Chemical Accident Investigation Report. http://www.epa.gov/oem/docs/chem/cterra.pdf.

[9] UK Health and Safety Executive, *Storing and Handling Ammonium Nitrate.* Sudbury, Suffolk, UK : November, 2004. p. 12. INDG230.

[10] Marlair, Guy, Marie-Astrid Kordek, and Christian Michot. High Challenge Warehouseing: Ammonium Nitrate as a Typical Case Study. *National Fire Protection Association web site.* [Online] February 16, 2010. [Cited: December 15, 2010.] http://www.nfpa.org/assets/files//PDF/Foundation%20proceedings/High_Challenge_Warehousing-Ammonium_Nitrate_as_a_Typical_Cas.pdf.

[11] *Safety of Ammonium Nitrate Fertilizers.* Shah, K.D. s.l. : The International Fertilizer Society, 10-Oct-1996. ISBN 0-85310-018-7.

[12] Keenan, A.G. and B.Dimitriades. *Mechanism for Chloride-Catalyzed Themal Decomposition of Ammonium Nitrate.* 8, 15-Oct-1962, Journal of Chemical Physics, Vol. 37, pp. 1583-1586.

[13] Rosser, W.A., S.H.Inami, and H.Wise. *Decomposition of Liquid Ammonium Nitrate Catalyzed by Chromium Compounds.* 1964, Trans Faraday Society, Vol. 60, pp. 1618-1625.

[14] Keenan, A.G., K.Notz, and N.B.Franco. *Synergistic Catalysis of Ammonium Nitrate Decomposition.* 12, June 4, 1969, Journal of the American Chemical Society, Vol. 91, pp. 3168-3171.

[15] Rosser, W.A., S.H.Inami, and H.Wise. *The Kinetics of Decomposition of Ammonium Nitrate.* 1963, Journal of Physical Chemistry, Vol. 67, pp. 1753-1757.

[16] Oxley, Jimmie C., S.M.Kaushick, and Nancy Gilson. *Thermal Stability and Compatibility of Ammonium Nitrate Explosives on a Large and Small Scale.* 1992, Thermochimica Acta, Vol. 212, pp. 77-85.

[17] Stephan, Clete R., P.E. Coal Dust Explosion Hazards. *Mine Safety and Health Administration web site.* [Online] March 27, 2009. [Cited: December 15, 2010.] http://www.msha.gov/S&HINFO/TECHRPT/P&T/COALDUST.pdf.

[18] Greiner, Maurice. Ammonium Nitrate Fertilizer - Exploding the Myth. *Emergency Film Group website.* [Online] 2009. [Cited: December 20, 2010.] http://www.efilmgroup.com/Exploding-the-Myth.html.

[19] Kilpatrick, Dan. *AN Detonability/Decomposition Studies and Plant History.* Tucson, AZ : s.n., 2002. ANPSG Conference Proceedings.

[20] Chemical Emergency Preparedness and Prevention Office. *Explosion Hazard from Ammonium Nitrate.* Washington, DC : U.S. Environmental Protection Agency, December, 1997. p. 5. EPA 550-F-97-002dR.

[21] Dowell III, Arthur M., et al. *Layer of Protection Analysis-Simplified Process Risk Assessment.* New York : American Institute of Chemical Engineers, 2001. ISBN 0-8169-0811-7.

[22] Frank, Walt and John Farquharson. *Guidelines for Developing Quantitative Safety Risk Criteria.* New York : Center for Chemical Process Safety of the American Institute of Chemical Engineers, 2009. p. 210. ISBN 978-0-470-26140-8.

An Analysis of CSB Investigation Reports for Inherent Safety Learnings

Paul R. Amyotte
Department of Process Engineering & Applied Science
Dalhousie University, Halifax, NS, Canada B3J 2X4
paul.amyotte@dal.ca

Dustin K. MacDonald
Department of Process Engineering & Applied Science
Dalhousie University, Halifax, NS, Canada B3J 2X4

Faisal I. Khan
Faculty of Engineering & Applied Science
Memorial University, St. John's, NL, Canada A1B 3X5

Keywords: inherent safety, inherently safer design, hierarchy of controls, process safety management, U.S. Chemical Safety and Hazard Investigation Board

Abstract

Sixty reports resulting from process incident investigations conducted by the U.S. Chemical Safety Board (CSB) have been analyzed for evidence of examples related to inherent safety, passive and active engineered safety, and procedural safety. These risk reduction measures, which collectively form the hierarchy of safety controls, were also analyzed for their contribution to both incident prevention and consequence mitigation, as well as their applicability to specific elements in a recognized process safety management (PSM) system.

Over two hundred examples from the hierarchy of controls were identified with the following breakdown: inherent safety (36 % of total overall examples), passive engineered safety (8 %), active engineered safety (14 %), and procedural safety (42 %). Numerous examples for both prevention and mitigation measures were found during the CSB report review. Inherently safer design (ISD) items were observed to be nearly equally split among the four primary principles of minimization (25 % of total ISD examples), substitution (22 %), moderation (25 %), and simplification (27 %). Widely applicable PSM elements (in terms of a link to the identified hierarchical safety measures) were determined to be process and equipment integrity, training and performance, process knowledge and documentation, capital project review and design procedures, and management of change.

The results from the current work are generally consistent both qualitatively and quantitatively with that of other workers who examined incidents investigated by the CSB and other agencies. CSB reports themselves afford multiple opportunities to learn and reinforce the concept of inherently safety design and its relationship to other available safety measures. The value of these reports in demonstrating lessons learned and in training/educational efforts is significant.

1. Introduction

The scope of the current paper is an analysis of various reports issued by the United States Chemical Safety and Hazard Investigation Board (CSB), primarily from the perspective of the potential for application of inherently safer design principles. Motivation for this work was provided by two factors: (i) the significant value of CSB reports in promoting lessons learned from process safety incidents, and (ii) a desire to encourage continued and increased usage of the terminology and principles of inherent safety in incident investigation. The specific objective in undertaking the analysis was to highlight the key role of inherent safety within both the hierarchy of risk control measures and process safety management systems.

To achieve this objective, the CSB reports cited in references [1-60] were analyzed within a threefold framework according to: (i) the individual inherent safety principles of minimization, substitution, moderation and simplification, (ii) the risk components of likelihood (i.e., prevention) and severity of consequences (i.e., mitigation), and (iii) the twelve elements of Process Safety Management (PSM) recommended by the Canadian Society for Chemical Engineering, which are based on the PSM structure developed by the Center for Chemical Process Safety prior to the recent introduction of risk-based PSM. As the research proceeded, the decision was made to extend the analysis using the above framework to include safety measures categorized as passive engineered, active engineered and procedural. The following sub-sections give definitions and a brief discussion of each of the key concepts identified in this Introduction.

1.1 Chemical Safety Board

As noted on its web site (*www.csb.gov*), the United States Chemical Safety and Hazard Investigation Board (*Chemical Safety Board* or *CSB*) is an independent, non-regulatory federal agency charged with investigating industrial chemical accidents. Such incidents are investigated by a team of CSB employees, and from the evidence collected, root and contributing causation factors are identified. With this information, the CSB creates sets of recommendations for various bodies such as facility managers, regulatory agencies and technical associations. Following a completed investigation, documentation in the form of a full investigation report, case study, safety bulletin or urgent recommendations is made available on the CSB web site. These documents often have accompanying video support and are widely recognized as valuable learning tools for improving safety in the process industries.

1.2 Inherent Safety and the Hierarchy of Controls

Inherent safety is a proactive approach in which hazards are eliminated or lessened so as to reduce risk without over-reliance on engineered (add-on) devices and procedural measures. The concepts of inherent safety (or *inherently safer design – ISD*) have been formalized in the process industries over the past thirty or so years, beginning with the pioneering work of Professor Trevor Kletz (largely in response to the cyclohexane explosion at Flixborough, UK in 1974). Several publications on ISD are available, including recent texts [61, 62] and a resource article [63].

Professor Kletz and others worldwide have formulated a number of principles or guidewords to facilitate inherent safety implementation in industry. Four basic principles have gained widespread acceptance: (i) minimization (or intensification), (ii) substitution, (iii) moderation (or attenuation), and (iv) simplification.

Minimization calls for the use of smaller quantities of hazardous materials when the use of such materials cannot be avoided. It may also involve performing a hazardous procedure as few times as possible when the procedure is unavoidable. *Substitution* calls for the replacement of a substance with a less hazardous material, or a processing route with one that does not involve hazardous material. It could also involve replacing a hazardous procedure with one that is less hazardous. *Moderation* implies the use of hazardous materials in their least hazardous forms, or the identification of processing options that involve less severe conditions – for example, a lower temperature, pressure or speed of rotation. *Simplification* requires the design of processes, processing equipment and procedures in a manner so as to eliminate opportunities for errors by eliminating excessive use of add-on safety features and protective devices.

These and other principles of inherent safety work in conjunction with other means of reducing risk – namely, *passive* and *active engineered safety*, and *procedural safety* – within a framework commonly known as the *hierarchy of controls* (or as some call it, the *priority of controls* or the *safety decision hierarchy*). Inherent safety, being the most effective and robust approach to risk reduction, sits at the top of the hierarchy; it is followed in order of decreasing effectiveness by passive engineered safety devices (such as explosion relief vents), then active engineered safety devices (such as automatic fire suppression systems), and finally procedural safety measures (such as ignition source control by hot-work permitting).

Inherent safety is not a stand-alone concept; as just described, ISD works through a hierarchical arrangement in concert with engineered and procedural safety to reduce risk. However, inherent safety is not necessarily a solution for all hazards and all risks. Nor does the hierarchy of controls invalidate the usefulness of engineered and procedural safety measures. Quite the opposite; the hierarchy of controls recognizes the importance of engineered and procedural safety by highlighting the need for careful examination of the reliability of both mechanical devices and human actions. Hendershot [63] has described the hierarchy of chemical process safety strategies as a *spectrum of options*.

1.3 Risk – Prevention and Mitigation

Functional definitions of *hazard* and *risk* according to Wilson and McCutcheon [64] are as follows:

Hazard: The potential of a machine, equipment, process, material or physical factor in the working environment to cause harm to people, environment, assets or production.

Risk: The possibility of injury, loss or environmental incident created by a hazard. The significance of risk is a function of the *probability* of an unwanted incident and the *severity* of its consequence.

Process safety efforts that address the probability (or likelihood) component of risk can therefore be classified as *prevention* measures. Similarly, efforts aimed at lessening consequence severity are termed *mitigation* (or *protection*) measures.

1.4 Process Safety Management

Employing an effective management system for process-related hazards (i.e., fires, explosions and releases of toxic materials) is a critical corporate objective in the chemical process industries. An approach widely adopted in the Canadian (and worldwide) chemical industries is

PSM, Process Safety Management (where PSM is defined as the application of management principles and systems to the identification, understanding and control of process hazards to prevent process-related injuries and accidents). The suite of PSM elements is shown in Table 1, taken from the Process Safety Management Guide of the Canadian Society for Chemical Engineering (CSCHE) [65]. As previously mentioned, the material in the CSChE PSM guide [65] is based on that developed by the Center for Chemical Process Safety (CCPS) of the American Institute of Chemical Engineers (AIChE), prior to the recent introduction by the CCPS of its 20-element risk-based process safety management system [66].

Table 1 Process Safety Management (PSM) elements [65].

Number	Element
1	Accountability: objectives and goals
2	Process knowledge and documentation
3	Capital project review and design procedures
4	Process risk management
5	Management of change
6	Process and equipment integrity
7	Human factors
8	Training and performance
9	Incident investigation
10	Company standards, codes and regulations
11	Audits and corrective actions
12	Enhancement of process safety knowledge

2. Explicit Inherent Safety Considerations in CSB Reports

There has been a welcome trend in recent CSB reports toward increased use of the terminology of inherently safer design. At the time of the full analysis described in the current paper, reference [60] was the most recent report reviewed. In this report released in 2010, the CSB identified a set of urgent recommendations arising from its investigation into the Kleen Energy natural gas explosion in Middletown, CT, which occurred during the natural gas purging of a pipeline and resulted in the death of six workers. In the recommendations made to the U.S. Occupational Safety and Health Administration [60], there is clear evidence of use of the hierarchy of controls. These recommendations incorporate first, elimination of the fire and explosion hazard by prohibiting the release of flammable gas to the atmosphere for the purpose of cleaning fuel gas piping (*inherent safety*), through to the use of combustible gas monitoring (*engineered safety*), and the development of flammable gas safety procedures and training (*procedural safety*).

Further, the Kleen Energy recommendations explicitly invoke the inherent safety principle of *substitution* in the case of replacing natural gas with a less hazardous (i.e., inherently safer) gas, such as air, in the cleaning of fuel gas piping. The principle of *simplification* is also relevant in this case in helping to ease the complex requirements for discharge design when using natural gas for pipeline cleaning, which are noted in the urgent recommendations. The issue of whether alternative cleaning methods are technically feasible is addressed in reference [60] by the comment that safer alternatives, such as air, are just as effective as natural gas for cleaning fuel gas piping.

The facts that ISD implementation is hazard-specific, and is not a cure for all safety issues, are also recognized in the Kleen Energy recommendations [60]. Clear reference is made to the removal of fire and explosion hazards by alternative cleaning methods such as pigging with air or nitrogen, using air, nitrogen or steam blows, and cleaning with water or chemicals. New hazards may be introduced, however, as noted in the recommendations by the identification of the asphyxiation hazard of nitrogen. This illustrates the importance of having an effective safety management system – with management of change, or MOC, as a key system element (as further indicated in a later section of the current paper).

The inherent safety concept and the principles of ISD are, therefore, explicitly referenced in the Kleen Energy urgent recommendations [60] and form an integral component of the investigation findings. The same conclusion can be drawn for the 2007 report on the BP Texas City incident [44], in which inherently safer approaches to the use of an atmospheric vent stack are discussed. Similarly, the 2008 report on the Valero McKee Refinery LPG fire [52] includes explicit ISD consideration in the sections dealing with management of dead-legs and substitution of sodium hypochlorite for chlorine.

Since the time of undertaking the full analysis reported here, and writing of the current paper, three additional reports [67-69] have appeared on the completed investigations page of the CSB web site. Of these new reports (which were not included in the quantitative analysis), two explicitly use the terminology of inherent safety, while the third makes an implicit reference to ISD. The ConAgra urgent recommendations [67] and the Xcel Energy investigation report [69], contain the following statements (with underlining added for emphasis):

Purging fuel gas piping to a safe location outside avoids the hazard of forming a flammable atmosphere inside an occupied building and is an inherently safer approach [67].

Using this method of hazard control, primary consideration is given to eliminating the hazard or using engineering controls such as substitution; for instance, using less hazardous, non-flammable substitute solvents for the highly flammable MEK. Another more effective safety approach within the hierarchy of controls would have been to conduct the work outside the confined space [69].

The Veolia Technical Solutions case study [68] contains the following implicit reference to inherent safety (specifically, process unit segregation to avoid domino or knock-on effects – via the *moderation* sub-principle of *limitation of effects*), again with underlining added for emphasis:

...include a section requiring a written engineering analysis to determine the safe separation distance for occupied buildings, control rooms, and operating areas [68].

3. Other Examples of Inherent Safety in CSB Reports

Most of the CSB reports analyzed in the current work do not explicitly incorporate ISD terminology. As discussed in the previous section, use of the language of inherent safety in these reports is a fairly recent, and most welcome, trend. Other organizations, as well as process industry companies, should be encouraged to follow the lead of the Chemical Safety Board in this regard.

Notwithstanding the comments in the previous paragraph, each CSB report typically contains one or more examples of the potential application of inherently safer design. In the analysis reported here, such examples were identified through consideration of whether a lack of application of inherent safety principles played a role in a given incident from the perspectives of both causation and propagation. Illustrative cases are presented in the remainder of this section, prior to the quantitative analysis presented in the next section.

The concept of *minimization* is prevalent in a number of CSB reports, particularly those dealing with explosions of combustible dusts [27, 28, 30, 40, 57]. Repeated calls to avoid and clean up dust accumulations have been made in these reports; housekeeping efforts in this regard can help to either prevent primary dust explosions or mitigate their consequences by prevention of secondary dust explosions [70]. A review of ISD applications in the field of dust explosion risk reduction [71] was recently completed by the present authors with, during discussion of the *minimization* principle, extensive reference to relevant CSB reports available at the time [27, 28, 30, 40]. Additionally, the case study chapter in Kletz and Amyotte [62] includes a section on CSB reports, which describes a methanol tank explosion and fire at a wastewater treatment plant [43]. Several inherent safety aspects of this incident are described in relation to both *minimization* and *substitution*.

Kaszniak [72] has examined a number of CSB investigation reports for lessons learned with respect to conducting process hazard analyses. Key among his findings is the difficulty of scaling-up chemical processes while at the same time ensuring an acceptable level of risk. Here one sees essentially a negation of the *minimization* principle in cases related to commercial-scale, first-batch production from laboratory- and pilot plant-scale data, and doubling of batch size to fill a customer order. The ensuing runaway reactions were aided by poor control of reactant addition, inadequate cooling capacity and undersized relief systems [72]. These examples also illustrate the inherently less safe features of batch reactors as compared to continuous flow reactors.

Another implicit example of ISD can be found in the CSB safety bulletin on positive material verification [38]. During shutdown of the plant in question, three elbows were removed from a high-pressure, high-temperature hydrogen line for cleaning. Due to different process conditions in the line, two of the elbows were composed of a stronger metal alloy, while the third was constructed of less expensive carbon steel. During reinstallation, the carbon steel elbow was incorrectly positioned and subsequently corroded, leading to a hydrogen release and fire that caused extensive property damage and minor injury to an employee. A key finding identified by the CSB in relation to this process incident is the following:
Piping systems can be designed such that incompatible components cannot be interchanged. All three elbows could have been made from the same low alloy steel material, even though this would have meant additional expense. Alternatively, elbow 1 could have been dimensionally different from elbow 2 and 3, although this would have meant additional construction costs [38].

The above quote provides an excellent example of *simplification* (in particular, the sub-principle of *making incorrect assembly impossible*). The same comment can be made for the CSB safety bulletin on the hazards of nitrogen asphyxiation [16]; this document describes how a cylinder of pure nitrogen, in addition to several oxygen cylinders, was accidentally delivered to a nursing home. The nitrogen cylinder was mistakenly connected to the oxygen supply system, resulting in the death of four residents and injury to six others. Procedural safeguards alone are inadequate to prevent such tragedies.

4. CSB Report Analysis

Quantitative analysis of the CSB reports [1-60] was undertaken initially by one of the authors (DKM) over a period of several months. The resulting data were then reviewed by one of the two remaining authors (PRA) with support provided by the other (FIK). This data review stage resulted in minor adjustments to some of the analysis categories, particularly the inherent safety/passive safety and prevention/mitigation classifications. On the matter of inherent safety and passive safety, Hendershot [63] has remarked that it is not uncommon for people to disagree on the category in which to place a particular safety approach. The more important consideration is whether a given measure is a good engineering design feature [63].

With respect to prevention and mitigation, it is to be expected that differences of opinion may arise in terms of the initiation point of an incident. For example, if one minimizes the amount of combustible dust in suspension to a level below the minimum explosible concentration, then a dust explosion has been *prevented* if event initiation is considered to be ignition of the dust cloud. However, if for the same scenario one considers event initiation to be the aerodynamic disturbance leading to formation of the dust cloud, then the consequences of the dust dispersion have been *mitigated.*

4.1 Inherent Safety

The analysis identified a total of 258 examples from the various categories in the hierarchy of risk control measures, of which 93 examples are related to inherent safety. Table 2 provides illustrative ISD examples according to the previously described threefold analytical framework.

Table 2 Inherent safety examples from CSB reports.

ISD Principle	Risk Component	PSM Element	CSB Reference	Example
Minimization	Prevention	5	48	Reactant amount increased, rendering heat removal capability ineffective.
Minimization	Mitigation	4	1	Excessive amount of explosives placed in storage.
Substitution	Prevention	8	46	Flammable gas detected by using open flames rather than portable gas detectors.
Substitution	Mitigation	2	41	Recommendation made to use chemistry with reduced toxicity.
Moderation	Prevention	6	31	Unit run at higher pressure than industry norm, resulting in increased rate of corrosion and amount of scaling.
Moderation	Mitigation	3	14	Warehouses located in close proximity, with offices breaching separation.
Simplification	Prevention	12	12	Chemical hazard ratings normalized for inappropriate purposes.
Simplification	Mitigation	2	56	Plant not designed to withstand effects of (unknown) runaway reaction.

Overall, there is a near equal split among the four basic principles of ISD: *minimization* (25 % of total ISD examples), *substitution* (22 %), *moderation* (25 %), and *simplification* (27 %). The fact that numerous examples for each principle were identified is consistent with the work of Kidam et al. [73], who reviewed 364 chemical process industry incident descriptions in the Failure Knowledge Database maintained on the Japan Science and Technology web site. These workers identified applicable ISD measures for the incidents in the data base and found the breakdown to be: *minimization* (8 % of total ISD examples), *substitution* (21 %), *moderation* (29 %), and *simplification* (42 %). (It should be noted that reference [73] considered the sub-principle of *limitation of effects* to be separate from *moderation*, and the sub-principle of *error tolerance* to be separate from *simplification*; the percentages given in the previous sentence have combined each of these sub-principles with its respective parent principle.) The analysis by Kidam et al. [73] resulted in percentages similar to the current work for both substitution and moderation, a lower percentage for minimization, and a higher percentage for simplification.

Approximately 70 % of the total ISD items were judged to be pertinent to *prevention* efforts and thus about 30 % are aimed at *mitigation*. From the perspective of individual principles, the simplification examples are heavily weighted toward prevention. Those for minimization and substitution are also weighted toward prevention (although not quite to the same extent as simplification), while moderation examples are equally split between prevention and mitigation. The higher percentage for prevention than for mitigation is likely due, in part, to the fact that the CSB reports are the result of incident investigations aimed at identifying root and contributing causes to facilitate *prevention* of potential future incidents. The general conclusion to be drawn from the current work is that the principles of inherent safety can be helpful in both the prevention and mitigation of process incidents. This comment is consistent with the title of Hendershot's paper [74] to be presented at the same congress as the current paper: *Inherently Safer Design – Not Only About Reducing Consequences!*

With respect to the Process Safety Management elements shown in Table 1, the leading elements in terms of number of ISD examples are *process and equipment integrity*, *process knowledge and documentation*, *capital project review and design procedures*, and *management of change*. The full breakdown by percentage of total ISD examples is given in Figure 1. Fully half of the element 3 examples (capital project review and design procedures) are related to facility siting issues as captured under the moderation principle (mitigation). This is a similar finding to one of the conclusions reached by Kraszniak [72] in his independent review of CSB investigation reports.

4.2 Hierarchy of Controls

With a total of 258 examples from the various classifications in the hierarchy of controls, and 93 of these being related to ISD, the remaining 165 examples are drawn from the categories of passive engineered, active engineered and procedural safety. Table 3 provides illustrative examples of these latter three hierarchy levels according to the applicable risk component and PSM element. With respect to the overall hierarchy (i.e., all four levels) the distribution among the total 258 examples is: *inherent safety* (36 % of total overall examples), *passive engineered safety* (8 %), *active engineered safety* (14 %), and *procedural safety* (42 %). The overall hierarchy split between prevention and mitigation measures is the same as for the ISD measures alone – i.e., 70 % *prevention* and 30 % *mitigation*.

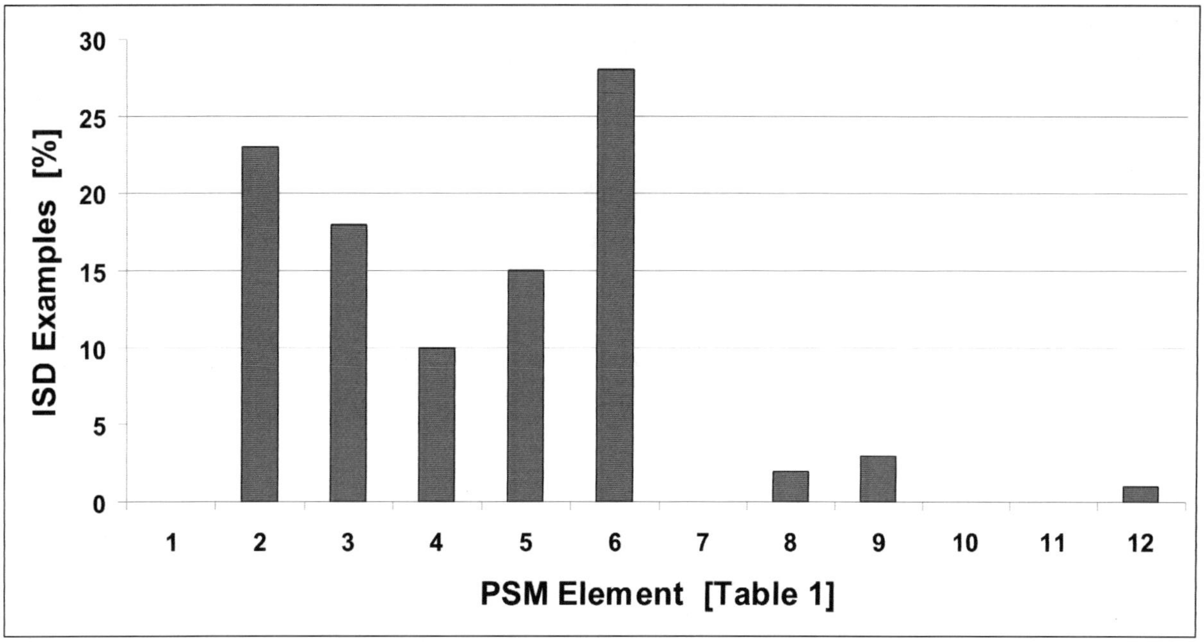

Figure 1 Breakdown of inherent safety examples according to applicable PSM elements.

Table 3 Passive, active and procedural safety examples from CSB reports.

Hierarchy Category	Risk Component	PSM Element	CSB Reference	Example
Passive Engineered	Prevention	6	8	Insulation not reinstalled on piping after maintenance work, resulting in line freezing and accumulation of flammables.
Passive Engineered	Mitigation	3	14	Diking around storage tank farm inadequately designed and constructed.
Active Engineered	Prevention	6	58	Audible alarm on level indicator malfunctioned.
Active Engineered	Mitigation	3	49	Lack of fire suppression systems allowed fire to propagate.
Procedural	Prevention	1	44	Company safety guidelines contravened by lack of both supervisory oversight and technically trained personnel during startup.
Procedural	Mitigation	8	37	Employees not prepared for large release of flammable chemical.

The fact that numerous examples in the remaining (non-ISD) hierarchy levels were identified in the present analysis is again consistent with the work of Kidam et al. [73]. The overall distribution of examples throughout the hierarchy of controls determined by the current authors is also relatively consistent with that found by Kidam et al. [73]: *inherent safety* (17 % of total overall examples), *passive engineered safety* (13 %), *active engineered safety* (15 %), and *procedural safety* (50%). (It is noted in reference [73] that 4 % of the total corrective actions were classified as unknown due to insufficient information; presumably the remaining 1 % is due to rounding). A preponderance of procedural safety measures can be observed in both analyses.

With respect to the Process Safety Management elements shown in Table 1, the leading element in terms of total hierarchy examples is element 6, *process and equipment integrity* (as with the ISD measures alone). The full breakdown by percentage of total overall hierarchy examples is given in Figure 2. (Note that Figure 2 gives the distribution for all four hierarchy categories, and therefore includes the examples shown in Figure 1.) The data trends in Figure 2 are similar to those shown in Figure 1 with the addition of a significant contribution from element 8, *training and performance*, because of the high procedural component in the total safety measure count (42 %).

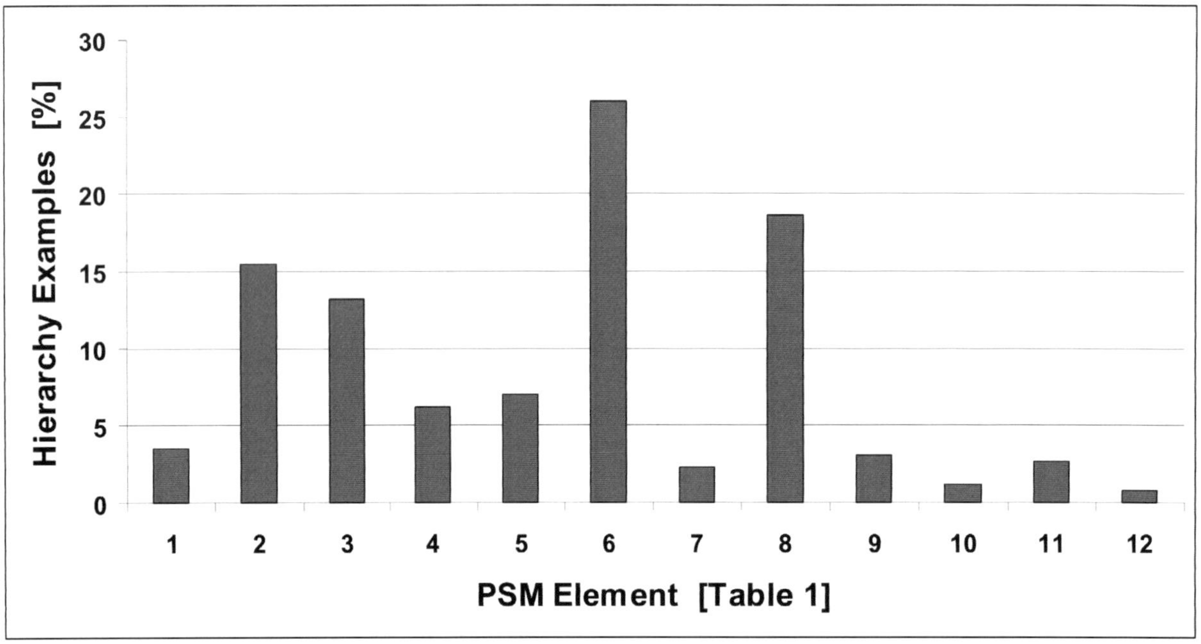

Figure 2 Breakdown of overall hierarchy examples according to applicable PSM elements.

The trends in Figures 1 and 2 can also be compared with those shown in Figure 3 from reference [75]. Figure 3 is a collection of incident causation data collected and analyzed over the indicated seven-year period by the Process Safety Management Committee of the then Canadian Chemical Producers' Association (CCPA) thorough a procedure known as PRIM (Process-Related Incidents Measure). (CCPA is now known as the Chemistry Industry Association of Canada.) A general observation can be made here that deficiencies in Process Safety Management elements are related in part to deficiencies in the hierarchical control measures aligned with specific elements. Further, a select group of PSM elements (i.e., those with the highest percentages in Figures 1-3) are particularly important in this regard.

5. Inherent Safety Learnings

What can be learned – or perhaps reinforced – with respect to inherent safety from the current analysis of CSB investigation reports? There are two parts to the answer to this question; the first deals with the usefulness of ISD in the process industries while the second relates to the intrinsic value of the CSB reports themselves. Simply put, inherent safety works – and the CSB reports show what can happen when the concept is not given its full measure of consideration.

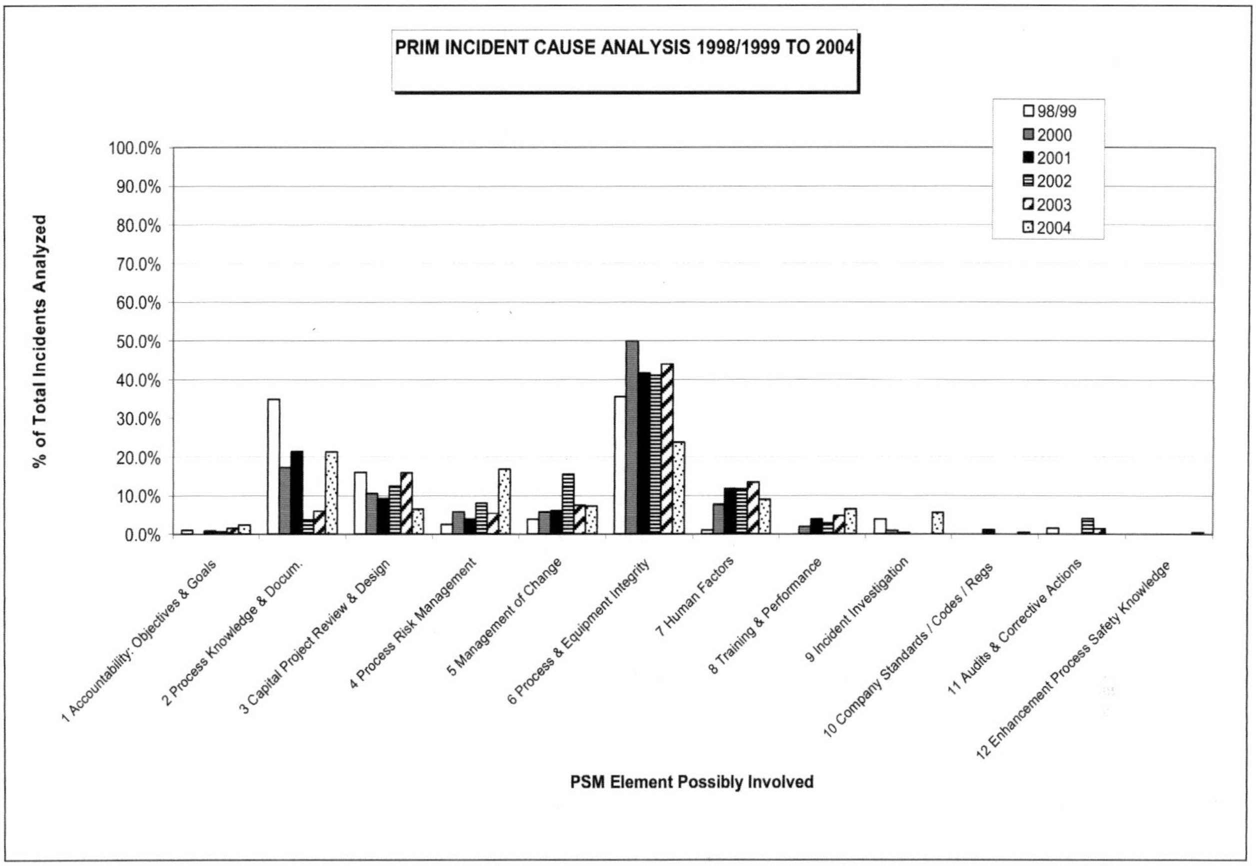

Figure 3 Incident causation by PSM element from 1998/99 to 2004 (PRIM data) [75].

The comments were made earlier that inherent safety is not a stand-alone concept and that its use within the hierarchy of controls highlights the need for caution with respect to the reliability of mechanical devices and human actions. The validity of these remarks has been demonstrated in the present work by extension of the analysis to the full hierarchy and by examination of the potential failings of passive and active safety devices as well as procedural measures such as safety guidelines and emergency procedures. A quantitative link has also been established between inherent safety and process safety management; previous work by the present authors [75] had demonstrated the qualitative link between these key process safety concepts.

The work of Kaszniak [72] has been referenced previously to indicate consistency of the current work with analyses conducted by other authors. Kaszniak's full set of findings related to process hazards analyses (PHAs) from his review of CSB investigations are as follows [72]:

- *No PHA conducted.*
- *Previous incidents not included.*
- *PHA recommendations not completed.*
- *Facility siting.*
- *Adequacy of controls, safeguards, and layers.*
- *Credible hazard scenarios not evaluated.*
- *Scale up is hard to do.*

In the work presented here, it has been demonstrated that the CSB reports contain inherent safety learnings related to three of the above items (facility siting, adequacy of controls, and scale-up).

Yang et al. [76] have recently analyzed a set of case histories (including a small subset of CSB investigations) for common lessons learned. Their findings, in the form of recommendations, are as follows [76]:

- *Perform a careful review of materials used in construction; create a record change notification.*
- *Maintenance and inspection should be performed periodically to ensure equipment reliability.*
- *Manufacturers should develop effective worker training programs.*
- *Major Hazard Installations (MHI) should be installed to minimize the impact of an explosion.*
- *Plant layout and siting should be designed far away from the residential area.*
- *Complete plan and appropriate protection system should be in place in case of emergency.*
- *Process hazard analysis must be implemented periodically.*
- *Actions must be taken to prevent potential hazards according to lessons learned from similar incidents in the past.*
- *Appropriate protection system should always be in place in the case of an emergency.*
- *Appropriate operation procedures must incorporate all safety procedures and standard industry practices.*

There are many commonalities between the current work and the comprehensive listing given by Yang et al. [76]. For example, inherent safety learnings are apparent with respect to materials of construction (*substitution*), and MHI installation, plant layout and siting (*moderation*). Maintenance and inspection to ensure equipment reliability are clearly related to the PSM element of *process and equipment integrity*. Training programs are critical to effective *procedural safety*, and emergency protection systems and appropriate operation procedures essentially complete the *hierarchy of controls*.

A common item in the lists compiled by Kaszniak [72] and Yang et al. [76] is the use of case histories, expressed in the following manner:

- *Previous incidents not included* [72].
- *Actions must be taken to prevent potential hazards according to lessons learned from similar incidents in the past* [76].

The current authors agree with this finding and would add that the reports provided by the CSB have immense value as case studies for multiple process safety purposes. With respect to the primary scope of the present paper, the following inherent safety uses for CSB reports are enthusiastically recommended:

- in the performance of PHAs, HAZOPs, etc. as a valuable data base of past incidents,
- at the initial design stage for guidance on facility layout considerations,
- for training exercises aimed at enhancing knowledge on the practical applications of ISD,
- in performing incident investigations with an objective of making recommendations to eliminate or reduce hazards,
- as evidence of the need to manage changes that may arise from ISD implementation, and
- by process safety educators in their lectures on inherent safety.

6. Conclusion

Investigation reports, case studies, safety bulletins and urgent recommendations produced by the U.S. Chemical Safety Board contain numerous examples where safety measures in all categories of the hierarchy of controls would be helpful in reducing risk in the process industries. The four main principles of inherent safety (minimization, substitution, moderation and simplification), as well as passive and active engineered safety devices and procedural safety measures, all have a role to play in the prevention and mitigation of process incidents such as those investigated by the CSB. There is also a strong connection evident in CSB investigations between available risk control measures and the various activities required to effectively manage process safety.

In addition to providing relevant industrial examples of inherent, engineered and procedural safety, CSB reports have multiple potential uses with respect to ISD implementation and process safety in general. Examples include evidence-based guidance on process hazard analyses, facility layout and siting, incident investigation, and management of change. The training usefulness and educational value of these reports are also of benefit to practitioners and educators alike.

Recent reports from the Chemical Safety Board have directly incorporated the language or terminology of inherent safety. This trend will hopefully continue, and in doing so, should enhance the adoption of explicit inherent safety terms (e.g., *inherently safer alternative*, *moderation of process conditions*, *minimization of hazardous inventories*, etc.) in process company investigation reports. In this manner, wider adoption of ISD principles themselves will be encouraged.

7. Acknowledgement

The authors gratefully acknowledge the financial support of the Natural Sciences and Engineering Research Council of Canada in the form of discovery and strategic grants and an undergraduate student research award.

8. References

[1] CSB, *Investigation Report – Explosives Manufacturing Incident – Sierra Chemical Company*, Report No. 98-001-I-NV, U.S. Chemical Safety and Hazard Investigation Board, Washington, DC (1998).

[2] CSB, *Summary Report – Nitrogen Asphyxiation – Union Carbide Corporation*, Report No. 98-05-I-LA, U.S. Chemical Safety and Hazard Investigation Board, Washington, DC (1999).

[3] CSB, *Investigation Report – Propane Tank Explosion – Herrig Brothers Feather Creek Farm*, Report No. 98-007-I-IA, U.S. Chemical Safety and Hazard Investigation Board, Washington, DC (1999).

[4] CSB, *Investigation Report – Chemical Manufacturing Incident – Morton International Inc.*, Report No. 1998-06-I-NJ, U.S. Chemical Safety and Hazard Investigation Board, Washington, DC (2000).

[5] CSB, *Investigation Report – Catastrophic Vessel Overpressurization – Sonat Exploration Company*, Report No. 1998-002-I-LA, U.S. Chemical Safety and Hazard Investigation Board, Washington, DC (2000).

[6] CSB, *Investigation Report – Refinery Fire Incident – Tosco Avon Refinery*, Report No. 99-014-I-CA, U.S. Chemical Safety and Hazard Investigation Board, Washington, DC (2001).

[7] CSB, *Safety Bulletin – Management of Change*, No. 2001-04-SB, U.S. Chemical Safety and Hazard Investigation Board, Washington, DC (2001).

[8] CSB, *Investigation Report – Steel Manufacturing Incident – Bethlehem Steel Corporation*, Report No. 2001-02-I-IN, U.S. Chemical Safety and Hazard Investigation Board, Washington, DC (2002).

[9] CSB, *Case Study – The Explosion at Concept Sciences: Hazards of Hydroxylamine – Concept Sciences, Inc.*, No. 1999-13-C-PA, U.S. Chemical Safety and Hazard Investigation Board, Washington, DC (2002).

[10] CSB, *Investigation Report – Thermal Decomposition Incident – BP Amoco Polymers, Inc.*, Report No. 2001-03-I-GA, U.S. Chemical Safety and Hazard Investigation Board, Washington, DC (2002).

[11] CSB, *Investigation Report – Refinery Incident – Motiva Enterprises LLC*, Report No. 2001-05-I-DE, U.S. Chemical Safety and Hazard Investigation Board, Washington, DC (2002).

[12] CSB, *Hazard Investigation – Improving Reactive Hazard Management*, Report No. 2001-01-H, U.S. Chemical Safety and Hazard Investigation Board, Washington, DC (2002).

[13] CSB, *Investigation Report – Hydrogen Sulfide Poisoning – Georgia-Pacific*, Report No. 2002-01-I-AL, U.S. Chemical Safety and Hazard Investigation Board, Washington, DC (2003).

[14] CSB, *Investigation Report – Petroleum Products Facility Incident – Third Coast Industries*, Report No. 2002-03-I-TX, U.S. Chemical Safety and Hazard Investigation Board, Washington, DC (2003).

[15] CSB, *Investigation Report – Chlorine Release – DPC Enterprises, L.P.*, Report No. 2002-04-I-MO, U.S. Chemical Safety and Hazard Investigation Board, Washington, DC (2003).

[16] CSB, *Safety Bulletin – Hazards of Nitrogen Asphyxiation*, No. 2003-10-B, U.S. Chemical Safety and Hazard Investigation Board, Washington, DC (2003).

[17] CSB, *Investigation Report – Vapor Cloud Deflagration and Fire – BLSR Operating, Ltd.*, Report No. 2003-06-I-TX, U.S. Chemical Safety and Hazard Investigation Board, Washington, DC (2003).

[18] CSB, *Case Study – Hydrogen Sulfide Exposure – Environmental Enterprises, Inc.*, No. 2003-02-C-OH, U.S. Chemical Safety and Hazard Investigation Board, Washington, DC (2003).

[19] CSB, *Investigation Report – Chemical Waste-Mixing Incident – Kaltech Industries Group, Inc.*, Report No. 2002-02-I-NY, U.S. Chemical Safety and Hazard Investigation Board, Washington, DC (2003).

[20] CSB, *Investigation Report – Explosion and Fire – First Chemical Corporation*, Report No. 2003-01-I-MS, U.S. Chemical Safety and Hazard Investigation Board, Washington, DC (2003).

[21] CSB, *Case Study – Fire and Explosion: Hazards of Benzoyl Peroxide – Catalyst Systems, Inc.*, No. 2003-03-C-OH, U.S. Chemical Safety and Hazard Investigation Board, Washington, DC (2003).

[22] CSB, *Investigation Report – Catastrophic Vessel Failure – D.D. Williamson & Co., Inc.*, Report No. 2003-11-I-KY, U.S. Chemical Safety and Hazard Investigation Board, Washington, DC (2004).

[23] CSB, *Safety Bulletin – Removal of Hazardous Material from Piping Systems*, No. 2004-03-B, U.S. Chemical Safety and Hazard Investigation Board, Washington, DC (2004).

[24] CSB, *Safety Bulletin – Sodium Hydrosulfide: Preventing Harm*, No. 2003-03-B, U.S. Chemical Safety and Hazard Investigation Board, Washington, DC (2004).

[25] CSB, *Investigation Report – Vent Collection System Explosion – Technic Inc.*, Report No. 2003-08-I-RI, U.S. Chemical Safety and Hazard Investigation Board, Washington, DC (2004).

[26] CSB, *Case Study – Explosion at Biochemical Facility: Liquid Nitric Oxide Release – Isotec*, No. 2003-15-C-OH, U.S. Chemical Safety and Hazard Investigation Board, Washington, DC (2004).

[27] CSB, *Investigation Report – Dust Explosion – West Pharmaceutical Services, Inc.*, Report No. 2003-07-I-NC, U.S. Chemical Safety and Hazard Investigation Board, Washington, DC (2004).

[28] CSB, *Investigation Report – Combustible Dust Fire and Explosions – CTA Acoustics, Inc.*, Report No. 2003-09-I-KY, U.S. Chemical Safety and Hazard Investigation Board, Washington, DC (2005).

[29] CSB, *Investigation Report – Chlorine Release, Contaminated Antimony Pentachloride Exposure, Hydrogen Fluoride Release – Honeywell International, Inc.*, Report No. 2003-13-I-LA, U.S. Chemical Safety and Hazard Investigation Board, Washington, DC (2005).

[30] CSB, *Investigation Report – Aluminum Dust Explosion – Hayes Lemmerz International-Huntington, Inc.*, Report No. 2004-01-I-IN, U.S. Chemical Safety and Hazard Investigation Board, Washington, DC (2005).

[31] CSB, *Case Study – Oil Refinery Fire and Explosion – Giant Industries' Ciniza Oil Refinery,* No. 2004-08-I-NM, U.S. Chemical Safety and Hazard Investigation Board, Washington, DC (2005).

[32] CSB, *Safety Bulletin – Explosion at ASCO: Dangers of Flammable Gas Accumulation*, No. 2006-01-B, U.S. Chemical Safety and Hazard Investigation Board, Washington, DC (2006).

[33] CSB, *Investigation Report – Sterigenics– Sterigenics*, Report No. 2004-11-I-CA, U.S. Chemical Safety and Hazard Investigation Board, Washington, DC (2006).

[34] CSB, *Investigation Report – Toxic Chemical Vapor Cloud Release – MFG Chemical, Inc.*, Report No. 2004-09-I-GA, U.S. Chemical Safety and Hazard Investigation Board, Washington, DC (2006).

[35] CSB, *Case Study – Polyethylene Wax Processing Facility Explosion and Fire – Marcus Oil and Chemical*, No. 2005-02-I-TX, U.S. Chemical Safety and Hazard Investigation Board, Washington, DC (2006).

[36] CSB, *Safety Bulletin – Fire at Praxair St. Louis – Dangers of Propylene Cylinders in High Temperatures*, No. 2005-05-B, U.S. Chemical Safety and Hazard Investigation Board, Washington, DC (2006).

[37] CSB, *Case Study – Fire at Formosa Plastics Corporation: Evaluating Process Hazards – Formosa Plastics Corporation*, No. 2006-01-I-TX, U.S. Chemical Safety and Hazard Investigation Board, Washington, DC (2006).

[38] CSB, *Safety Bulletin – BP Texas City, TX Refinery Fire – Positive Material Verification: Prevent Errors During Alloy Steel Systems Maintenance*, No. 2005-04-B, U.S. Chemical Safety and Hazard Investigation Board, Washington, DC (2006).

[39] CSB, *Case Study – Confined Space Entry - Worker and Would-be Rescuer Asphyxiated – Valero Energy Corporation Refinery*, No. 2006-02-I-DE, U.S. Chemical Safety and Hazard Investigation Board, Washington, DC (2006).

[40] CSB, *Investigation Report – Combustible Dust Hazard Study*, Report No. 2006-H-1, U.S. Chemical Safety and Hazard Investigation Board, Washington, DC (2006).

[41] CSB, *Investigation Report – Chlorine Release – DPC Enterprises, L.P.*, Report No. 2004-02-I-AZ, U.S. Chemical Safety and Hazard Investigation Board, Washington, DC (2007).

[42] CSB, *Investigation Report – Vinyl Chloride Monomer Explosion – Formosa Plastics Corp.*, Report No. 2004-10-I-IL, U.S. Chemical Safety and Hazard Investigation Board, Washington, DC (2007).

[43] CSB, *Investigation Report – Methanol Tank Explosion and Fire – Bethune Point Wastewater Treatment Plant*, Report No. 2006-03-I-FL, U.S. Chemical Safety and Hazard Investigation Board, Washington, DC (2007).

[44] CSB, *Investigation Report – Refinery Explosion and Fire – BP*, Report No. 2005-04-I-TX, U.S. Chemical Safety and Hazard Investigation Board, Washington, DC (2007).

[45] CSB, *Case Study – Mixing and Heating a Flammable Liquid in an Open Top Tank – Universal Form Clamp, Inc.*, Investigation No. 2006-08-I-IL, U.S. Chemical Safety and Hazard Investigation Board, Washington, DC (2007).

[46] CSB, *Case Study – Hot Work Control and Safe Work Practices at Oil and Gas Production Wells – Partridge-Raleigh Smith County Oilfield.*, Report No. 2006-07-I-MS, U.S. Chemical Safety and Hazard Investigation Board, Washington, DC (2007).

[47] CSB, *Safety Bulletin – Emergency Shutdown Systems for Chlorine Transfer*, No. 2005-06-I-LA, U.S. Chemical Safety and Hazard Investigation Board, Washington, DC (2007).

[48] CSB, *Case Study – Runaway Chemical Reaction and Vapor Cloud Explosion – Synthron, LLC*, No. 2006-04-I-NC, U.S. Chemical Safety and Hazard Investigation Board, Washington, DC (2007).

[49] CSB, *Case Study – Fire and Community Evacuation in Apex, North Carolina – Environmental Quality Company*, 2007-01-I-NC, U.S. Chemical Safety and Hazard Investigation Board, Washington, DC (2008).

[50] CSB, *Investigation Report – Confined Vapor Cloud Explosion – CAI, Inc. and Arnel Company, Inc.*, Report No. 2007-03-I-MA, U.S. Chemical Safety and Hazard Investigation Board, Washington, DC (2008).

[51] CSB, *Case Study – Barton Solvents – Static Spark Ignites Explosion Inside Flammable Liquid Storage Tank*, No. 2007-06-I-KS, U.S. Chemical Safety and Hazard Investigation Board, Washington, DC (2008).

[52] CSB, *Investigation Report – LPG Fire at Valero-McKee Refinery – Valero Energy Corporation*, Report No. 2007-05-I-TX, U.S. Chemical Safety and Hazard Investigation Board, Washington, DC (2008).

[53] CSB, *Case Study – Static Spark Ignites Flammable Liquid During Portable Tank Filling Operation*, No. 2008-02-I-IA, U.S. Chemical Safety and Hazard Investigation Board, Washington, DC (2008).

[54] CSB, *Investigation Report – Little General Store - Propane Explosion – Little General Store, Inc.*, Report No. 2007-04-I-WV, U.S. Chemical Safety and Hazard Investigation Board, Washington, DC (2008).

[55] CSB, *Investigation Report – Allied Terminals, Inc. - Catastrophic Tank Collapse – Allied Terminals, Inc.*, Report No. 2009-03-I-VA, U.S. Chemical Safety and Hazard Investigation Board, Washington, DC (2009).

[56] CSB, *Investigation Report – T2 Laboratories, Inc. – Runaway Reaction*, Report No. 2008-03-I-FL, U.S. Chemical Safety and Hazard Investigation Board, Washington, DC (2009).

[57] CSB, *Investigation Report – Sugar Dust Explosion and Fire – Imperial Sugar Company*, Report No. 2008-05-I-GA, U.S. Chemical Safety and Hazard Investigation Board, Washington, DC (2009).

[58] CSB, *Case Study – Uncontrolled Oleum Release - Petrolia, Pennsylvania*, 2009-01-I-PA, U.S. Chemical Safety and Hazard Investigation Board, Washington, DC (2009).

[59] CSB, *Safety Bulletin – Seven Key Lessons to Prevent Worker Deaths During Hot Work In and Around Tanks – Effective Hazard Assessment and Use of Combustible Gas Monitoring Will Save Lives*, No. 2009-01-SB, U.S. Chemical Safety and Hazard Investigation Board, Washington, DC (2010).

[60] CSB, *Urgent Recommendations from Kleen Energy Investigation*, U.S. Chemical Safety and Hazard Investigation Board, Washington, DC (2010).

[61] Center for Chemical Process Safety, *Inherently Safer Chemical Processes. A Life Cycle Approach*, 2nd edition, John Wiley & Sons, Hoboken, NJ (2009).

[62] Kletz, T. and Amyotte, P., *Process Plants. A Handbook for Inherently Safer Design*, 2nd edition, CRC Press/Taylor & Francis Group, Boca Raton, FL (2010).

[63] Hendershot, D.C., *A Summary of Inherently Safer Technology*, Process Safety Progress, **29** (4), 389-392 (2010).

[64] Wilson, L. and McCutcheon, D., *Industrial Safety and Risk Management*, University of Alberta Press, Edmonton, AB (2003).

[65] Canadian Society for Chemical Engineering, *Process Safety Management*, 3rd edition, Canadian Society for Chemical Engineering, Ottawa, ON (2002).

[66] Center for Chemical Process Safety, *Guidelines for Risk Based Process Safety*, John Wiley & Sons, Hoboken, NJ (2007).

[67] CSB, *Urgent Recommendations from ConAgra Investigation*, U.S. Chemical Safety and Hazard Investigation Board, Washington, DC (2010).

[68] CSB, *Case Study – Explosion and Fire in West Carrollton, Ohio – Veolia Technical Solutions, LLC*, 2009-10-I-OH, U.S. Chemical Safety and Hazard Investigation Board, Washington, DC (2010).

[69] CSB, *Investigation Report – Xcel Energy Hydroelectric Plant Penstock Fire – Cabin Creek*, Report No. 2008-01-I-CO, U.S. Chemical Safety and Hazard Investigation Board, Washington, DC (2010).

[70] Amyotte, P.R. and Eckhoff, R.K., *Dust Explosion Causation, Prevention and Mitigation: An Overview*, Journal of Chemical Health and Safety, **17** (1), 15-28 (2010).

[71] Amyotte, P.R., Pegg, M.J. and Khan, F.I., *Application of Inherent Safety Principles to Dust Explosion Prevention and Mitigation*, Process Safety and Environmental Protection, **87** (1), 35-39 (2009).

[72] Kaszniak, M., *Oversights and Omissions in Process Hazard Analyses: Lessons Learned from CSB Investigations*. Process Safety Progress, **29** (3), 264–269 (2010).

[73] Kidam, K., Hurme, M. and Hassim, M.H., *Inherent Safety Based Corrective Actions in Accident Prevention*, Proceedings of 13th International Symposium on Loss Prevention and Safety Promotion in the Process Industries, Volume 02, Brugge, Belgium (June 6-9, 2010), pp. 447-450.

[74] Hendershot, D.C., *Inherently Safer Design – Not Only About Reducing Consequences!*, 13th Process Plant Safety Symposium (7th Global Congress on Process Safety), Chicago, IL (March 13-16, 2011).

[75] Amyotte, P.R., Goraya, A.U., Hendershot, D.C. and Khan, F.I., *Incorporation of Inherent Safety Principles in Process Safety Management*, Center for Chemical Process Safety, American Institute of Chemical Engineers, Proceedings of 21st Annual International Conference – Process Safety Challenges in a Global Economy, Orlando, FL (April 23-27, 2006), pp. 175-207.

[76] Yang, X., Dinh, L., Castellanos, D., Osorio, C., Ng, D. and Mannan, M.S., *Common Lessons Learned from an Analysis of Multiple Case Histories*, Hazards XXI, IChemE Symposium Series No. 155, Manchester, UK (November 10-12, 2009), pp. 43-50.

Inherent Safety in Front End Engineering

Ian Sutton
AMEC Paragon
10777 Clay Road
Houston, TX 77041
ian.sutton@amec.com

Keywords: Inherent Safety

Abstract

The best time to apply the principles of inherent safety is during the early stages of a design. In particular it is at the early stages that the "Eliminate" option is best applied. For example, in the early stages of one design the process engineers had come up with a system of three knockout vessels in series. The vessels contained very high pressure gas. Serious consideration was given to eliminating two of the vessels and combining their functions in the first knockout vessel. Doing so led to a dramatic reduction in inventory of gas and also reduced problems to do with finding sufficient vertical space for the three vessel set-up.

The application of other inherent safety guidewords such as "Substitute" or "Minimize" is also best achieved during front end design. It is relatively easy to change the design to allow for a less toxic chemical to substitute for one that is more hazardous. However, it is much more difficult to make such as change once the final design has been authorized, or once the system is in operation.

This paper describes experiences with the application of the principles of inherent safety in the early design stages of a large project.

An early lesson learned was that most lead engineers do not have much training or exposure to the concepts of inherent safety, and they tend "not to think that way". Just telling them to apply the principles of inherent safety is not sufficient. In response to this concern a requirement was made that all technical documents, regardless of the discipline concerned, should start off with a short preamble showing how the author had considered the elements of inherent safety in his or her design.

1. Introduction

The topic of Inherent Safety (IS) in the process industries is well established and has been thoroughly described in many publications. Moreover, Inherent Safety is now being included in some process safety regulations, such as that from the New Jersey Department of Environmental Protection.

> *Facilities regulated by the TCPA program must perform Inherently Safer Technology reviews.*

(NJDEP 2010)

It is recognized that removing hazards, or at least reducing their magnitude, is generally a more effective way of improving safety than adding additional layers of instruments and/or improved operator training. Furthermore, hazard elimination or reduction is accomplished by means that are inherent to the design and are thus permanent and inseparable from the process. Add-on safety devices can always fail, but inherent safety is part and parcel of the process itself.

Yet, in spite of its obvious value, the application of Inherent Safety concepts to actual projects poses some challenges. In particular, because Inherent Safety often represents a new way of thinking, management can find it difficult to get engineers and other technical specialists to adjust the manner in which they approach the design of equipment and process systems.

The purpose of this paper is to discuss some of the practical issues and challenges that management faces when implementing Inherent Safety on their projects. The paper was written during the course of a project to do with the initial design, procurement, construction and commissioning of an offshore deepwater drilling and production platform. Client management had emphasized that Inherent Safety was a critical project driver, so the Technical Safety team was required to develop a program for implementing IS as the project moved from the Conceptual Stage into Front End Engineering Design (FEED) and on into detailed design. The team goal was to, "Bake Inherent Safety into the Design Pie".

Based on the lessons learned on this project, this paper provides practical suggestion and guidance as to how the principles of Inherent Safety can be applied to a project, with a particular focus on how to get project team members to "think that way".

TIMING

Although the principles of Inherent Safety can be applied to a facility at any time, even after it is in operation (Edwards and Chosnek 2011), it best used during the early stages of a design. It is at this time that fundamental changes can be made quickly and easily. Once detailed design is underway, such changes are harder to make, and they become even harder as a project enters the procurement, construction and commissioning phases. For example, it is relatively easy to change to a less toxic chemical before the Process Flow Diagrams (PFDs) have been finalized; it is much more difficult to do so once the final design has been authorized, or after the system is in operation.

The project described above was a good one on which to apply the principles of Inherent Safety because it was at the concept phase, which meant that most of the design decisions either had not been made or could be changed. Preliminary PFDs and Piping & Instrument Diagrams (P&IDs) had been prepared, but they were far from being finalized. Similarly, a general layout had been prepared, but the design team still had a good degree of freedom to make changes.

CHOICE OF GUIDEWORDS

Inherent Safety programs are usually structured around a set of guidewords. Different authorities and authors use different sets of words. For example, The Center for Chemical Process Safety (CCPS 1996) suggests the following guidewords.

1. Minimize
2. Substitute
3. Moderate
4. Simplify

One of the early decisions made on this project was to allow the team to select its own guidewords. Much of the Inherent Safety literature is written for the chemical industries, yet the safety environment on an offshore platform differs from that of a chemical plant. For example, highly toxic chemicals are not generally used on an offshore platform (although the presence of hydrogen sulfide and carbon dioxide can be a major hazard), but the risks associated with dropped objects and blocked emergency escape routes are high. Therefore guidance that is appropriate for the onshore process industries is valuable, but also limited in usefulness.

For this project, the team selected the guidewords shown below.

1. Eliminate
2. Minimize
3. Substitute
4. Moderate
5. Simplify

The hierarchy is significant; the words at the top of the list were considered to have a greater effect than those lower down. For example, completely removing a pump from the process (*Eliminate*) is likely to have a much greater impact on safety than merely making the pump easier to use (*Simplify*).

Eliminate

Generally, the best way of achieving Inherent Safety is to eliminate the hazard altogether, as canbe seen from the following risk equation.

$$Risk_{Hazard} = Consequence * Predicted\ Frequency$$

Risk is estimated and then reduced by first identifying the hazards, generally through the use of a HAZID or HAZOP. Each hazard has associated consequences (safety, environmental, economic), and also a predicted frequency rate.

The equation shows that, as long as hazards are present, the total risk can never be zero — a truth not always grasped by members of the general public or the news media. Hazards are always present within all industrial facilities. Those hazards always have undesirable consequences, and their likelihood of occurrence is always finite. The consequence and likelihood terms can be reduced in size, but they can never be eliminated. The only way to achieve a truly risk-free operation is to remove the hazards altogether, *i.e.,* to apply the Inherent Safety *Eliminate* guideword.

Overton and King (2006) provide an example in which the liquefaction step in a chlorination process allowed for the elimination of at storage tank containing 750 tons of chlorine.

For the offshore platform project, attention was paid to opportunities for eliminating pumps and valves (although the latter often provide a major safety function). For example, one proposed process design consisted of a system of three knockout vessels in series. The vessels contained very high pressure gas. Serious consideration was given to eliminating two of the vessels and combining their functions in the first knockout vessel. Doing so led to a sizeable reduction in inventory of gas and also reduced problems to do with finding sufficient vertical space for the three vessel set-up (space limitations are always a problem on offshore platforms).

Minimize

'Minimization' generally refers to the use of smaller quantities of hazardous substances. The Bhopal tragedy of 1984 was a missed opportunity for application of this guideword. Had the inventory of methyl isocyanate been less than it was, far fewer people would have died.

For an offshore platform there are not as many opportunities to reduce inventories as there are onshore. Platforms generally have little or no storage (oil, gas and water are treated and immediately sent to an onshore facility). Nor do they generally have significant quantities of highly hazardous chemicals on board.

One minimization initiative considered on the project was to connect piping sections with welds rather than flanges. Doing so reduces the number of potential leak sources. (It does, however, make general maintenance more difficult and thereby increase risk.)

Substitute

'Substitution' also tends to be more applicable to chemical plants than it is to an offshore facility. Although chemicals such as glycol are used to removed hydrates from subsea piping, these chemicals are usually relatively benign. The same can be said of many of the miscellaneous chemicals that are used for activities such as corrosion control.

Moderate

'Moderation' generally refers to the use of a use a less hazardous condition. In many cases, the team will determine if it is possible to operate at lower temperatures or pressures. On this project much of the electrical equipment operated at high voltage. Therefore careful consideration was given to 'moderating' high voltage with low voltage.

Simplify

Simplification is the lowest of the levels of Inherent Safety considered by the team. It often relates to making equipment easy to use and tolerant of operating error. Simplification also makes operation of the facility during an emergency less prone to error.

With regard to the offshore project, the design carefully considered the 'Simplify' guideword during the procurement stage, particularly with respect to rotating equipment. If one vendor offered a design that used fewer moving parts, for example, then that product was considered favorably.

The project was at too early a stage for simplification to do with human factors and operating procedures to be considered.

Removing People

Related to the concept of hazard elimination, that was discussed above, is the idea of removing people from the site of a potential incident. The term, "What you don't have, can't leak" was used by Kletz (1978). A derivative of that phrase would be, "If a man's not there, he can't be killed" (Sutton 2010).

A good example of the application of this concept can be seen with respect to hurricanes in the Gulf of Mexico (GoM). During the period 2003 - 2005 some 164 offshore platforms were either lost or seriously damaged due to hurricanes. Although the economic loss was high, the number of fatalities and serious injuries associated with the storms was zero; the reason being that, whenever a hurricane is brewing and heading toward a platform the crew is evacuated.

In the long-term, one of the best means of improving safety is to develop systems that are so automated that very few humans are required to be in the vicinity of operating equipment so that they are not exposed to hazards. With regard to the offshore project, the platform will have a substantial crew (more than 70 people) during the drilling phase, but will be totally unmanned once drilling ceases and production is the only activity. Control of the operations would be from an onshore facility.

In general manning levels can be reduced by the following actions:

- Simplification of the design;
- Robust design;
- Use of passive rather than active corrosion management;
- Designing out the need for local intervention; and
- Specification of proven, high-reliability equipment.

IMPLEMENTATION PROGRAM

Some of the techniques used to implement Inherent Safety on the project are described in this section. As already noted, the overall project aim was to "Bake Inherent Safety into the Design Pie". The idea behind this phrase was that Inherent Safety would become part of all project activities, and that engineers and other technical specialists would follow its principles at all times. Inherent Safety would not be an add-on, or something directed by the Technical Safety group.

The Inherent Safety goal was integrated into three other, compatible drivers. They were:

- No injuries to people at any stage of the project or when the facility was in operation.
- A flawless startup.
- No surprises regarding the project budget or schedule.

Team Effort

The first and most crucial decision was to make the application of Inherent Safety a multi-discipline team effort. The Inherent Safety work was managed by the Technical Safety lead, but all decisions of significance, including the choice of guidewords, were a team decision.

Not an "Add On"

It was made clear to the team that Inherent Safety be considered throughout the design — in no way should the concept be an "add on" or as something as being just in the domain of Technical and Process Safety.

Safety Moment

It is commonplace for companies to start formal meetings with a Safety Moment, in which someone describes an event or lesson learned that can help the other people better understand and improve their own safety performance. For this project it was decided that many of the Safety Moments would be to do with Inherent Safety. The person responsible for the Safety Moment would discuss some activity that had been carried out or that was planned that showed how Inherent Safety was being used on the project.

Technical Documents

An early lesson learned was that most lead engineers do not have much training or exposure to the concepts of Inherent Safety, and they tend "not to think that way". Just telling them to apply the principles of Inherent Safety is not sufficient. In response to this concern a requirement was made that *all* technical documents, regardless of the discipline concerned, should start off with a short preamble showing how the author had considered the elements of Inherent Safety when preparing that document.

Examples included the following:

- An electrical engineer issuing a specification for a compressor motor would demonstrate how he considered the possibility of using lower voltage electricity ("Moderate").
- A mechanical engineer responsible for designing the facility crane would show how consideration was given to minimizing the maximum weight load so as to reduce the impact of a dropped object ("Minimize").
- A systems engineer would demonstrate that consideration was given to importing electrical power rather than generating it locally ("Eliminate").

In practice, many of the lead engineers were not entirely comfortable with the above approach; it represented a new way of thinking. In these cases an alternative approach was used. All the technical leads would gather in a meeting to carry out a traditional squad check of the document. However, in addition to carrying out the normal engineering review, a technical safety expert would guide the team through a discussion of the Inherent Safety guidewords as they applied to that document.

Team meetings led to a second benefit. The application of the principles of Inherent Safety can lead to unanticipated and unwanted results (the Law of Undesirable Consequences, discussed below). In the case of the three pressure vessel train, for example, merging their functions into one vessel meant that solids removal became more difficult and potentially exposed maintenance workers to greater risk. A team discussion is more likely to generate an understanding of potential increased risks than is a single-person review.

Hazard Analyses

The principles of Inherent Safety were also applied during the normal HAZID (Hazard Identification) process and in the early HAZOPs. HAZIDs traditionally aim to identify hazards such as high level in a vessel, total loss of power or a spill of a toxic chemical. The team then considers the consequences and likelihood of such events and, if necessary, generates findings and recommendations. On this project the Inherent Safety guidewords listed above were explicitly called out during the HAZID.

Implementing the principles of Inherent Safety during the routine HAZIDs helped make the final design safer. However, there was a second, and more subtle, benefit that resulted from this activity. As the project progressed most of the engineers and other technical specialists started to "get" the principles of Inherent Safety, and start to apply them in all situations, and at all times.

Such a way of thinking is ultimately what Inherent Safety is all about, and is analogous to what occurred as HAZOPs were introduced. The immediate benefit of a HAZOP is to identify and correct high risk hazards. However, a more subtle, and ultimately much more important, role of a HAZOP is to inculcate a way of thinking that is baked into the way all personnel act and work. People learn to "think the unthinkable". So it is with Inherent Safety; use of the techniques discussed above help make the topic "part of the furniture".

CHALLENGES

Although the concepts of Inherent Safety were accepted intellectually without much difficulty by the project team members; the practical application of these concepts was more of a challenge. Some of the issues faced included the following:

- Simple design
- Need for imagination
- Engineering standards;
- Focus on process engineering;
- Inherent and passive safety; and
- Value of backup equipment.

Simple Design

Although large, the offshore platform that was being designed was quite simple in its overall design, and it was quite similar to other platforms of the same type. Therefore, it was difficult at times to come up with ideas for the application of the Inherent Safety guidewords. The lack of opportunity for finding new ideas tended to dampen the enthusiasm of the team members.

Imagination

On September 25th 2001, just days after the 9/11 attacks, the New York Times columnist Thomas Friedman wrote,

The World Trade Center is not the place where our intelligence agencies failed. It is the place where our imaginations failed.

Prior to that attack very few people had even conceived that terrorists could take over airplanes and use them to hit large buildings. They lacked imagination. The same problem can apply to the use of Inherent Safety.

By its very nature the technique requires that people come up with new ideas, that they "think the unthinkable" — but not that "they think out of the box" (anyone who uses the phrase "thinking out of the box" shows by the staleness of their imagery that they are well inside the box). Simply put, thinking is hard work — yet such thinking is at the heart of any Inherent Safety program. In the words of one of the Monte Python Gumbys, "My brain hurts!"

Most people prefer to work in a well-established system using previous designs and proven engineering standards. Yet Inherent Safety requires that people use their imagination, particularly with regard to the 'Eliminate' guideword. This is difficult.

In addition, many processes — including offshore platforms — are quite similar to one another. There is considerable pressure on the team simply to repeat previous designs on the basis of, "It it's not broken, don't fix it." Moreover, the strict discipline that engineers and technical specialists must follow with regard to the application of industrial standards can discourage imaginative thinking.

A further difficulty is that a person who is skilled and experienced at leading a standard risk management program such as a HAZOP analysis, may not be as effective in having team members think creatively. It is relatively easy to define the hazards, estimate their consequences and likelihood, come up with an assessment of overall risk, and then generate some recommendations. A different patter of thinking is needed when applying concepts such as eliminate or simplify, and the HAZOP team leader may not be the best choice of person to lead the activity.

Rather than having a good knowledge of risk management, as most HAZOP leaders do, the person leading an Inherent Safety analysis needs a detailed understanding of the different engineering disciplines.

Engineering Standards

Engineers organize their work around the hundreds of standards from bodies such as the API, ASME, NFPA and IEEE. Since these standards (and their regular updates) have typically been in service for many years engineers tend to instinctively follow them, rather than look for fresh ideas. For example, when faced with a pressure control problem on a vessel, an engineer designing an offshore platform will instinctively tend to reach out for API Recommended Practice 14C rather than think through whether the instrumentation in its present form is even needed.

Focus on Process Engineering

Much of the literature to do with Inherent Safety is published in chemical process journals, and therefore tends to look at process-related issues, such as the replacement of a hazardous chemical with one that is more benign. Other industries have a different set of safety concerns. As already discussed, offshore platforms and drill rigs do not typically handle highly toxic materials, and their process steps are relatively simple when compared to a chemical plant or refinery. However, they have serious concerns with dropped objects (usually when items are being transferred from a service boat by crane). Also, if there is an emergency offshore, escape can be a major problem. Movement on the deck could easily be obscured by smoke, and finding a safe location can be a challenge. The Inherent Safety program for an offshore drill rig or platform needs to reflect these realities.

Inherent and Passive Safety

Hendershot (2006) places safety strategies into one of four categories:

- Inherent
- Passive
- Active
- Procedural

It was found on the offshore project that the team members readily understood the philosophical differences between Inherent and Active Safety. For example, removing a vessel from a process is likely to be more effective in improving safety than adding instruments to prevent incidents to do with high pressure and high level.

However, the distinction between Inherent and Passive Safety is more subtle. Using an example from a typical onshore operation, it is standard practice to put a containment (berm or bund) wall around a tank that stores a hazardous chemical. This concept is illustrated for Tank, T-100, in the sketch below.

The wall is often an earthern berm, although concrete is sometimes used. The capacity of the containment system is generally 110% of the capacity of the largest tank. This means that, if the tank leaks for reasons such as corrosion or impact with a vehicle, the liquid will be contained.

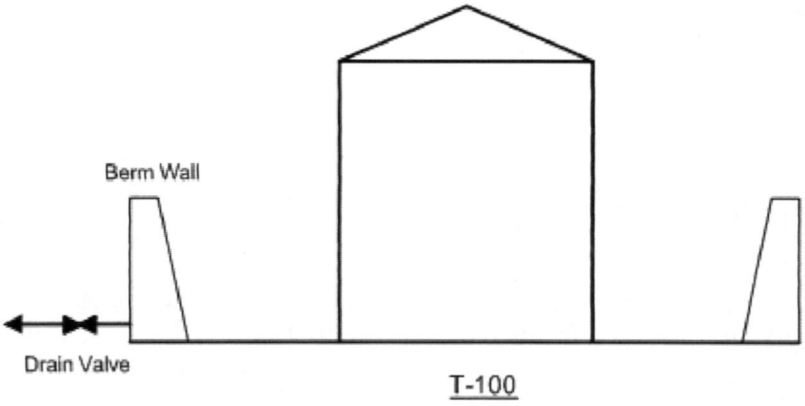

This system is *passively* safe, *i.e.,* it does not rely on the active response of equipment items such as a level gauge on the tank, or a pump to remove spilled liquid. The berm wall will remain in existence regardless of what is going on around it, and will work effectively, even if all the other control systems have failed. However, the system is not *inherently* safe. The spilled liquid may be contained, but it could still be hazardous (for example, it may catch fire or emit toxic fumes). Also, there is usually a drain valve placed in the wall so that rain water can be removed. This drain valve could be inadvertently left open. Indeed, it is possible that a large explosion at the facility could destroy the wall itself. The use of a containment wall has reduced risk both by reducing the likelihood of a spill to the general environment, and by mitigating the risk in case something does go awry. However this system is not "inherently safe" — it is just much less risky. An inherently safe system would be one in which the process was redesigned such that there was no need for a tank at all.

Inventory Management

The 'Moderate' guideword would suggest that small quantities of hazardous materials be used wherever possible. Then, if there were to be a leak, the consequences would be less severe. However, the use of small quantities may actually reduce safety. For example, a facility that uses a toxic chemical in liquid form may have the choice of either bringing in one large shipment every week, or multiple small deliveries ten times a week. The latter would appear to be the inherently safer of the two operations. However, the latter operation involves many more deliveries, and opportunities for error during the connection and disconnection operations. In other words, the risk reduction associated with having lower consequence events may be erased due to the higher likelihood of an event taking place.

A detailed discussion to do with the tradeoff between storage size and the number of loading/unloading activities is provided by Englund (1991).

Value of Backup Equipment

Many equipment items are provided with an on-line spare or backup. On the offshore project, for example, critical pumps were spared so that, if the primary pump were to fail, the second pump could be brought on line quickly.

From an Inherent Safety point of view, the spare item poses an unnecessary hazard — it could leak or otherwise cause a problem, and it will require routine maintenance, which always involves a risk to the maintenance workers. However, if it is removed, and the primary pump fails then a new set of safety problems can arise. These could include the possibility of high level or low level in associated vessels and the need to blowdown equipment to the flare.

The "Law of Unintended Consequences" is a term that is generally used somewhat ironically or tongue in cheek, but that should be considered when applying the principles of inherent safety. The basic idea behind this "law" is that human intervention in complex systems tends to produce a range of unexpected outcomes, most of which are undesirable, and which could readily lead to losses which negate the benefits many times over. For example, the rabbit was introduced into Australia for food, but eventually became a highly destructive pest.

Law of Unintended Consequences

The unintended consequences generally fall into one of three general categories.

- *The unexpected outcome is desirable.* Such a consequence is sometimes referred to as serendipity. An example could be to do with a pumping system. The original design may call for two pumps: one operating and the other serving as a backup. If the pump can be redesigned such that it is much more reliable, then the spare pump can be removed, thus improving Inherent Safety. But it may also be found that the new design results in much less maintenance, thus reducing the number of times maintenance workers have to be put at risk. Or it may be found that the reduced pump downtime results in increased catalyst life in a downstream reactor because it is subject to fewer swings in flow rate and temperature.

- *The unexpected outcome is undesirable.* Using the same example to do with the pumps, it may be found that, because the workers now have to work on the item less frequently they are less experienced, and so, when they have to do so they therefore are more likely to make a mistake and increase their chances of being hurt.

- *The outcome is one that makes the original situation worse.* Once more using the pump example, it may be found that having a spare pump makes the operations personnel more willing to operate the system closer to the safety limits.

Conclusions

The concept of Inherent Safety is well understood and has been extensively discussed in the process safety literature. Its application to actual projects can certainly lead to reduced risk, and improved safety in particular. However there are challenges, of which probably the biggest is that it can be difficult to persuade senior engineers and other highly qualified professionals to adopt a new way of thinking.

Citations

Center for Chemical Process Safety (CCPS). Inherently safer chemical processes: A life cycle approach. D.A. Crowl (editor). American Institute of Chemical Engineers. New York, 1996.

Edwards, V.H., J Chosnek. Making Existing Process Plants Inherently Safer. 7th Global Congress of Process Safety. March 2011.

Englund, S.M. Design and Operate Plants for Inherent Safety - Part 1. Chemical Engineering Progress. March 1991.

Hendershot, Dennis, C. An Overview of Inherently Safety Design. Process Safety Progress. June 2006.

Kletz, T. A. What you don't have can't leak. Chem Ind. May 1978.

NJDEP. Rules and Regulations. http://www.state.nj.us/dep/rpp/brp/tcpa/tcpadown.htm. December 12th 2010.

Overton, T., G.M. King. Inherently Safer Technology: An Evolutionary Approach. Process Safety Progress. June 2006.

Sutton, Ian S. Process Risk and Reliability Management. Elsevier, Oxford. 2010.

Not the abstract again. Perhaps background information. Why did you do what was reported. A general outline of the paper.

2. Main section 1 heading

2.1 sub-section title

sub-section text

2.1.1 sub-sub section title

sub-sub-section text

2.1.2 sub-sub section title

sub-sub-section text

example of equation inclusion:

$$\mathbf{x} = f(\mathbf{y}) \quad \text{[Eq. 1]}$$

2.2 sub-section title

sub-section text

2.2.1 sub-sub section title

sub-sub-section text

2.2.2 sub-sub section title

including an example table

Table 1. Title of Table

Column Title	Column Title	Column Title

Use a 'justified' paragraph format so that each side of paragraphs are to the margin. Use a 'justified' paragraph format so that each side of paragraphs are to the margin. Use a 'justified' paragraph format so that each side of paragraphs are to the margin. Use a 'justified' paragraph format so that each side of paragraphs are to the margin. Use a 'justified' paragraph format so that each side of paragraphs are to the margin. Use a 'justified' paragraph format so that each side of paragraphs are to the margin.

Do not use very narrow margins. 1" around is good. (Ultimately the paper is in a back-to-back format in a manual. Narrow margins cause the text to be too close to the spine of the book.)

Most of the time it is advantageous to stick to one type of font. Times New Roman is good. And it is good not to over-use emphasized text (underlined, italicized, semi-quoted text; Used sparingly, the right things get emphasized. Used widely so that everything is emphasized and nothing ends up really getting emphasized.) Text size: suggest 12 for basic paragraph text, table and figure titles, 10 for the actual data within tables and a slow progression upward in size (from 12) for subheadings, headings, all the way up to the paper title. Be consistent and do not over-do changing text size. Your paper may eventually be published in a periodical.

Continued text: When including a figure such as Figure 2 below, generally it is best to use color figures but print it out in black and white and make sure that the figure colors and styles make sense in B&W. Use of Figures with various colored (dark only) lines that have data points with x's and triangles and circles, etc. on the various lines of different style will make them readable in B&W.

A numbered list:

1. Text 1
2. Text 2

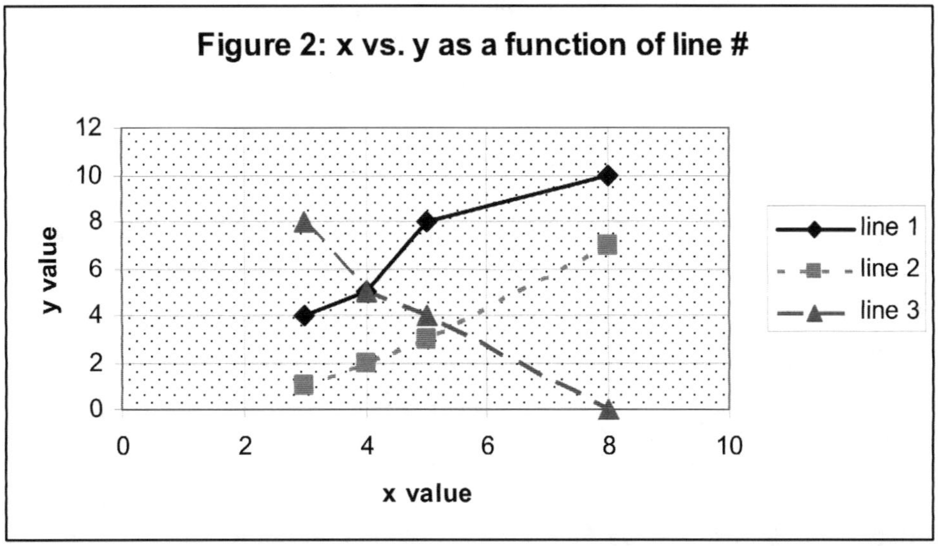

Burger dot example

- Main word/phrase - Text.
- Main word/phrase - Text.
- Main word/phrase - Text.

Continued text. When you are at the 50%, 80%, and 99% completed levels, have a good friend read it, with a 'good friend' defined as one who will be brutally honest with you and make constructive comments and criticisms that, when acted by you, will improve your paper both technically and readability-wise. There is only one thing worse than not sharing your information with others that could benefit from it. That one thing is sharing information in a way that emphasizes not what you have to say but emphasizes mostly your lack of writing skills.

You have several months to write this paper. It will likely take you 2 hours per page overall, on average, from start to finish. If you start now on a 10 page paper it will take you a total of 20 hours of your time to write it. If you start much later on a 10 page paper it will take you a total of 20 hours of your time to write it. The same time. But if this 20 hours is spread over a several month time period and not a several week or several day time period just prior to the date it absolutely, positively, has to be done by, you will gain think time and time for it to be proofed by a 'friend' several times and time to re-write it applying the suggestions and correcting errors. The quality of the paper that your name will be attached to will be far greater if you leave time for think time, review time, reorganization time, correction time. Start now with a goal to have a 1st draft in 2 to 3 weeks.

X Conclusion

The conclusion. Tie in that which was mentioned in the introduction. Use similar and familiar terms. If a reader reads the abstract, the introduction and the conclusion ... does he/she know, while not the details, the important aspects of what you want to convey?

X+1. References (examples given to show format)

[1] Announcement- Pharmaceutical cGMPs Initiative, www.fda.gov/oc/guidance/announce (accessed Oct. 16, 2003)

[2] Kolpin, Dana W., "Pharmaceuticals, Hormones, and Other Organic Wastewater Contaminants in U.S. Streams 1999-2000: A National Reconnaissance," *Environmental Science & Technology*, Volume 36, NO. 6, 2002

Additional references not cited:

[3] "Guidance for Industry Q7A Good Mfg Practice Guidance for Active Pharmaceutical Ingredients (FDA)," Aug 2001, www.fda.gov/cder/guidance/4286fnl.htm (accessed Oct. 26, 2004

[4] "Compliance Program Guidance Manual Program (FDA)," 7356.002, Form FDA 2438, 30 June 2001,www.fda.gov/ora/cpgm/default.htm#drugs (accessed Oct. 26, 2004)

[5] "GMPs for the New Millennium: A Risk Based Approach," www.fda.gov/oc/guidance/gmp (accessed Oct. 16, 2003)

OTHER SUGGESTIONS

Layout

- Paper Size 8.5" x 11", letter
- Do not format for A4 paper
- Left and right margins: 1.25 inches
- Top and bottom margins: 1.0 inch
- Gutter: 0
- Header and footer from edge: 0.5 inches

Graphics, pictures, and equations

- Equations: Use MS Word equation editor if possible. If not possible, insert the equation into the document by Copy – Paste Special – As Picture
- Graphs: Do not insert the Excel data that goes with the graph. Instead, insert the graph by selecting the graph and Copy – Paste Special – As Picture
- Photos and illustrations: when inserting pictures preferably start with files in JPG format and 200 dpi resolution. Reduce the native photo size to that more in line with the size required in your paper using an outside program. Just shrinking the file to fit the available space in Word does not reduce the file size. Native size might be 8 x 10 in but that shown is about 2 x 2.
- The graphics that you insert should have a minimum resolution of 200 dpi for grayscale files and 600 dpi for bitmap files.
- If you wish to have graphics wrap around your graphics as overlain here, you may do so by right clicking the picture and clicking Format Picture – Layout – Square. Otherwise, right click the picture and click Format Picture – Layout – In line with text.

Checklist before submitting your paper:

- ☐ Check to be sure that all editing and redline changes have been accepted.
- ☐ Be sure your documents are saved as Word (preferred) or RTF files.

- Be sure all graphics have been inserted into the document as pictures
- Send to another machine as a test to see if the document opens up in good order.
- Convert Mac format papers to PC format
- Remove all page numbering

Inherently Safer Design – Not Only About Reducing Consequences!

Dennis C. Hendershot
Process Safety Consultant
930 Greenhouse Drive
Bethlehem, PA 18017
dennis.hendershot@gmail.com

Keywords: Inherently safer design, risk analysis, risk management

Abstract

Process risk is a function of both the likelihood of occurrence of an incident, and of the consequences of the incident. There is a common perception that inherently safer design focuses solely on reducing or eliminating the consequences. However, inherently safer design can also focus on the frequency part of the risk equation. An inherently safer design can make an incident inherently less likely to occur, although the consequences of the incident should it occur would be unchanged. In event tree or Layer of Protection (LOPA) terms, this may be thought of as a design modification which reduces the frequency of the initiating event, rather than improving the reliability or increasing the number of layers of protection. Several examples of inherently safer designs which reduce the likelihood of an incident will be discussed.

1. Introduction

Inherently safer design (ISD), also known as inherently safer technology (IST), has been recognized as an important tool for process safety management since the concept was first developed by Trevor Kletz in the late 1970s [1]. Recently, second editions of two important books on ISD have been published: *Inherently Safer Chemical Processes: A Life Cycle Approach* by CCPS [2], and *Process Plants: A Handbook for Inherently Safer Design* by Trevor Kletz and Paul Amyotte [3].

The United States Department of Homeland Security (DHS), concerned about the potential for intentional release of toxic chemicals caused by a terrorist attack, has also been interested in ISD. DHS has listed 360 chemicals of interest in Appendix A of its Chemical Facility Anti-Terrorism Standards (CFATS) (6 CFR 27). The Chemical Security Analysis Center (CSAC) of DHS has initiated a program to enhance the safety and security of these chemicals. ISD/T is often proposed as a strategy to achieve safety and security. In 2010, CSAC asked the Center for Chemical Process Safety (CCPS) to develop a formal scientific definition of IST. CCPS completed this project in mid-2010, and a final report including the definition and additional explanations has been published [4]. The definition is intended to express the concept of IST clearly, and to consider the full life cycle and footprint of the chemical process, including manufacture, use, storage, and transportation.

2. CCPS/DHS Definition of IST/ISD

Here is the definition of IST/ISD which was developed by the 2010 CCPS/DHS project [4]:

> Inherently Safer Technology (IST), also known as Inherently Safer Design (ISD), permanently eliminates or reduces hazards to avoid or reduce the consequences of incidents. IST is a philosophy, applied to the design and operation life cycle, including manufacture, transport, storage, use, and disposal. IST is an iterative process that considers options, including eliminating a hazard, reducing a hazard, substituting a less hazardous material, using less hazardous process conditions, and designing a process to reduce the potential for, or consequences of, human error, equipment failure, or intentional harm. Overall safe design and operation options cover a spectrum from inherent through passive, active and procedural risk management strategies. There is no clear boundary between IST and other strategies.
>
> IST's are relative. A technology can only be described as inherently safer when compared to a different technology, including a description of the hazard or set of hazards being considered, their location, and the potentially affected population. A technology may be inherently safer than another with respect to some hazards while being inherently less safe with respect to others, and may not be safe enough to meet societal expectations.
>
> IST's are based on an informed decision process. Because an option may be inherently safer with regard to some hazards and inherently less safe with regard to others, decisions about the optimum strategy for managing risks from all hazards are required. The decision process must consider the entire life cycle, the full spectrum of hazards and risks, and the potential for transfer of risk from one impacted population to another. Technical and economic feasibility of options must also be considered.

3. ISD and Process Risk

Risk has been defined as a measure of human injury, environmental damage or economic loss expressed in terms both of the incident likelihood and the magnitude of the injury, damage or loss. Risk can therefore be considered to be a function of the existence of a hazard, the frequency of occurrence of an incident associated with the hazard and the consequence or impact of the incident should it occur:

Risk = f (incident, frequency, consequence)

The risk function can be quite complex, and many numerically different measures of risk can be calculated from the same set of incident, frequency and consequence data [5].

Note that the first paragraph of the CCPS/DHS definition of ISD/IST in Section 2 above includes the following statement in the eighth line:

> "...designing a process to ***reduce the potential for, or consequences of***, human error, equipment failure, or intentional harm."

In the course of the definition project, during extensive discussions with process safety management experts, it was often mentioned that many people consider ISD to be focused only on the consequence part of the risk function. The project reached the conclusion that ISD/IST actually applies to both frequency (likelihood) and consequence. A number of examples will be used to illustrate that this is true.

4. Examples of ISD to Reduce Frequency (Likelihood)

4.1 Example 1: Inherently reducing the likelihood of reverse flow

There are many situations where reverse flow can cause a significant hazard. For example, consider a reactor or other process vessel which contains a highly water reactive material, perhaps a material such as phosphorus trichloride which reacts with water to form large quantities of non-condensable HCl gas. If the vent of the vessel containing the water reactive material is connected to a vessel containing water (for example, a scrubber containing an aqueous treatment solution), it is possible that a downstream failure could cause backflow from the downstream vessel, resulting in an exothermic reaction and perhaps generation high pressure from the non-condensable gas. This can be prevented using active safety devices such as check valves, and perhaps using level, flow, or pressure alarms to close automatic valves and stop backflow. However, Kletz [6] suggests an inherently safer solution which may work in some situations and which eliminates the need to install and maintain all of the active safety systems. As shown in Figure 1, if the connecting pipe is routed to a high elevation, with a vent in the top to prevent siphoning, it may be possible to make backflow impossible. I have seen a similar system in which the vent pipe connecting a reactor to a scrubber was routed in excess of 32 feet in the air, making backflow from the downstream tank, which was open to the atmosphere, impossible even without a vent because it was too high for a siphon to occur (Figure 2).

This is clearly an example of an inherently safer design reducing the likelihood of an incident, not the consequence. If backflow were to occur, the consequences would be the same with the proposed inherently safer system as they would for the system using active safety devices. Whatever reaction and potential overpressure would occur in the upstream vessel would be the same if backflow were to occur somehow. The advantage of the inherently safer design is that it eliminates the need for active devices such as check valves, sensors, or interlocks and automatic shutoff valves. The inherently safer design relies on physical principles of elevation, pressure, and flow, and has no moving parts. This makes the likelihood of backflow inherently less likely. But, backflow is not impossible – for example, in Figure 1, the vent could be blocked. If this happens, the consequence of backflow is the same, so the consequence part of the risk function has not been changed.

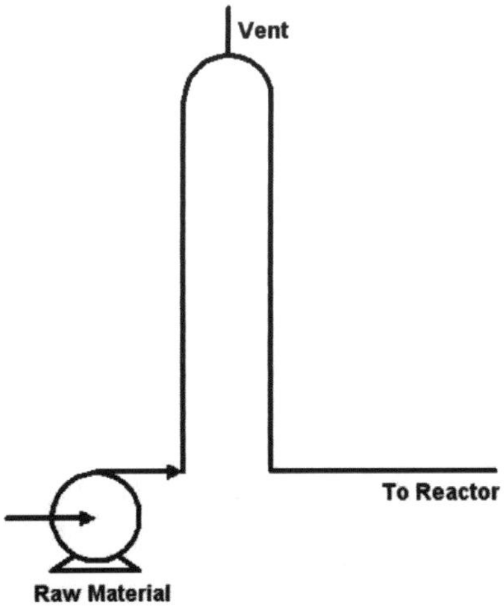

Figure 1. An inherently safer backflow prevention design [6]

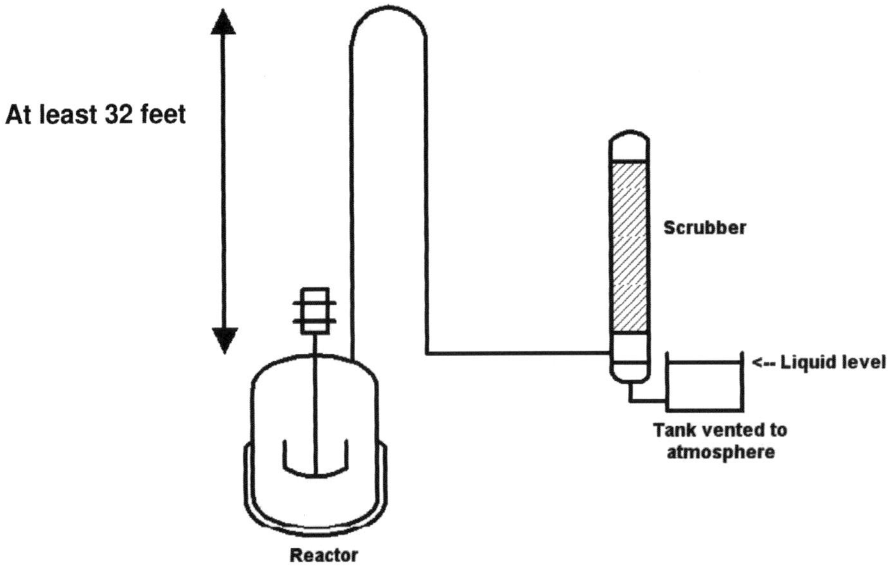

Figure 2. A similar system to prevent backflow from a scrubber to a reactor

4.2 Example 2: Inherently reducing the likelihood excess reactant charge or feed rate to a semi-batch reactor

This example is described in Section 11.3 of Reference 2. A highly exothermic reaction is being carried out in a semi-batch reactor using a process in which a limiting reactant is gradually added to the reactor at a rate determined by the heat removal capability of the reactor. For this process, the total amount of the gradual addition reactant is also important – excess reactant will not react as desired, but if it builds up it can react exothermically with the reaction solvent and there is a runaway reaction potential. So, it is important to control both the rate of reactant addition, and also the total amount charged. The original design used gravity feed for the limiting reactant from a feed tank to the reactor, controlling the feed control valve based on reactant temperature. Various flow, temperature, pressure, and agitation interlocks were used to stop the feed in case of process upset. The reactant was fed from a feed tank which was filled from the top and drained from the bottom (Figure 3). This design relies on active controls to monitor and control the feed rate to prevent either of two possible runaway reactions – from feeding the reactant too fast or with insufficient cooling, or the reaction with solvent if the total amount of reactant is excessive.

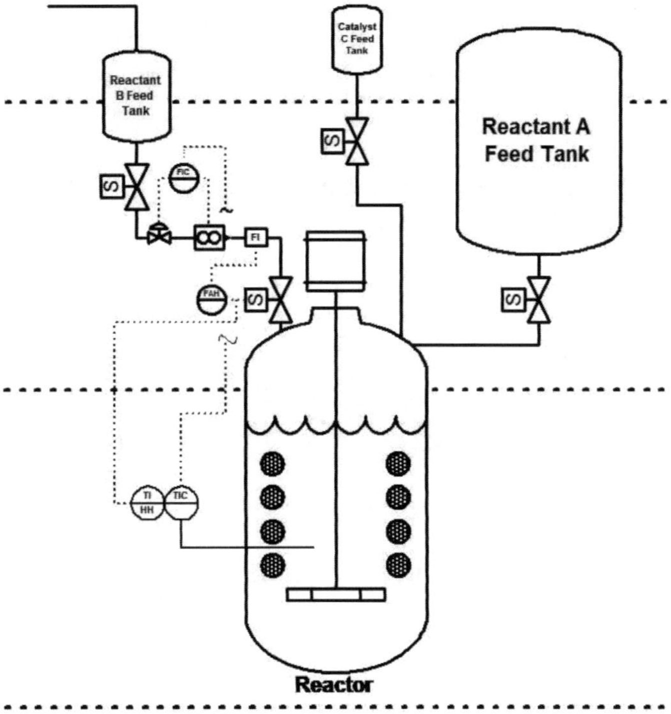

Figure 3. Original design of the semi-batch reactor

An inherently safer design was developed as shown in Figure 4. Flow rate of the reactant is limited by the capacity of a metering pump, so that it is not possible to exceed the cooling capacity of the reactor. Also, the feed is stopped by stopping the pump as well as closing the valves – in the original design gravity is always on (!) so a stuck or leaking valve could result in continuing feed of reactant. The total amount of possible feed is limited by the size of the feed tank, which holds exactly one batch charge. The 3-way valve on the bottom of the tank does not permit direct flow from the reactant storage tank in the tank farm – the only way to feed the reactor is to fill and then empty the feed tank (Figure 5).

Once again, this inherently safer design addresses the likelihood part of the risk function. The reactor is the same size, so, if there is a runaway reaction, the magnitude of the potential explosion is exactly the same with either design. The inherently safer design makes it much more difficult for the runaway reaction to occur, and it accomplishes this by equipment and piping design features and the location of the equipment, not by using active safety devices, alarms, interlocks, and procedures.

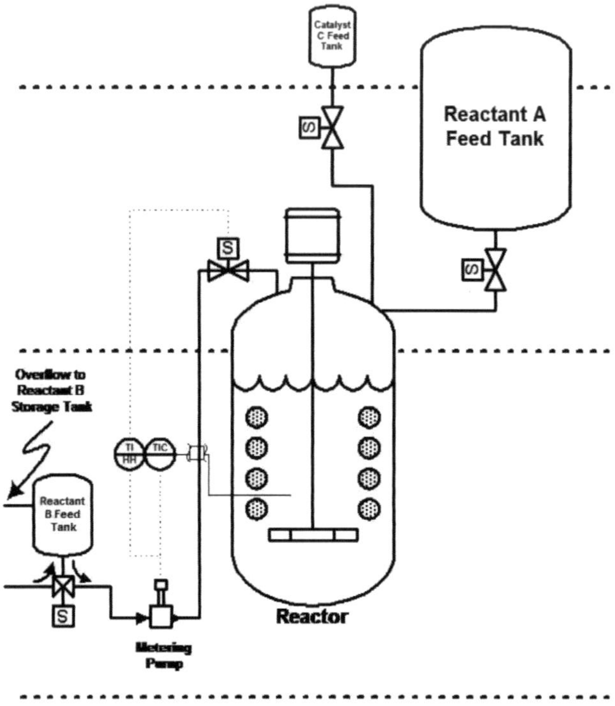

Figure 4. Inherently safer semi-batch reactor design

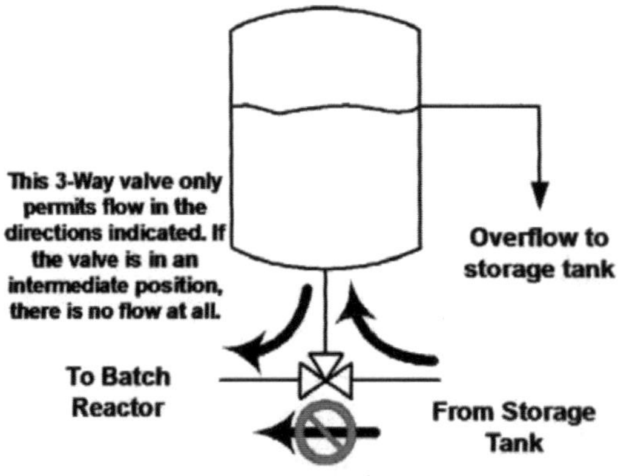

Figure 5. Feed tank design to limit maximum charge

4.3 Example 3: Using a Non-reactive Heat Transfer Fluid

It is common to use a non-reactive heat transfer fluid for many water-reactive process materials where there is concern for a hazardous reaction in case of a leak from heat exchanger tubes, reactor coils, or other heat transfer equipment. This also focuses on the likelihood part of the risk equation. In most cases, the vessel containing the water reactive material is essentially the same size regardless of the heat transfer fluid being used. In fact, it may even be necessary to use a larger heat exchanger if the non-reactive heat transfer fluid is less efficient than water or steam. Therefore, the consequence of water contamination is the same, but use of the non-reactive heat transfer fluid makes the likelihood lower – you have made it very difficult for water to leak into the reactive material. But, you have not made it impossible – the non-reactive heat transfer fluid could be contaminated with water, or somebody could connect up a hose to use water if there is a failure of the heat transfer system.

4.4 Simplification and "User Friendly Plants"

Many of the published examples of the ISD strategy "Simplify", and of designing facilities to be "user friendly" focus on the likelihood part of the risk function. In many of these cases, the inherently safer design makes it less likely that people will make an error, either in operation or maintenance. However, if somebody makes the error, the consequence is not changed. Here are some examples:

- Confusing controls on a stove (Figure 6) [2]: An inherently safer design would use a more logical layout, making it less likely that a person would turn on the wrong burner. However, the consequence of turning on the wrong burner is no different.
- Poor assignment of equipment identification numbers (Figure 7) [2]: A more logical arrangement would make it less likely that a person would select the wrong pump, for example, for maintenance. But, the consequence of selecting the wrong pump remains the same.
- Another example of confusing equipment numbering (Figure 8): Once again, using a more logical system of equipment identification makes it less likely that a person will make an error, but the consequence of making the error does not change.
- Confusing disconnect between the control room layout and the plant layout (Figure 9): In this case, a more consistent arrangement makes error less likely, but does not change the consequence of the error.
- Nitrogen supply lines to a vessel routed across manway so they must be disconnected to enter the vessel (Figure 10) [2]: This makes it less likely that nitrogen will be connected to the vessel when the manway is open for vessel entry, but does not change the consequence of entering the vessel when it contains an atmosphere which is hazardous.

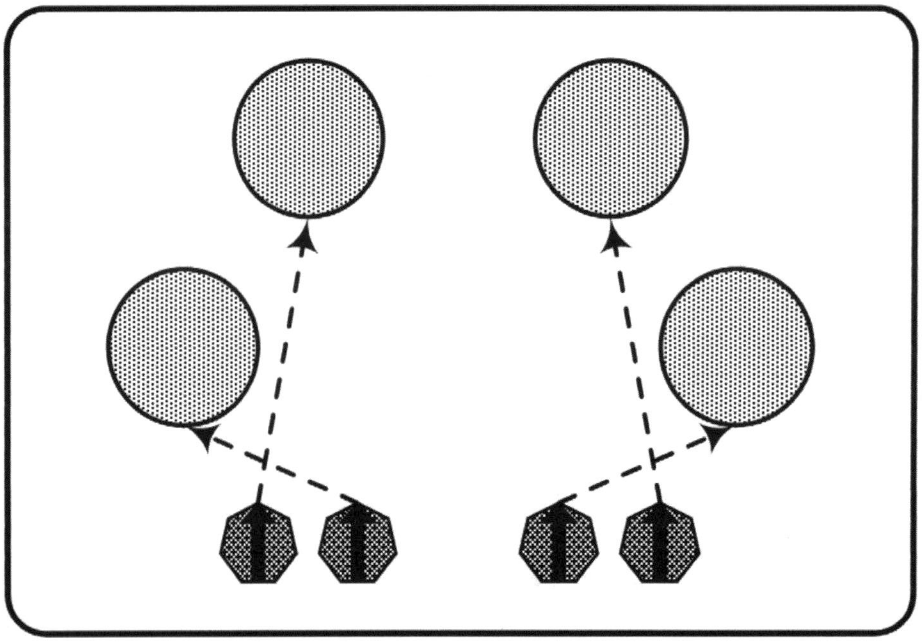

Figure 6. Confusing controls on a stove [2]

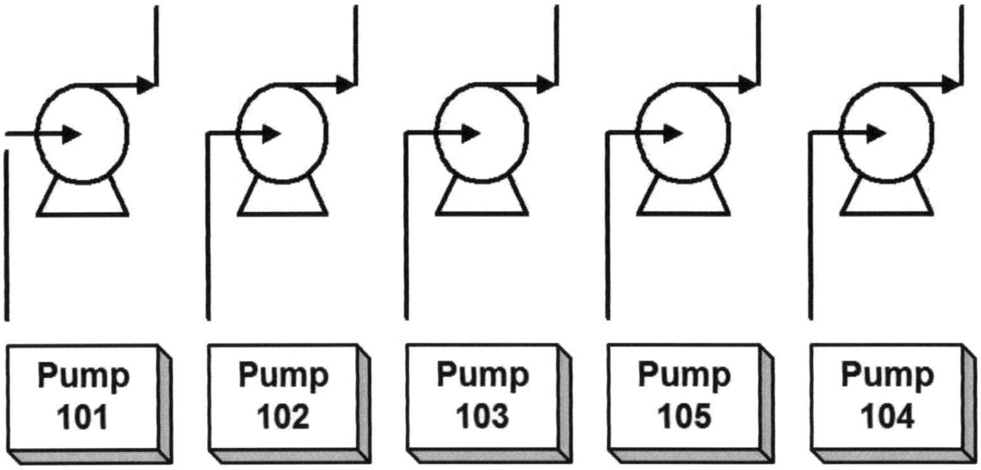

Figure 7. Confusing pump numbers [2]

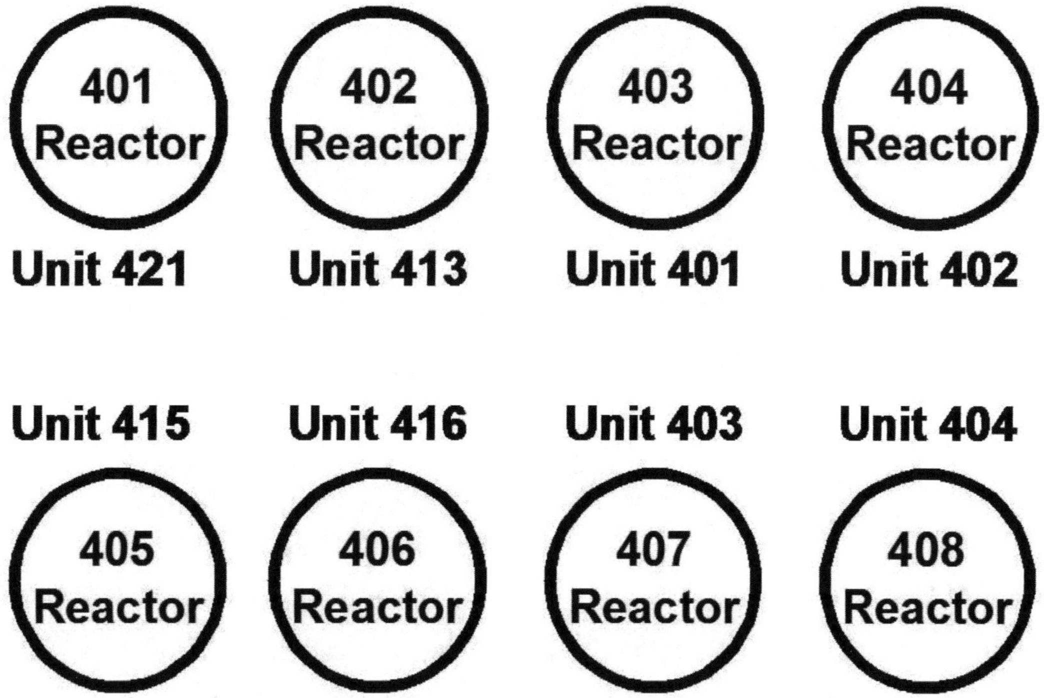

Figure 8. Confusing equipment designations

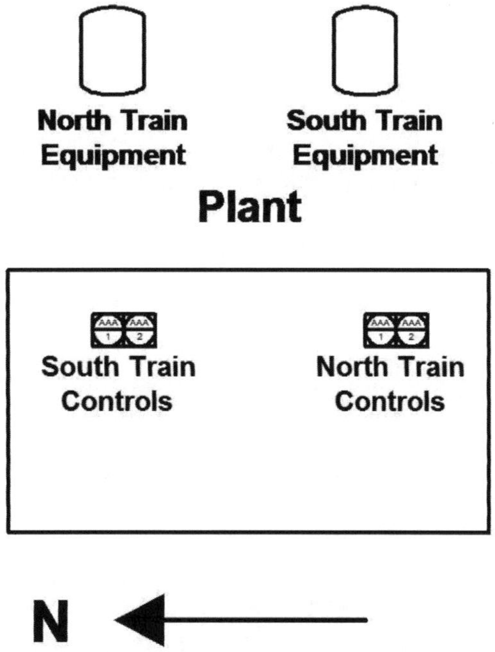

Figure 9. Confusing control system layout compared to plant layout

Figure 10. All nitrogen connections to this vessel go through the hoses positioned so they must be disconnected to open the manway [2]

5. Using Event Tree Analysis (ETA) and Layer of Protection Analysis (LOPA) to Understand ISD/IST

Event tree analysis (ETA) and Layer of Protection Analysis (LOPA) can be used to help visualize how IST/ISD options impact the process risk function. These two process safety management tools are closely related. LOPA can be described as "one path (typically we choose the path to the worst consequences) through an event tree" [7], as shown in Figure 11. We can use this model to understand how IST/ISD impacts process risk.

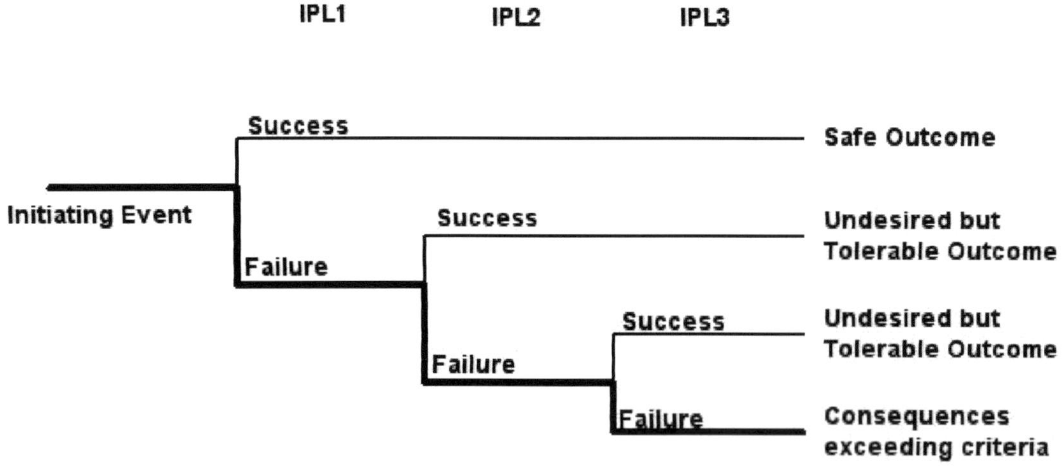

Figure 11. Event tree analysis and LOPA [7]
Note: IPL – Independent Protective Layer

5.1 IST/ISD as a consequence reduction tool

Figures 12-14 illustrate how IST/ISD can reduce the consequence portion of the risk function. Figure 12 shows the initial process, which has a single layer of protection, and the result is a risk which does not meet tolerance criteria.

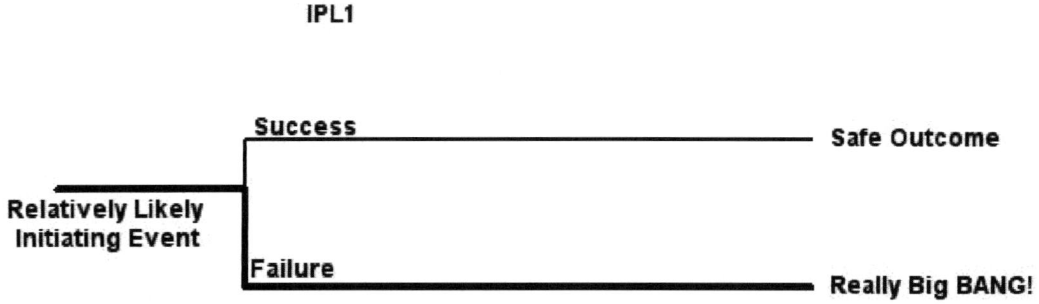

Figure 12. Initial LOPA event tree – intolerable risk

A traditional approach to risk management would add additional layers of protection, and perhaps also increase the reliability of some of those layers of protection, as shown in Figure 13, with the magnitude of the consequence remaining the same.

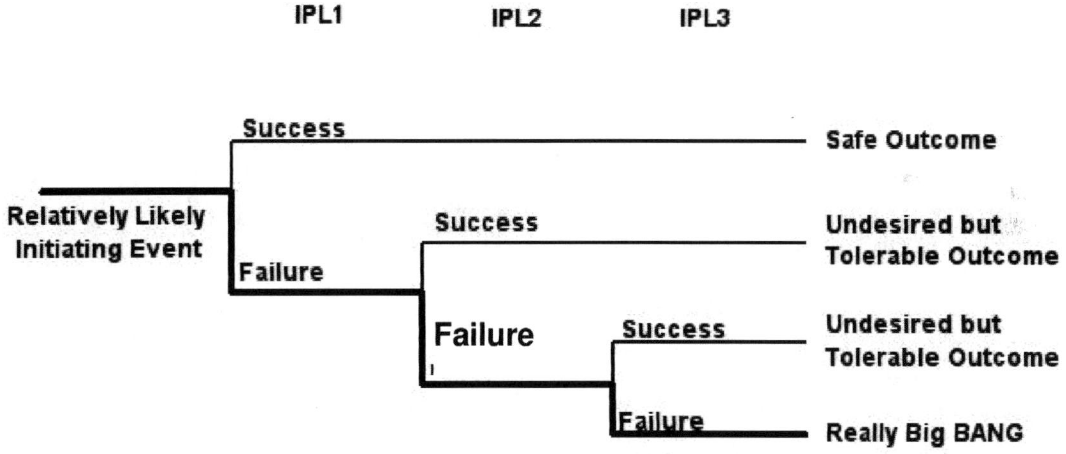

Figure 13. Risk management by additional layers of protection, resulting in tolerable risk

An IST/ISD approach would be to reduce the magnitude of the consequence (Really Big BANG!), as illustrated in Figure 14. This addresses the consequence part of the risk function, and fewer layers of protection are needed to meet risk tolerance criteria. If the consequence is reduced enough, so that the event is not capable of causing harm, perhaps no layers of protection would be needed.

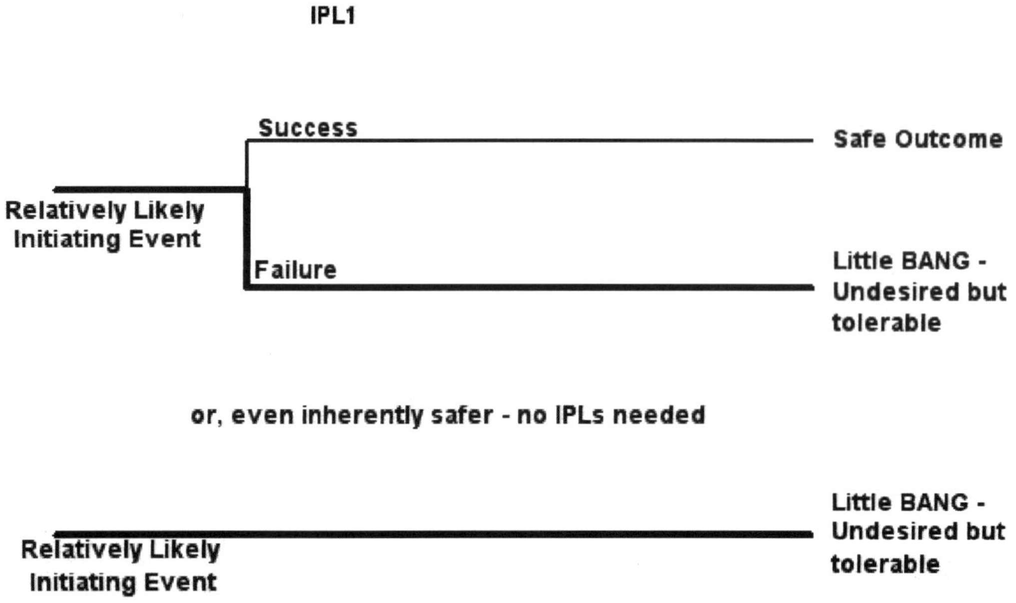

Figure 14. Risk management by IST/ISD – reducing the magnitude of the consequence so fewer, or perhaps no, layers of protection are needed

5.2 IST/ISD as a frequency reduction tool

Figures 15-17 illustrate how IST/ISD can reduce the frequency portion of the risk function. Figure 15 shows the initial process, and the risk is too high (exceeds tolerance criteria) despite two layers of protection.

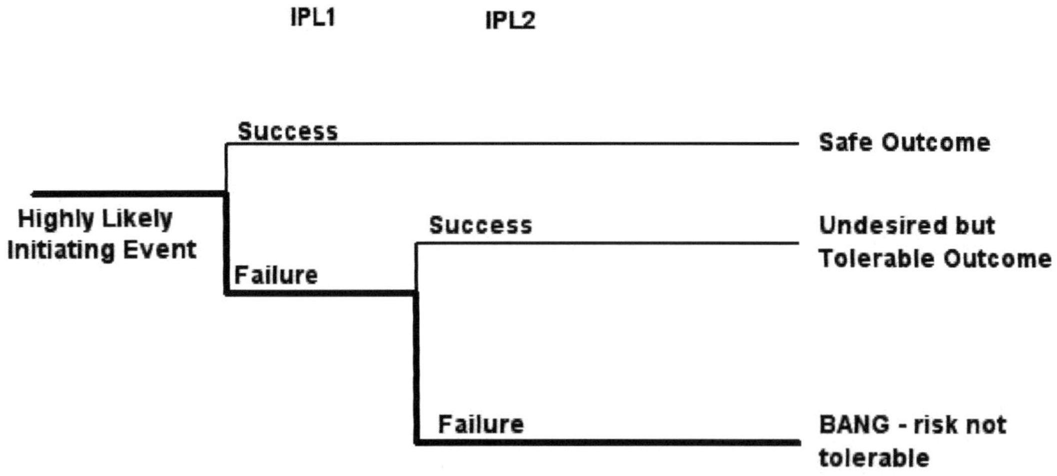

Figure 15. Initial LOPA event tree – intolerable risk

A traditional approach to risk management might be to add an additional layer of protection – IPL3 in Figure 16 – to reduce the risk to a value that meets risk tolerance criteria. Alternatively, the reliability of the existing layers of protection could be increased.

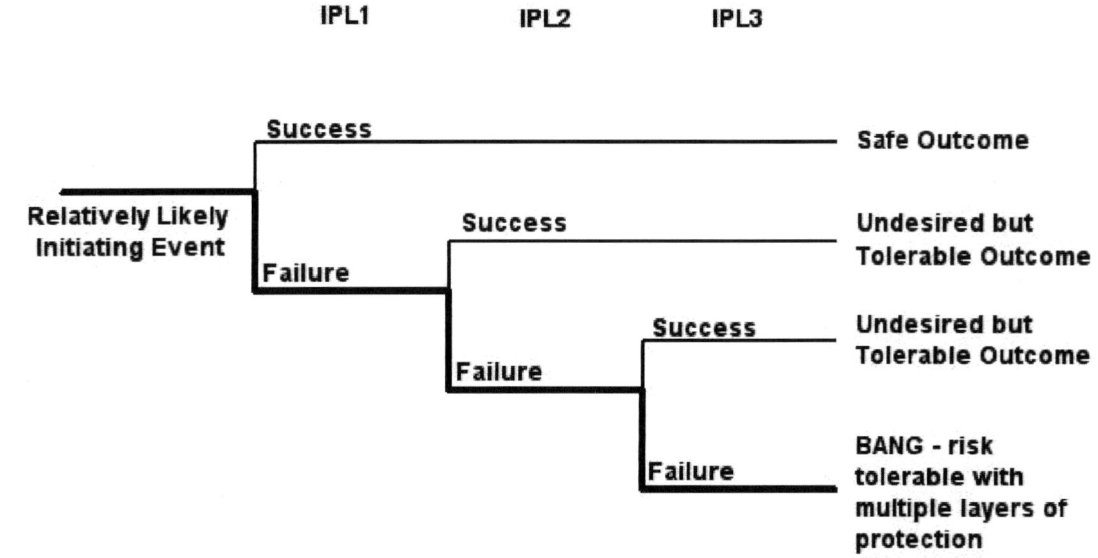

Figure 16. Risk management by adding an additional layer of protection

An IST/ISD approach might be to modify the design to make it extremely difficult for the initiating event to occur, as shown in Figure 17. Many of the examples in this paper illustrate some specific instances of this approach. If the initiating event likelihood is low enough, it may be possible to attain risk tolerance criteria with fewer layers of protection.

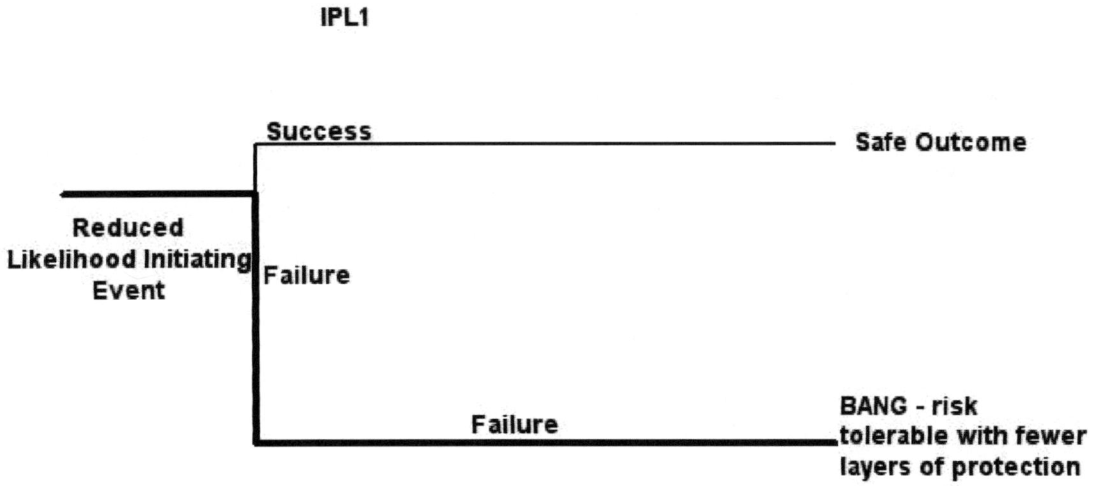

Figure 17. An ISD/IST risk management approach by reducing the likelihood of the initiating event

6. Conclusion

The examples in this paper clearly show that the inherently safer design philosophy is not solely based on reducing the consequences of a potential incident. IST/ISD also applies to the frequency part of the risk function, by making an incident inherently less likely to occur. Event/LOPA trees can be useful in understanding how inherently safer design options impact the risk function.

7. References

[1] Kletz, T. A.. "What you don't have, can't leak." *Chemistry and Industry*, 287-292, (6 May 1978).

[2] Center for Chemical Process Safety, *Inherently Safer Chemical Processes: A Life Cycle Approach*, 2nd Edition, American Institute of Chemical Engineers, New York, NY, and John Wiley & Sons, Hoboken, NJ, (2009).

[3] Kletz, T. A. and P. Amyotte, *Process Plants: A Handbook for Inherently Safer Design*, 2nd Edition, Taylor CRC Press, Boca Raton, FL (2010).

[4] Center for Chemical Process Safety of the American Institute of Chemical Engineers, *Final Report: Definition for Inherently Safer Technology in Production, Transportation, Storage, and Use*, U. S. Department of Homeland Security, Science & Technology, Chemical Security Analysis Center, Aberdeen Proving Ground, MD 21010-5424, July 19, 2010, available for download from: http://www.aiche.org/Conferences/Specialty/GCPS/IST.aspx

[5] Center for Chemical Process Safety. *Guidelines for Chemical Process Quantitative Risk Analysis.* 2nd Edition. American Institute of Chemical Engineers, New York, (2000).

[6] Kletz, T. A. *Process Plants: A Handbook for Inherently Safer Design*, Taylor & Francis, Philadelphia, PA (1998).

[7] Center for Chemical Process Safety, *Layer of Protection Analysis: Simplified Process Risk Assessment*, American Institute of Chemical Engineers, New York, NY, (2001).

Make Existing Process Plants Inherently Safer

Victor H. Edwards, Ph. D., P. E.
Aker Solutions Americas Inc.
3010 Briarpark Drive
Houston, TX 77042
Vic.edwards@akersolutions.com
&
Jack Chosnek, Ph. D., P. E.
Knowledge*One*
P.O. Box 580465
Houston, TX 77258
jc@knowledge1.net
www.knowledge1.net

Keywords: design review, hazard, inherently safer, ISD, IST, process plants, risk

Abstract

Inherently safer design (ISD) methods have matured during the past decade. Making process plants inherently safer has obvious advantages and is easiest to do when designing a new plant. Many older plants currently in operation were designed and built without the benefits of ISD. Inherently safer design reviews of these existing plants may identify opportunities for improvements, permit life cycle cost-benefit evaluations of the potential improvements, and sometimes allow improvements to be made during scheduled maintenance.

This paper outlines methods to conduct an inherently safer design review of an existing plant, and concludes with two examples of successful modifications of existing plants.

1. Introduction

An inherently safer design (ISD) will totally eliminate some hazards and reduce others resulting in a safer plant (CCPS, 2009; Kletz and Amyotte, 2010). But is it appropriate or economical? When a hazard has been eliminated or its frequency of occurrence reduced, as intended by ISD, the instrumentation and procedures used to control the hazard are no longer necessary. Depending on the nature of these controls, their life cycle costs can be significant in terms of expertise and time required for maintenance. Besides, even the best designed and maintained controls can fail, potentially leading to an incident and the resulting loss.

2. Why Consider Inherently Safer Design for Existing Plants?

Most currently operating plants were designed and built prior to the 1990s. In those days engineers tended to design with more fat because instrumentation and controls weren't that sophisticated, computer modeling was less prevalent, and instrumentation costs were a smaller percentage of total installed cost than they are today. This overdesign turned out to be a boon in the years following the initial design because plants could increase capacity with little investment.

Unfortunately, the end result was that plants started running close to their design limits, reducing safety margins and increasing the probabilities of process failures. Process hazards analyses (PHAs) have uncovered many of these increased risks. In many cases, this has led to increasing controls with added complexity and cost. The complexity itself can represent a risk. An inherently safer design could replace these designs with less costly maintenance and increased ease of operation while at the same time reducing risks.

3. Inherently Safer Design Methods Have Matured

Many of today's operating plants did not incorporate inherently safer process technology in their design. In addition, inherently safer process design methods have progressed significantly in the three decades since they were first proposed by Trevor Kletz (1978).

Because it is much easier and less costly to incorporate ISD when designing a new process plant than it is to modify an existing facility, much of the ISD literature has focused on incorporation of ISD during the research, development, and design of new process plants. The time is ripe for ISD reviews for existing plants because most have undergone significant upgrades in process safety information during the past several decades. Many of the related process safety advances of the past several decades facilitate the ISD evaluation of existing plants.

Although ISD can enhance the safety of some existing process plants, it is either impractical or not cost effective in others. This paper will outline methods to evaluate the potential value that ISD can bring to an existing plant and will illustrate them with several examples.

4. Could Your Plant Benefit from ISD?

An ISD review could discover opportunities to reduce risks and lower costs. Has new process technology been developed that offers both ISD benefits and economic benefits? Then life cycle cost/benefit analysis may make ISD of the new process, with subsequent replacement of the existing process, an obvious choice.

Four choices with the ISD of new process technology are whether to:

(1) Modernize the existing plant with the new process technology.

(2) Build a new plant based on the new process technology.
(3) Continue to operate the existing plant with no changes.
(4) Or shut down the existing plant.

In either of the first two choices, it is essential that ISD methods be integrated into the associated design work (Edwards, 2009; Sutton, 2011). If the decision is made to continue to operate the existing plant, there may still be less costly ISD improvements that can be made that offer potential benefits. This will be addressed later in this paper.

5. Is Your Plant's Process Safety Information an Adequate Basis for an ISD Review?

Regulatory requirements, as well as recognized and generally accepted good engineering practice have led to the creation and updating of the process safety information of many plants in the US, Europe, Japan, and other parts of the world. This process safety information includes adequate and up-to-date documentation of the design and of the plant and safety instrumented systems. This information is supplemented by process hazards reviews, consequence modeling, and facility siting studies.

In many cases, ISD may have been considered during initial design and/or baseline PHAs or during subsequent cyclic PHAs. If ISD has been done well, then further consideration of ISD is less likely to result in improvements.

If the existing process safety information for the plant is inadequate, then the first task in an ISD review is to upgrade the plant's process safety information. If upgrading the process safety information includes the conducting of a process hazards analysis (PHA) and risk assessment, then the ISD review should be integrated into the PHA and risk assessment (Kletz and Amyotte, 2010; Sutton, 2011).

If the process safety information is adequate and if significant hazards exist but ISD does not appear to have been applied, then two ISD reviews are recommended.

6. There are Two Levels of ISD Improvements and Two Types of ISD Reviews

- ***ISD Screening Review:*** Major changes in chemistry and/or process configuration (The changes are more costly with existing plants, but big potential improvements may result in both safety benefits and lower operating costs). An example is replacement of a highly toxic process route with a new and much less toxic alternative process with lower operating costs.
- ***ISD Detailed Review:*** Minor changes in equipment, operating conditions, and process configuration (Easiest to do with existing plants, but still potentially significant improvements) For example, upgrades and standardization in gaskets

may reduce leaks as well as preventing selection of the wrong gasket from the plant warehouse.

7. Conduct an ISD Screening Review

The ISD Screening Review is intended to identify potentially major process changes with major benefits.

The ISD Screening Review Team should first identify the major process hazards and rank the risks.

This review requires the maximum imagination and creativity, so a Guided What If method using appropriate guidewords selected for the process (Table 1 is one example). The ISD review leader should keep the focus on major changes during this review, focusing on brain-storming broad new approaches, and capturing but not exploring minor issues when they arise. ISD methodology has been clearly documented (CCPS, 2009; Kletz and Amyotte, 2010; Sutton, 2011).

The ISD Review leader should have expertise in process safety and at least one of the following areas: technology of the current process and products; process research and development; process engineering and plant design, plant operations. The ISD Screening Review Team should have experience typical of a revalidation or cyclic PHA team, except that in addition, expertise in research and development and in engineering design is very beneficial (Table 2). The best choice of the ISD Screening Review Team is based on the process technology and on the experience and personalities of the team members.

If one or more major opportunities are identified then each of these should be explored in more detail by the team after the initial brainstorming phase is complete. If potentially practical ISD approaches are identified, then these should be documented. These potentially practical ISD approaches should then be evaluated by a design team created after the ISD Screening Review.

Table 1 – Example of What If Guide Words for ISD Screening Review

- Problems with the Current Technology
- Intensification or Minimization
- Substitution
- Supplementation
- Attenuation or Moderation
- Limitation of Effects
- Simplification
- Ensuring Dynamic Stability
- Avoiding Knock-on Effects

Table 2 – Experience that May Be Needed in a ISD Screening Review
- ISD Review Leader: Process Safety plus at least one other relevant expertise
- Process Technology
- Process Research and Development
- Process Engineering and Plant Design
- Plant Operations
- Process Dynamics and Control
- Mechanical Engineering

8. Conduct an ISD Detailed Review

The second type of the ISD Review, the ISD Detailed Review, looks for smaller but potentially worthwhile improvements in the detailed design of the existing facility. It may be desirable to change the team composition somewhat for this detailed review. For example, it will be essential to have at least some of the following disciplines present: plant operations, instrumentation, electrical, mechanical, maintenance and quantitative risk assessment expertise. The ideal team composition will depend on the nature of the existing plant.

Here a What If/Checklist approach is recommended using a checklist of detailed changes that might offer improvements (Table 3 shows several example guide words and questions). Again, after the ISD Team has identified potentially promising ways to make the process inherently safer and/or more user friendly, these approaches should be turned over to plant management to create one or more design teams to develop and implement the upgrades.

Table 3 – Example of What If Guide Words and Questions for Detailed ISD Review
- Problems in the Existing Plant
- Making Incorrect Assembly Impossible
- Making Status Clear
- Tolerance of Misuse
- Ease of Control
- Ensuring Dynamic Stability
- Can More Frequent Testing of the SIS Improve Safety
- Can Modification of Equipment and/or Vessels Improve Safety

9. Example of ISD Success with an Existing Plant

Inherently safer designs can be applied to operating plants as evidenced by the work done by Celanese in the area of ethylene oxide manufacture. Ethylene oxide is produced in the vapor phase reaction of ethylene and oxygen. The gaseous product is quenched in a water absorber and then taken to a distillation column where the water is separated and taken out the bottom ("heavy ends column") and the overhead product is taken to a "light ends column" where the pure ethylene oxide is taken at the bottom and the light ends removed at the top (see Fig. 1).

Since ethylene oxide is a very hazardous chemical, having flammable limits from ~3% to 100% (it has its own oxygen), and being a decomposable and polymerizable chemical through very exothermic reactions, it's no wonder that explosions have occurred in the light ends column where pure ethylene oxide is reboiled. Such explosions happened in a BP Chemicals plant in Antwerp, Belgium, in 1987 where 14 people were injured (Melin, 1991), and in a Union Carbide plant in Seadrift, Texas, in 1991 where 1 person was killed and 32 injured, *Chem. Eng. News* (1991). In both cases, destruction was extensive with total damage to the unit and damage to neighboring units, and some damage even outside the plant. Investigation of the latter explosion indicated that in the presence of rust compounds ethylene oxide can violently decompose at lower temperatures than previously estimated.

In order to eliminate the contact of pure ethylene oxide with the hot surfaces provided by a reboiler the two columns were consolidated into one (see Fig. 2). The pure ethylene oxide is removed as a mid stream and the light ends are removed overhead. Thus, only water and heavy ends are in contact with the reboiler eliminating the potential of ethylene oxide decomposition. In addition, if the column were to lose its heat source and have all the liquid concentrate in the bottom, there would be just a dilute mixture of water and ethylene oxide that wouldn't present any danger of decomposition on restart of the column. By eliminating the hazard of runaway exothermic decomposition an inherently safer design was created.

A patent has issued on this inherently safer process: Coffey (1996)

This is an example of attenuation or moderation, in which the highly reactive ethylene oxide product is handled at a lower temperature by an inherently safer design that avoids contact of concentrated product with a reboiler.

10. Another Novel Increase in Inherent Safety

In December, 1984, the Bhopal tragedy took place in India. A release of Methyl Isocyanate (MIC) killed approximately 3000 people and injured tens of thousands more. Here is a description of a proactive ISD response by DuPont (C. Curtis Clements, 2005):

"In December of that year, another infamous process safety incident would further the application of ISP (Inherently Safer Processes) within DuPont. The tragic event in Bhopal, India provided the driving force. At the time, a major business in DuPont's

agrichemicals portfolio was critically dependent upon methyl isocyanate, or MIC, as a key ingredient in the manufacture of a very profitable insecticide. At the time, DuPont was purchasing MIC from its West Virginia-based supplier and shipping it in railcars half way across the country to a plant located just outside one of the most populous metropolitan areas in the United States. Routinely several million pounds of MIC were maintained in storage on the site at any given time. The Bhopal tragedy immediately put this entire business at risk, as concerns about the safety of making, shipping and storing MIC threatened the supply chain for this product."

"The answer was provided by the Company's research and development organization in the form of a totally new process, one that utilized the ISP principles of substitution and minimization. Specifically, an alternative chemistry was developed that allowed for use of a significantly less hazardous material than methyl isocyanate as a starting raw material, thus eliminating the risk of shipping and storing MIC. Alternative reactor technology was then utilized to convert the new raw to MIC in situ. The MIC was then immediately reacted and consumed as part of the next reaction step, limiting the total amount of MIC in in-process inventory at any point in time to something on the order of 10 to 20 pounds. Thus, the need to ship and store MIC was completely eliminated and a roughly five order of magnitude reduction in hazardous material inventory was achieved. From a business impact perspective, this is still perhaps the single most significant application of ISP for the Company to date."

11. Conclusions

There have been numerous recent applications of inherently safer design, resulting in inherently safer processes. However, a significant number of plants operating today may not have considered ISD principles during their creation. The modification of existing plants to become inherently safer can be much more difficult and costly than applying ISD during the conception and design of a new plant. However, there are still opportunities for cost effective application of ISD to some existing plants. This paper has outlined methods to apply ISD principles to evaluate potential application to existing plants.

12. References

Center for Chemical Process Safety, "Inherently Safer Chemical Processes – A Life Cycle Approach", 2nd Edition, American Institute of Chemical Engineers, New York, NY (2009).

Chemical and Engineering News, **69** (11) page 6 (1991).

C. Curtis Clements, "Application of Inherently Safer Process Concepts in DuPont", paper presented at the Process Plant Safety Symposium and Spring National AIChE Meeting, Atlanta, GA (2005).

Freylon Coffey, Improved Process for Ethylene Oxide Recovery, United States Patent 5,529,667 (1996).

Victor H. Edwards, "Develop and Design Inherently Safer Process Plants", Paper presented at the Mary Kay O'Connor International Symposium, Texas A & M University, College Station, Texas, (October 27-28, 2009) (accepted for publication in *Chemical Engineering*).

Trevor A. Kletz, "What You Don't Have Can't Leak", *Chemistry and Industry*, 287-292 (May 1978).

Trevor Kletz and Paul Amyotte, "Process Plants – A Handbook for Inherently Safer Design", 2nd Edition, CRC Press, Taylor & Francis Group, Boca Raton, FL (2010).

B. E. Melin, "Ethylene Oxide Plant Explosion, 3 July 1987, BP Chemicals, Antwerp, Belgium", Loss Prevention Bulletin 100, Institution of Chemical Engineers, London, UK (August 1991).

Ian Sutton, "Inherent Safety in Front End Engineering", paper accepted for presentation at the Process Plant Safety Symposium and Spring National AIChE Meeting, Chicago, IL (March 2011).

Figure 1 – Original Process

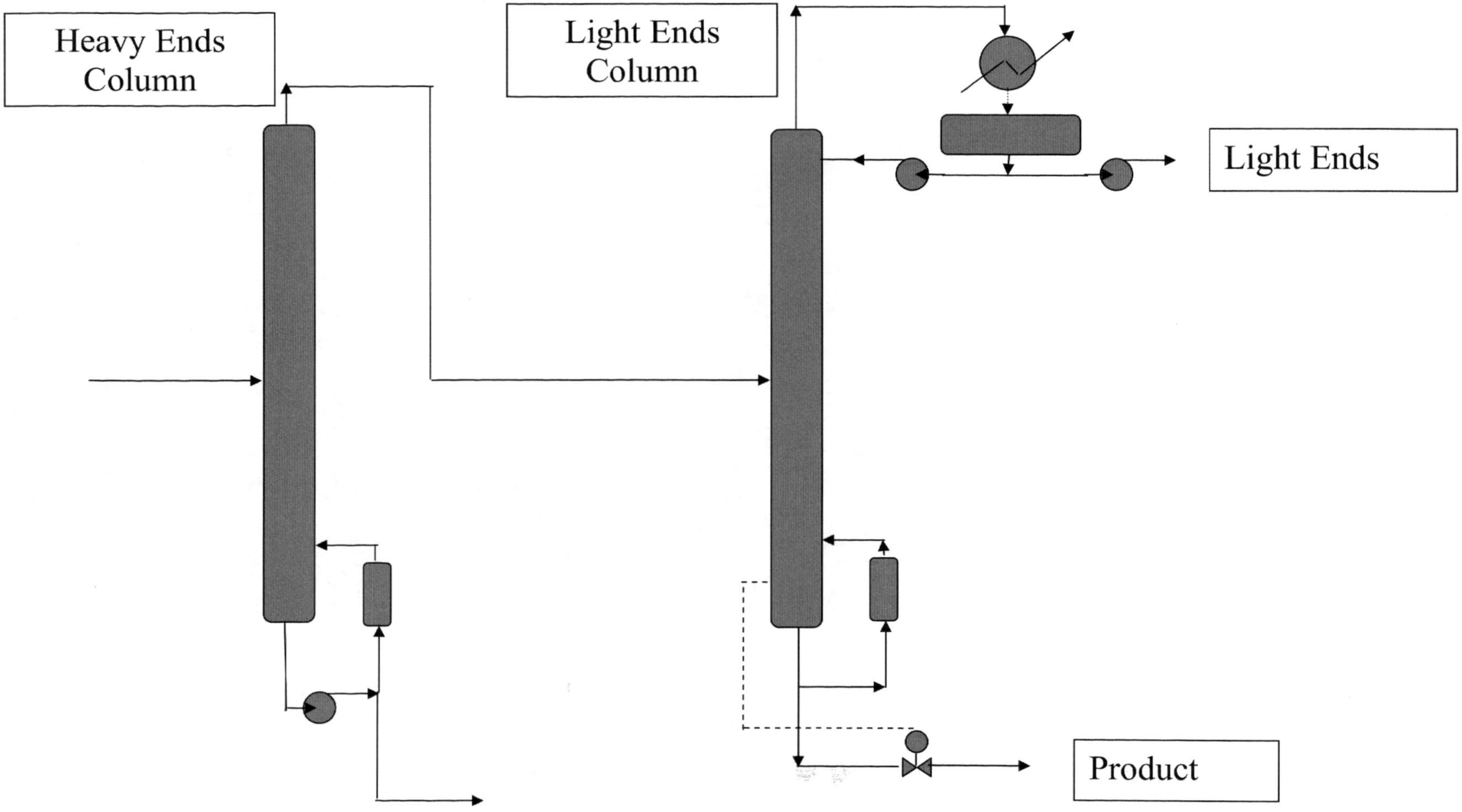

Figure 2 – Inherently Safer Process

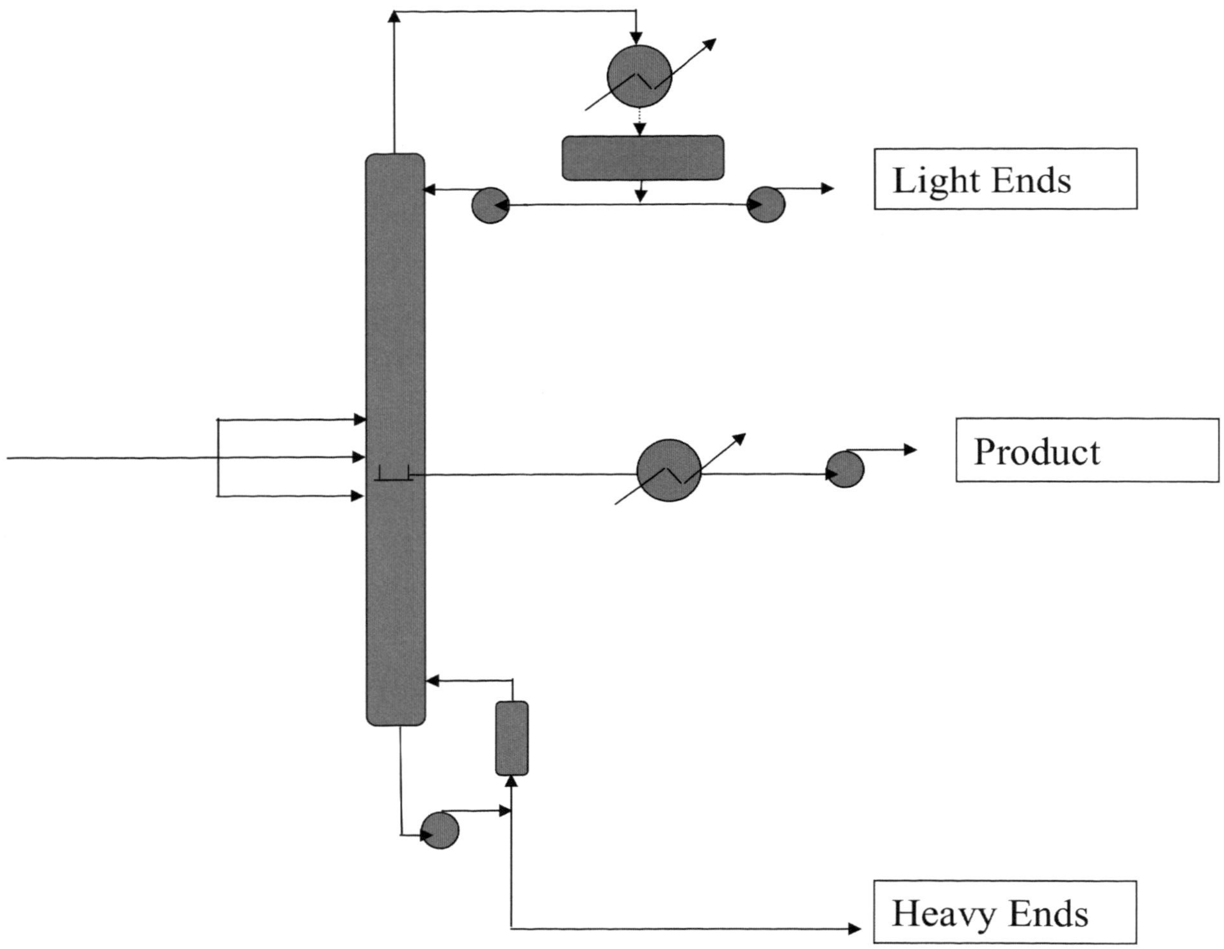

Inherently Safer Acids: A Case Study on the Application of Polyelectrolytes

Elliot Wolf
NOVA Chemicals Inc.
1555 Coraopolis Heights Road
Moon Township, PA 15108
wolfelli@novachem.com

Keywords: Inherent Safety, Polyelectrolyte, Acid, Corrosion

ABSTRACT

Strong acids, such as aqueous hydrochloric acid (HCl), are commonly used in the chemical process industry to purify streams or remove particulates from raw material, product, and waste streams. The benefits of HCl are its' low cost, large dissociation constant, and stability over time. However, the hazards associated with handling and processing HCl must be seriously considered when designing plants and specifying materials of construction and personal protective equipment. This typically results in higher capital, maintenance, and replacement costs to minimize the potential for personnel injury, equipment damage, and associated business interruption. This paper examines one such facility where aqueous HCl had been in use for many years. A number of inherently safer process design (ISPD) principles were evaluated against the existing operation, in order to identify opportunities for risk reduction. As a result, the inherent safety of the process was enhanced by converting an HCl system for dissolving inorganic particulates to a sequestration process utilizing polyelectrolytes. Application of this inherently safer process reduces maintenance costs, capital replacement costs, and most importantly, the potential for employee exposure. A review of the process chemistry and examples of process technologies that ultimately led to this inherently safer design are provided.

1. INTRODUCTION

Modifying an existing production facility to be inherently safer tends to require an initial capital investment and can often be more costly to operate than the existing, less inherently safe process. When confronted with alternative process designs, it can be difficult to optimally balance inherent safety, profitability, and operability under capital budget constraints and existing equipment configurations. However, equipping R&D scientists and process development engineers with the fundamentals of inherently safer design strategies can lead to innovative solutions that can often satisfy the above objectives. The four major strategies of inherently safer process design are [1]:

- Minimize – Use smaller quantities of hazardous substances
- Substitute – Replace a material with a less hazardous substance
- Moderate – Use less hazardous conditions, a less hazardous form of a material, or facilities that minimize the impact of a release of hazardous material or energy
- Simplify – Design facilities which eliminate unnecessary complexity and make operating errors less likely, and which are forgiving of errors that are made

This study is an example that demonstrates that it is never too late to apply the principles of inherent safety. Even after decades of plant operation, innovations in related industries may provide opportunities for risk reduction. This application of an alternative process chemistry culminated through a combination of colloidal science and an inherently safer way of thinking. The project demonstrates an inherently safer design of a process unit that reduces maintenance costs, capital replacement costs, and most importantly, the potential for employee exposure. From the beginning, it wasn't obvious that a more expensive and complex chemistry would ultimately lead to a more economical process. The alternative chemistry belongs to a family of water soluble polymers, polyelectrolytes, which have broad application in water treatment, oil recovery, concrete, and coatings. The process reported in this study utilizes anionic polyelectrolytes, which are effective sequestration agents, as a substitute for HCl to treat polymer pellets.

2. PROCESS DESCRIPTION

The process unit consists of three steps: (1) dispersion, (2) purification and (3) separation. Initially, resin is delivered to a dispersion vessel maintained at elevated temperature to achieve the desired mixing and reaction conditions. Suspending agents are added to the resin and water slurry to stabilize the particles and prevent agglomeration. The dispersion vessel operates above the softening point of the polymer, therefore, surfactants and insoluble inorganic partitioning agents are used to prevent resin particles from sticking together. Typical inorganic materials may consist of barium sulfate, calcium carbonate, or magnesium phosphate to name a few. The suspension slurry consisting of pellets, water, surfactants, and insoluble inorganics is transferred to a purification vessel at a lower temperature for removal of the suspending agents. The purification step utilizes HCl to dissolve the inorganic component. In this step, the pH of the slurry decreases from neutral to approximately three (3.0) due to dissociation of the strong acid. The last step consists of separating the pellets from the low pH aqueous suspension. Equipment in the downstream separation process not in direct contact with acidic slurry water but primarily wet resin is constructed of 316L stainless steel, whereas, the upstream purification system and associated piping is constructed of either Hastelloy® C or internally lined with a corrosion-resistant coating.

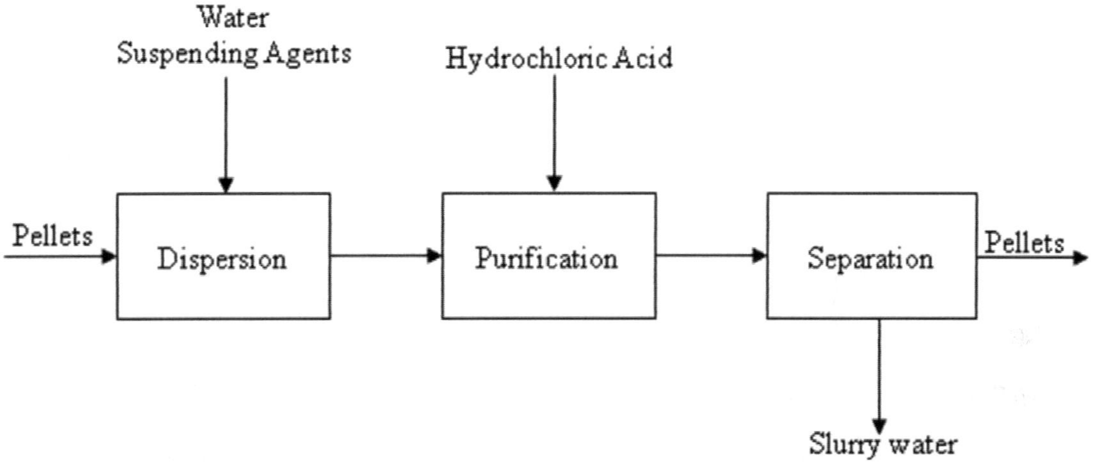

Figure 1. Diagram of the resin treatment process

3. HISTORICAL PERFORMANCE

Historically, the process unit experienced one-to-two corrosion-related incidents per year often leading to lost production time, maintenance costs, and capital replacement costs. After two debottlenecking efforts which significantly increased production throughput and proportionally the acid consumption, the process experienced accelerated rates of corrosion specifically in the downstream separation equipment constructed of 316L stainless steel. At this stage, the stream is greater than 99% pellets with trace amounts of acidic moisture on the surface. The failure mechanism in the equipment was confirmed to be chloride stress corrosion cracking based on expert metallurgical analysis. At the request of the business unit, the R&D group launched a project to investigate alternative equipment and process chemistries that would eliminate the use of HCl.

4. APPLICATION OF INHERENTLY SAFER PROCESS DESIGN

The project was led by a process development engineer that formerly received ISPD training while in a manufacturing engineer role. Exposure to the principles of inherent safety, guided the thinking of the project leader enabling the team to consider unconventional alternatives. Each of the ISPD principles considered in the brainstorming stage focused on reducing risk, maintaining operability, and minimizing cost. The following alternative processes and corresponding ISPD strategies were considered in the study.

4.1. *Minimize:* Reduce the quantity of suspending agent and acid requirement

To reduce the risk of an HCl release, the simplest approach would be to reduce the quantity of acid that needs to be stored and distributed in the process in order to meet the product quality specifications. In this process, a minimum ratio of HCl must be applied to the surface of the product to sufficiently dissolve the inorganic suspending agent. Complexometric titration is a common technique used to determine the residual suspending agent concentration by targeting cationic substances. Any excess suspending agent above product specifications, as measured by the technique, will have deleterious effects on the final product properties. Therefore, the only possible approach to reduce the amount of HCl, without compromising product quality, is to first reduce the level of suspending agent.

Typical suspending agent loadings for this product range from 0.25 to 5 parts per hundred depending on the product type. For every ten pounds of suspending agent, approximately one gallon of aqueous HCl (36 wt% purity) is needed to sufficiently reduce the concentration of suspending agent on the surface of the pellet. Historically, the suspending agent levels were kept conservatively high due to previous incidents with particle agglomeration. However, pilot scale studies under this project demonstrated that reduced levels, as much as 50% reductions, could be achieved without compromising the integrity of the suspension system. Consequently, these reductions and corresponding lower levels of acid did not significantly moderate the corrosion observed in the downstream equipment. Thus, minimization of the suspending agent and acid levels, although more economical and inherently safer, was not sufficient to mitigate corrosion failures and the risk of employee exposure.

4.2. *Moderate:* Apply an internal coating to enhance corrosion resistance

A less preferred approach, albeit, effective in terms of moderating the consequences of a release, was to modify the equipment with a fluoropolymer coating capable of withstanding acid attack. The combination of low pH and chloride ion concentration enabled conditions that were extremely corrosive to stainless steel. Processing equipment and transfer piping in acidic slurry service was constructed of either stainless steel with a fluoropolymer coating or Hastelloy® C. Vent piping from equipment to scrubbers was constructed of fiberglass. Equipment that was most susceptible and prone to corrosion was in the downstream size separation equipment that handled pellets and vapor fumes. Coating the separation equipment provided an immediate corrective safeguard against loss of containment but it was not economically sustainable due to the high cost of replacement. In contrast to minimizing the suspending agent and acid throughput which had an associated cost savings, extending the service life of new equipment with an internal coating was expensive and exceeded budget constraints. In order to meet production goals, spare equipment was coated and put into service until a permanent solution was identified.

4.3. Simplify: Mechanical Removal of Suspending Agent

Several physical removal designs were considered that would modify the process from chemical treatment to mechanical removal of the particulates, thus reducing the chemical requirements and simplifying the number of processing steps. Alternatives deemed feasible for removal of particulates consisted of centrifugation and ultrasonic cleaning. Both methods were demonstrated in the laboratory with moderate success but further pilot scale development was necessary to increase the probability of success on a commercial scale.

Laboratory experiments were performed in a sonication device. The ultrasonic removal method in a batch vessel was a factor of three less efficient than traditional HCl removal, thus it was not considered a viable alternative. A multi-stage ultrasonic processing train was proposed but that was not pursued because of the high capital cost. Furthermore, since ultrasonic technology had not been practically demonstrated on a commercial scale, the concept was abandoned.

On the contrary, centrifugation is a very common technology used for separation of solids of varying diameters and specific gravities. Screen bowl style centrifuges which combine the benefits of clarification and screening had been demonstrated in similar processes. Successful implementation of the centrifuge process would simplify the process by consolidating the purification and separation steps, in addition to eliminating HCl and the associated equipment. However, typical installations with centrifuges are effective when the suspending agent levels are below one part per hundred of polymer. Since this process consists of loadings that are 5 to 10 times greater, it was not possible to remove the suspending agent to the level required by product quality specifications. A feasible process at these higher loadings would likely require multiple centrifuges with separate washing steps in conjunction with the centrifuges. Again, the capital investment costs associated with this process were higher than coating all the equipment. Therefore, other inherently safer alternatives were pursued.

4.4.1. Substitute: Replacement of Hydrochloric Acid with Phosphoric Acid

Phosphoric acid was considered as an inherently safer substitute since it is less corrosive to stainless steel and the risk of employee exposure is lower. It is a weaker acid, absent of chloride ion, and has a lower acute oral toxicity. The reported lethal dosage (LD_{50}) is higher for phosphoric acid (1530 mg/kg) [2], compared to hydrochloric acid (700 mg/kg) [3]. Studies were performed to determine if the efficiency of phosphoric acid was comparable to that of hydrochloric and how this may impact usage quantities and material cost.

On a mass basis, approximately 0.7 pounds of HCl are needed for every 1 pound of inorganic to completely dissolve and purify the pellet surface. Solubility measurements (cloud point determination) of phosphoric acid with the inorganic component showed that approximately 2 pounds of phosphoric acid were needed per pound of inorganic to achieve an equal degree of removal. Given that the cost of phosphoric acid was about twice that of HCl and it was about three times less efficient, implementing this alternative would increase cost by a factor of six. Even considering the minimization of the suspending agent, the overall cost was still higher than the option to internally coat the equipment.

4.4.2. *Substitute:* Replacement of Hydrochloric Acid with Polyelectrolytes

Polyelectrolytes are polymers with charged functional groups attached to the chain and are generally water soluble. These polymers are typically classified as either polyanions (negative charges) or polycations (positive charges). There are extensive applications of polyelectrolytes in flocculation, adhesion, stabilization, compatibilization, viscosification, and suspension to name a few [4]. Table 1 provides a partial list of typical polyelectrolytes, manufacturers, and applications.

Table 1. Selected list of polyelectrolyte manufacturers [5]

Chemical Name	**Manufacturers**	**Applications**
Maleic Copolymers	Alco Chemical Co. BASF Corp. Rhom and Haas Co.	Industrial cleaners Sequestrant
Poly(2-acrylamido-2-methyl-1-propanesulfonic acid)	Alco Chemical Co.	Superplasticizer Scale inhibitor Drag reducer
Poly(acrylic acid)	Alco Chemical Co. BASF Corp. Rhodia, Inc. Rhom and Haas Co.	Dispersant Scale inhibitor Sequestrant
Poly(acrylate)	Alco Chemical Co. BASF Corp. Jacam Chemicals, L.L.C.	Chelate Sequestrant
Poly(aspartate)	NanoChem Solutions Inc.	Corrosion inhibitor Dispersant Scale inhibitor Fertilizer enhancer
Poly(methacrylate)	Alco Chemical Co.	Mineral dispersant Sequestrant Anti-deposition agent
Sulfonated Copolymers	Alco Chemical Co. Rhom and Haas Co.	Superplasticizer in concrete Ion exchange membranes

The fundamental problem of the application was to remove and/or solublize inorganic particles with a less hazardous process. Traditionally, this had been achieved with a very efficient and inexpensive acid posing significant corrosion potential and employee exposure risks. A weaker acid, such as, phosphoric acid still posed a health hazard, albeit accompanied by lower risk, and a cost about twice that of HCl. Alternatively, the cost of the synthetic polymers were about ten times higher than HCl on an active basis. Thus, they were not initially considered to be economical replacements. Another disadvantage initially considered with the polymer solution was the freezing point (0 °C) compared to HCl (-32 °C) which would require investment in a new storage and delivery system equipped with freeze protection. Table 2 provides a comparison of the acid and polyelectrolyte properties.

Table 2. Comparison of material properties

Property	**Hydrochloric Acid**	**Phosphoric Acid**	**Representative Anionic Polymer**
Concentration, wt%	36	75	50
Freezing point, °C	-34 [6]	-17.5 [7]	0
Odor	Pungent	Odorless	Slight
Health Effects	Corrosive, burns	Corrosive, burns	Slight/mild irritant
Toxicity (Oral LD_{50}), mg/kg	700 [3]	1530 [2]	>5,000
OSHA PEL, mg/m^3	7 (Ceiling) [8]	1 (TWA) [9]	Not Applicable

Upon detailed experimental investigation, however, research studies confirmed the extraordinarily high sequestration efficiency and ability to stabilize particles in the slurry. Analogous to the mechanism of scale inhibition in water treatment systems, anionic polyelectrolytes proved to be capable of removing chemical intermediates from the surface of polymeric resins. Figure 2 depicts the sequestration mechanism of the anionic polymer. Studies showed that on a pure basis, anionic polymers were able to sequester surface particles a factor of ten more efficiently that HCl removal. Despite the higher cost of the synthetic polymer, the increased removal efficiency resulted in a net cost savings. In addition to reducing the risk of exposure, implementing the sequestration process reduced chemical storage quantities and improved slurry processing rates. The small storage requirements were amenable to a tote delivery system that could be maintained indoors without freeze protection, thereby, eliminating the need for an outdoor unloading system and storage tank.

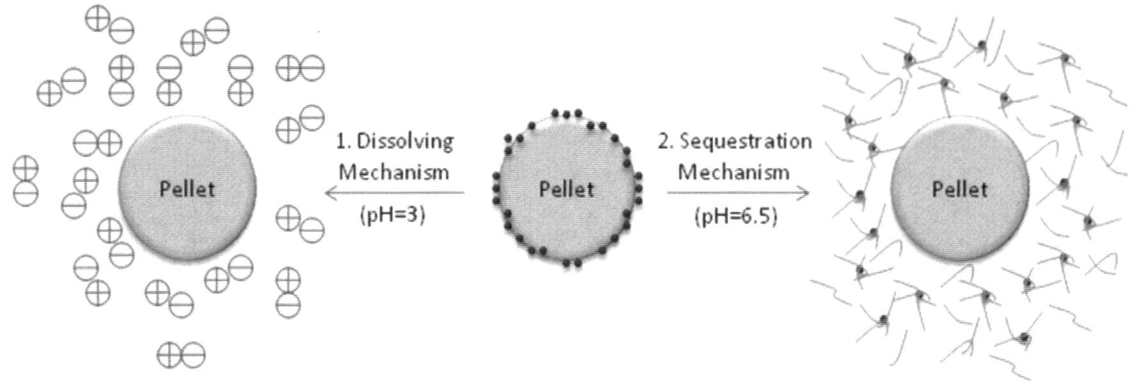

Figure 2. Diagram comparing acid dissolving and polymer sequestration mechanism

5. CONCLUSIONS

An existing process was modified to be inherently safer by substitution of aqueous hydrochloric acid with a less hazardous chemical technology, sequestration with polyelectrolytes. Consequently, the corrosion mechanism observed in the stainless steel separation equipment was eliminated, thus reducing the risk of employee exposure and improving plant operability. Not only did the inherently safer design reduce risk and improve operations, but it unexpectedly lowered operating costs due to increased particle removal efficiency. Application of ISPD principles in research and development led to a safer and more economical process demonstrating that it really is never too late to adopt an inherently safer way of thinking.

6. REFERENCES

1. Center for Chemical Process Safety (CCPS), Inherently Safer Chemicals Processes, American Institute of Chemical Engineers (AIChE), Wiley, Second Edition, pg. 27
2. Rhodia Material Safety Data Sheet, Phosphoric acid, December, 2000
3. Univar Material Safety Data Sheet, Hydrochloric acid, September 24, 2009
4. Water Soluble Polymers, Encyclopedia of Polymer Science and Technology, John Wiley & Sons, 2010
5. McCutcheon's, Functional Materials, North American Edition, MC Publishing, 2007
6. Hydrogen Chloride, Kirk-Othmer Encyclopedia of Chemical Technology, John Wiley & Sons, 2010
7. Phosphoric Acids and Phosphates, Kirk-Othmer Encyclopedia of Chemical Technology, John Wiley & Sons, 2010
8. OSHA, Hydrochloric Acid Exposure Limits, **http://www.osha.gov/dts/chemicalsampling/toc/chmn_H.html**
9. OSHA, Phosphoric Acid, Exposure Limits, **http://www.osha.gov/dts/chemicalsampling/data/CH_262500.html**

Shutdown Systems and Their Effect on Commissioning

Doug Cicerone
Cicerone & Associates, LLC
7111 W 151st Street, Suite 327
Overland Park, KS 66223 USA
doug.cicerone@gmail.com

Keywords: Shutdown Systems, Burner Management Systems, Sulfur Recovery, Commissioning

Abstract

As all process industries strive for inherently safer plant designs, shutdown systems are becoming more complex. For projects in the oil and gas industries, the process licensor often provides the initial design for the shutdown system (sometimes called Safety Instrumented System). During the detailed design phase, additional layers of complexity are often added as the various parties involved in a project (owner, licensor, EPC contractor, etc.) add devices and/or logic based on their past experiences. While these additions are made with the best of intentions, they are often the result of a lack of understanding of the project history or the process itself. As the level of complexity increases, so does the potential for difficulties during plant commissioning and initial operations. These difficulties can result in a significant number of plant shutdowns and subsequent restarts. Since a large percentage of incidents involving personnel safety and equipment damage occur during plant startups and shutdowns, the argument can be made that more complex shutdown systems can lead to a decrease in overall plant safety rather than the intended increase. This paper discusses the author's experiences with shutdown systems during the commissioning and initial operation of sulfur recovery units in overseas oil refineries and gas plants.

1. Introduction

Most process design projects in the oil and gas industries involve several companies. At a minimum, a project team is typically made up of personnel from a process licensor, an EPC contractor, and an owner. Each company brings with it experience and expertise based on previous design projects and previous operating experience. But ultimately, the project team is made up of individuals. And each individual has a different background and level of experience which they incorporate into their piece of the design.

Because of the long duration of a typical project, the project team members often change as the project and company needs change. Consequently, there is a high probability that the people working on it at the end of the project were not involved at the beginning. This creates the possibility of decisions being made by people who are not familiar with the project's history. Sometimes these decisions have negative consequences during the final stages of the project.

One area that can receive a lot of attention during the process design, particularly during design reviews and HAZOPs, is the shutdown system. This paper will focus on the shutdown systems of sulfur recovery units, and how decisions made during the design process can affect the unit during commissioning and operations. A brief background of sulfur recovery will first be given, followed by a discussion of specific safety concerns. The evolution of a shutdown system will be discussed followed by specific examples of real-world situations where uninformed design decisions created problems during startup and initial operations.

2. Sulfur Recovery

Because this paper focuses on shutdown systems for sulfur recovery units (SRU), a brief discussion of the sulfur recovery process follows. Although many sulfur recovery and tail gas cleaning technologies exist, only a standard straight-through Claus process will be discussed. (See reference [1] for more details on the sulfur recovery process.)

2.1 Process Chemistry

The modern Modified Claus process for the conversion of hydrogen sulfide (H_2S) to elemental sulfur (S) was first used in 1936. The process is divided into a high temperature thermal conversion section and a lower temperature catalytic conversion section. Although many reactions take place in the thermal section, the majority of the thermal conversion occurs according to the following two reactions:

$$\mathbf{H_2S + 1/2\ O_2 \rightarrow SO_2 + H_2O} \text{ [Eq. 1]}$$

$$\mathbf{2H_2S + SO_2 \rightarrow 3/x\ S_x + 2H_2O} \text{ [Eq. 2]}$$

In the catalytic section, the conversion of hydrogen sulfide occurs according to Equation 2 (also known as the Claus reaction) over a catalyst that is typically alumina or titania based.

2.2 Process Description

Figure 1 below shows a simplified process flow diagram for a typical sulfur recovery unit with two catalyst beds.

Figure 1: Sulfur Recovery Unit Simplified Process Flow Diagram.

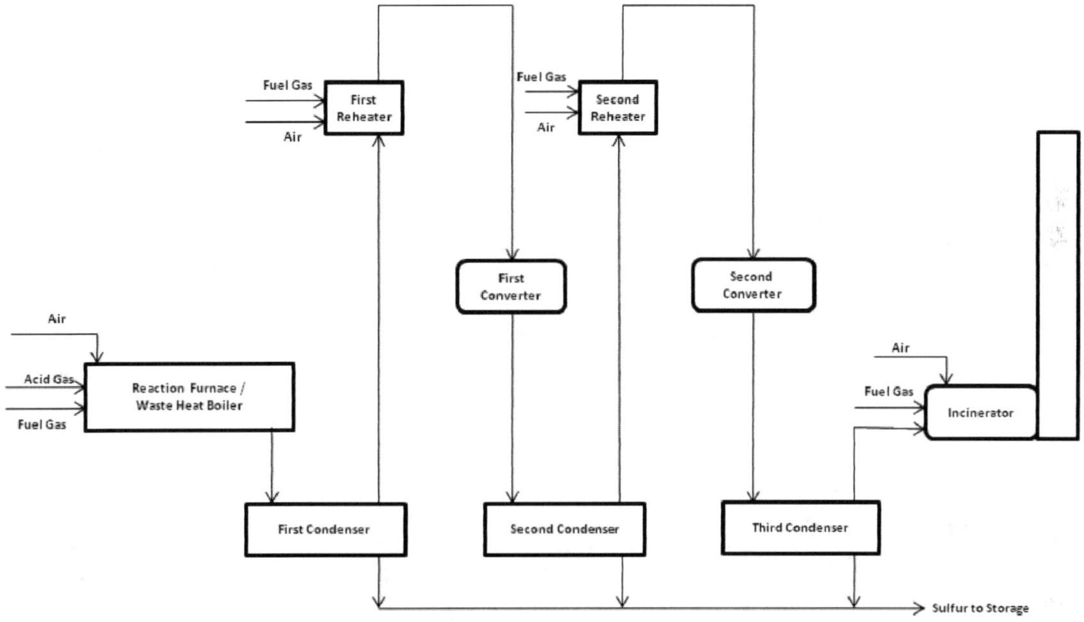

The H_2S bearing feed gas (acid gas) flows to a Reaction Furnace where it is combusted with air. The air is metered such that one-third of the H_2S is burned to sulfur dioxide (SO_2). The unburned H_2S reacts with the SO_2 according to Equation 2 and forms elemental sulfur. Temperatures in the Reaction Furnace range from 1,500°F to 2,500°F depending on acid gas composition. Heat is recovered in the Waste Heat Boiler, typically by generating high pressure steam.

The process gas flows to a Condenser where the sulfur vapor is condensed to liquid sulfur, which is sent to storage. Heat is recovered in the Condenser, typically by generating low pressure steam. The process gas exits the condenser at its sulfur dewpoint and passes through a Reheater. This Reheater raises the process gas temperature above the dewpoint to prevent sulfur from condensing on the downstream catalyst. Sulfur condensation causes a reduction in conversion by blocking active catalyst sites. The heat for the Reheater is typically supplied by high pressure steam or by the direct firing of fuel gas or acid gas. For the purposes of this paper, direct fired Reheaters (using fuel gas) will be considered.

The heated process gas flows to the Converter where Equation 2 takes place across a catalyst bed. Temperatures in the Converter are typically in the range of 525°F to 625°F. The sulfur vapor is condensed in the downstream Condenser, and the Reheat, Conversion and Condensing steps are repeated.

The process gas from the Final Condenser (tail gas) flows to the Incinerator where any remaining H_2S (along with any other reduced sulfur compounds) is oxidized to SO_2 before exiting to the atmosphere through a stack.

During startup, the plant is heated to normal operating temperatures by firing fuel gas in the Reaction Furnace, Reheaters and Incinerator. Once the catalyst has reached normal operating temperatures and the boilers have reached normal operating pressure, acid gas is introduced to the Reaction Furnace, the fuel gas is removed and sulfur production begins. The Reheaters and Incinerator continue to fire fuel gas during normal operation.

2.3 Process Safety

Note that this section is not in any way meant to be a discussion of all SRU safety concerns, and is only meant to give the background necessary for the remainder of the paper.

The principle safety concern in a sulfur recovery unit is personnel exposure to and environmental release of H_2S. The Threshold Limit Value (TLV) for H_2S is 10 ppm, and it is lethal in concentrations of 1,000 ppm and above. Although sulfur recovery units operate at low pressures (5 to 15 psig), H_2S concentrations are very high. Typical concentrations in the front end of an SRU are in the range of 50% to 90%, and in the back end of the unit they range from 1% to 2%. Consequently, any loss of containment of the process gas has the potential to cause injury or death.

The SRU shown in Figure 1 has four burners: the Reaction Furnace, two Reheaters and the Incinerator. Each burner has the potential to cause loss of containment due to damage to the vessel shell. The following three scenarios can lead to such damage:

Scenario #1: High temperature leading to refractory damage followed by vessel shell damage. The high temperature is typically due to a process upset or operator error.

Scenario #2: Deflagration leading to vessel shell damage. The deflagration is typically due to improper purging prior to light-off or process upset.

Scenario #3: Improper flame shape leading to refractory damage and vessel shell damage. The improper flame shape is typically due to burner damage.

These three scenarios are the focus of the Shutdown System examples given in Section 4.

3. Evolution of a Shutdown System

Many grass-roots or expansion projects for sulfur recovery units in the oil and gas industries follow these phases:

1. Initial Design
2. Engineering / Procurement / Construction (EPC)
3. Precommissioning / Commissioning / Initial Operations

Different companies often work on the Initial Design and the EPC phases. The owner's personnel assigned to the project will also often change over the course of the project. This means there are many different people involved with the various phases, and each person has a different background and set of experiences.

Based on input from the various parties involved, the design is enhanced as it progresses from the first phase to the second. In some cases, the design is further refined during the third phase when actual operating conditions do not match predicted conditions. The Shutdown System is one piece of the design that gets enhanced throughout all three phases.

The majority of the Shutdown System in an SRU is dedicated to the startup and shutdown of the fired equipment. This part of the Shutdown System is also known as the Burner Management System (BMS) which ensures the burners are started up and shut down in a safe manner through a combination of permissive and shutdown devices. The fired equipment in most modern sulfur recovery units employ a BMS which follows the guidelines set forth in NFPA 85 [2]. For increased reliability, the BMS typically resides in a Programmable Logic Controller (PLC) system which is separate from the plant Distributed Control System (DCS). And the BMS instruments (transmitters and final elements) are typically separate from the instruments used for DCS control. The progression of the BMS is outlined in the following sections.

3.1 Initial Design

The initial design (also known as Front End Engineering Design, or FEED package) is typically performed by a company who owns the process license for the selected technology. The FEED package normally contains a Cause & Effect Diagram which lists each instrument that has an input into the BMS, and the action caused by each instrument when it reaches its trip setpoint. The FEED package may also contain a write-up for the BMS which gives more detail for subjects like the purge requirements and startup bypasses. The BMS documents are prepared based on the Licensor's experience on previous projects and industry standard practices.

The Owner typically participates in design reviews and a HAZOP during the FEED effort. In this way the owner can include their own recommendations (or in some cases, demands) to the BMS.

3.2 Engineering / Procurement / Construction

During the EPC phase of the project, the equipment is purchased and the design enhanced as necessary based on input from the equipment vendors. For the BMS, the focus is on the vendors selected to provide the burners. Different vendors have different requirements for the BMS in order to ensure their equipment is protected. For example, some burner vendors require low air flow shutdowns to ensure the flame is maintained in a proper position in front of the burner face (preventing flame impingement on the burner itself). Other vendors use a low fuel flow shutdown to ensure the proper flame position. In all cases, the trip setpoints for each transmitter are determined based on vendor and/or process requirements.

The Cause & Effect diagram is updated as necessary to reflect the input from the vendors. A detailed BMS description is written to be given to the programmers who will develop the PLC logic. In some cases, logic diagrams are also prepared as a supplement to the Cause & Effect Diagrams.

As with the FEED stage, the EPC stage has design reviews and a HAZOP. Once the PLC has been programmed, a Factory Acceptance Test (FAT) typically occurs which allows the program to be tested. The Licensor, EPC Contractor and Owner are usually involved with all of these activities. And there is no guarantee that the various representatives will have been involved with the FEED stage. Inputs from the various parties may lead to additional instrumentation, changing the location of existing instrumentation and/or modification to the BMS logic itself.

3.3 Precommissioning / Commissioning

The line dividing the end of Construction and the beginning of Precommissioning / Commissioning is very blurry and many activities can fall on either side. For the BMS, this phase usually entails loop checks of all the BMS instrumentation and a Site Acceptance Test (SAT).

The loop checks ensure the BMS transmitters are communicating with the PLC, the PLC is communicating with the final elements (usually valves), and the PLC is communicating with any local control panels and/or the DCS. Once these are complete, an SAT is usually performed to prove that the BMS performs as per the design. This test is usually witnessed by representatives from the Licensor, the EPC Contractor and the Owner. As with previous stages, there is no guarantee the representatives are aware of the project history. At this time, changes to the BMS should be minimal, however changes can still occur based on input from the various parties.

4. Examples of BMS Changes

The following sections discuss specific examples of changes to a BMS and the effect these changes had during commissioning and initial operations. Each of these changes and consequences occurred during actual projects, however, they did not all occur on the same project.

4.1 Reaction Furnace High Temperature Trip

During the initial design of a new for an existing oil refinery, the BMS for the Reaction Furnace did not include a high temperature trip. This decision was based on the designer's previous experience with similar units and due to the difficulty of measuring the Reaction Furnace temperature.

When the Reaction Furnace is firing acid gas during normal operation, the operating temperature is determined by the acid gas composition. Control of the combustion air is set by the Claus reaction stoichiometry to burn 1/3 of the H_2S in the acid gas. As the H_2S concentration in the acid gas increases, so does the Reaction Furnace operating temperature. While varying acid gas CO_2 concentrations can cause Reaction Furnace temperature fluctuations in gas plant applications, such fluctuations are rare in oil refinery applications.

High operating temperatures are possible during the heat-up of the unit when fuel gas is fired in the Reaction Furnace. The refractory is designed for a maximum hot-face temperature of 2700°F. A stoichiometric fuel gas flame temperature is approximately 3600°F, so misoperation during fuel gas firing can lead to temperatures greater than the refractory design temperature. While the refractory can handle short-term temperature excursions, damage will occur in less than an hour at 3600 °F. In order to control the temperature during fuel gas firing, a tempering steam connection is provided to the Reaction Furnace burner. This injection of low pressure steam moderates the flame temperature and allows the furnace to operate at stoichiometric conditions while firing fuel gas. As part of the initial design for this project, operating procedures were written which explained the use of the tempering steam injection.

Measuring the Reaction Furnace temperature is difficult for a variety of reasons. The refractory system typically has at least two layers which can expand and contract at different rates. This can cause shearing of thermocouple elements if they are not installed correctly. The high temperatures necessitate precious metal thermocouple elements. During normal operation, the Reaction Furnace has high concentrations of hydrogen which can damage the elements. While the thermocouples are typically purged, loss of the purge medium for even a short time can cause permanent failure of the element. And because of the temperatures and toxic gases involved with Reaction Furnace operation, on-line replacement of the thermocouple elements is not possible.

Because high temperatures are likely only during startup and shutdown, periods when more attention is typically paid to unit operation, and because of the difficulties in reliably measuring the Reaction Furnace temperature, the decision was made to alarm, but not shutdown, on high Reaction Furnace temperature.

During the detailed design HAZOP (during the EPC phase), the EPC process engineer stated that the Reaction Furnace must shut down on high temperature. The FEED designers explained their reasons for not shutting down on high temperature, but the EPC process engineer insisted. The high temperature alarm was then changed to a high temperature shutdown which was maintained throughout the rest of the project.

Within the first few weeks of operation, the Reaction Furnace thermocouple purge regulator became choked with weld slag and the purge flow stopped. The thermocouple failed within a few hours, and it failed full-scale. This caused the Reaction Furnace to shut down. Luckily there was a spare thermocouple in the warehouse. However, it took five days to cool the Reaction Furnace, replace the thermocouple, reheat the Reaction Furnace and catalyst, and reintroduce acid gas. During this time, the refinery was forced to reduce throughput to stay within its emissions limits. The resulting loss of production was estimated to be greater than $150,000.

4.2 Reheater Loss of Flame Trip

The FEED designers of an SRU to be built in an existing gas plant decided to use direct-fired Reheaters to reheat the process gas to each Converter. Two flame scanners were provided on each Reheater to detect the flame. During the BMS development it was decided that the loss of flame in a Reheater would cause the fuel gas and air automated block valves associated with that Reheater to close. Although shutting down the entire SRU for loss of flame in the Reheaters was discussed, the original decision was maintained because shutting down the Reheater does not have an immediate negative effect on the overall SRU operation, the Reheater is constantly purged by an inert stream during normal operation, and because of the difficulty of getting reliable flame signals on SRU Reheaters.

When a Reheater is shut down during SRU operation, the process gas to the downstream catalyst bed cools to its sulfur dewpoint. This causes sulfur vapors to condense in the catalyst pores, which reduces the activity of the catalyst. This takes several hours to have an effect on overall plant operation, which gives operators the time necessary to relight the Reheater.

Because the Reaction Furnace is supplied with only enough air to burn 1/3 of the H_2S contained in the acid gas, the process gas downstream of the Reaction Furnace is oxygen-free. While it is common practice to have a pre-startup purge for any piece of fired equipment, Reheaters are constantly purged by an inert process gas during normal operation.

Reheater burners are typically small in diameter because the required heat duty is low. The burners typically have a common fuel and air nozzle, two flame scanner nozzles and one sight port nozzle (for visual inspection of the flame). The small diameter of the burner along with the required number of nozzles limits the size of the scanner nozzles, often to only 1" or 2". This leaves a very small viewing area for the flame, so small obstructions in the nozzle can affect the flame scanner signal. Reheaters are full of sulfur vapor and it is common to have sulfur deposit in the nozzles. The flame scanner nozzles are usually purged (in this case with nitrogen). However, fluctuations in the purge flow can lead to sulfur deposits, which in turn can cause a false loss-of-flame indication.

Based on the long time period to impact operation, the constant purge of the Reheater and the difficulties with maintaining flame indication, the decision was made to isolate only the fuel and air to the individual Reheater on loss of flame indication.

During the EPC phase, the EPC contractor sent documents to the FEED Licensor for review, including the Cause & Effect Diagrams. The personnel involved with the original FEED package had all been assigned to new projects, so an engineer who was new to the project reviewed the documents. This was the engineer's first job with direct-fired Reheaters. However he knew from his experience with Reaction Furnaces that the furnace is always isolated from all feed sources and purged with nitrogen before relighting after a loss of flame. Based on his prior experience, he marked the Cause & Effect diagram to show a full SRU shutdown upon loss of flame in any Reheater.

The EPC contractor also had limited experience with direct-fired Reheaters, so they did not question the Licensor's comment and incorporated it into the design. The new cause for a complete SRU shutdown was maintained through the remainder of the plant design.

Within the first few days after startup, plant management started a campaign to reduce nitrogen consumption. Operators followed their instructions and closed the valves on the nitrogen purge lines to the Reheater nozzles. This caused the nozzles to plug with sulfur and the First Reheater shut down soon after on the loss of flame signal which then caused a shutdown of the entire SRU.

The nozzle was unplugged relatively quickly by injecting steam. However, during the restart of the plant, an inexperienced operator fired the Reaction Furnace pilot incorrectly at too low of a fuel and air flowrate. The tip of the pilot quickly melted as the flame burned back inside the pilot and the melting steel caused damage to the Reaction Furnace burner.

Although the burner was designed to be removable from outside the Reaction Furnace through a nozzle, a fabrication error caused the burner diameter to be slightly larger than the removal nozzle diameter. It took one month for the Reaction Furnace to be cooled and readied for entry, the old burner tip cut off, the new burner tip and pilot installed, and the SRU restarted. In this case, the gas plant had spare sulfur recovery capacity, so production was not lost. However, the estimated cost of the repair work exceeded $100,000.

4.3 Incinerator Fuel Gas Low Pressure Trip

The Initial Design of an SRU revamp in an existing refinery included the replacement of several pieces of equipment, including the Incinerator. For the FEED package documents, the Incinerator, Stack, Burner Management System and associated instrumentation were shown as a vendor-supplied package. Within the vendor scope of supply, typical instrumentation was shown on the Piping & Instrumentation Diagrams (P&IDs) only to give the EPC contractor an idea of the instrumentation Input and Output scope necessary for the upgrade of the DCS.

The instrumentation included the following typical shutdown devices: high temperature, high fuel gas pressure, low fuel gas pressure, and loss of flame indication. The high and low fuel gas pressure transmitters were both shown just upstream of the burner on the fuel gas line, downstream of all valves. During the EPC phase, the Licensor confirmed that the low fuel gas pressure shutdown is provided to prevent burner damage caused by firing the burner at too low a fuel gas flow rate. Such burner damage can cause direct flame impingement on the refractory which in turn can lead to refractory failure, shell damage and toxic gas release.

The Incinerator vendor was selected and they submitted their own P&ID for review and HAZOP. As part of their detailed design development, they moved the low fuel gas pressure shutdown upstream of the fuel gas control valve, consistent with NFPA 85 guidelines [2]. The vendor confirmed that for their design, maintaining a minimum air flow prevented burner damage, and there was no minimum required fuel gas flow.

During the detailed design HAZOP, an owner representative strongly insisted that the low fuel gas pressure be reinstated to its original position just upstream of the burner. The vendor representative at the HAZOP explained the reason for moving the pressure transmitter. He also explained that the Incinerator will sometimes operate at a slight vacuum and sometimes run at slight positive pressure depending on the operating scenario. Locating the pressure transmitter just upstream of the burner meant that at low fuel gas flow rates, the pressure transmitter would read close to the Incinerator pressure (because the differential pressure across the burner would be small). Therefore, calculating a setpoint for this low fuel gas pressure shutdown appropriate for all operating scenarios would be difficult. The owner representative's opinion did not change, and in the interest of making progress in the HAZOP, it was agreed to move the pressure transmitter back to its original location.

The Incinerator vendor calculated a low fuel gas pressure shutdown setpoint of a slightly positive gauge pressure. However, they did not accurately predict the draft of the stack at high operating temperatures. During initial operations, a minor plant upset caused the Incinerator temperature to increase by 200°F. The resulting increase in the draft of the stack caused the Incinerator to shut down on low fuel gas pressure. The Incinerator shutdown caused an entire SRU shutdown.

During the restart of the SRU, fuel gas was fired in the Reaction Furnace to reheat the catalyst to normal operating temperatures. One of the combustion air flowmeters had lost its calibration and extra combustion air was sent to the Reaction Furnace. This excess oxygen ignited sulfur that was still contained in the catalyst causing fires in both catalyst beds. The excessive heat burned holes in the downstream piping and Condensers causing the ground-level release of significant quantities of SO_2. No personnel were injured, but the plant was shut down for six weeks to repair the piping and equipment. The refinery had excess sulfur recovery capacity, so production was not lost. But the repair costs were estimated to be more than $200,000.

5. Conclusion

Decisions made during the various phases of a design project can have adverse affects on the startup and initial operation of a plant. Often these decisions are made with the best of intentions, but they can be based on an incomplete knowledge of the history of the project, or a lack of understanding of the process. The design changes resulting from these uninformed decisions are then often approved by the design team because they seem simple and are inexpensive.

When uninformed decisions and changes occur with a shutdown system, it creates the possibility of an increased frequency of shutdowns and restarts during commissioning. And because of the complexity of the processes involved, every shutdown and restart has a potential for personnel safety and equipment damage issues.

The examples discussed in this paper show a clear relationship between bad decisions and loss of production and/or equipment damage. This illustrates a need for better communication of a project's history as project team personnel change, so the team can make more informed decisions that help prevent future commissioning and initial operation incidents.

6. References

[1] Engineering Data Book, 10th Edition, Gas Processors Suppliers Association, Tulsa, OK, 1994.

[2] *NFPA 85 Boiler and Combustion Systems Hazard Code*, National Fire Protection Agency, 2004 Edition, Quncy, MA, 2004.

The Obvious-to-Obscure Process Safety Checklist for Plant Engineers

Mr. John T. Perez, P.E.
Cognascents Consulting Group, Inc.
11777-A Katy Fwy Ste 438
Houston, TX 77079
john.perez@cognascents.com

Keywords: Process Safety Management (PSM), Management of Change (MOC), Process Hazards Analysis (PHA), Human Factors (HF), Mechanical Integrity (MI).

Abstract

The recent process safety performance of the energy industry has led to scrutiny of industry and company-specific process safety standards, guidelines, and best practices. Several elements of process safety focus directly on the knowledge-base and experience of a facility's workforce. Plant engineers are required to participate in and lead various elements of a facility's process safety management program (e.g. management of change reviews, process hazards analyses, and mechanical integrity program implementation). Although plant engineers have an undergraduate degree in their technical field of choice, few have any formal training or recognized credential in the field of process safety. Furthermore, few companies have implemented a formal competency assessment and technical authorization program. Without such training and competency verification requirements, the author contend that a greater likelihood exists for human error in the administration of a facility's process safety program.

This paper provides a safe operating window for engineers of all experience levels on process safety protocols and pitfalls. The advice and information shared in this paper are direct learnings from assignments as a plant engineer, technical consultant, and business owner. The content is divided into the following four categories:

1. The Obvious - information, knowledge, and responsibilities conveyed through schooling and company's technical onboarding process.
2. The Potentially Obvious - information, knowledge, and responsibilities conveyed through training from entry-level to mid-level engineer.
3. The Potentially Obscure - information, knowledge, and responsibilities conveyed through training from mid-level to senior-level engineer.
4. The Obscure - information, knowledge, and responsibilities conveyed through advanced/expert training, application, and experience.

The target audience for this paper includes plant managers, business unit managers, project managers, plant engineers, EH&S managers, PSM coordinators, and operators; however, anyone involved with plant operations and/or service delivery to the energy industry may benefit from this paper.

1. Process Safety Is Under Fire, Again

I have worked in the energy industry for fourteen years. I have seen many changes, but I have also seen many things remain the same. I have heard people in companies across the energy spectrum argue whether the industry has improved overall with respect to process safety. I do not believe that enough information is available for review to answer the question definitively; however, I am hopeful that current benchmarking and reporting initiatives will take root and soon provide this valuable information. Unfortunately for now, the lack of information is a problem because it leaves the industry unarmed to critics and regulators.

It is December 2010 as I write this paper and process safety (along with other significant business functions) is under fire at the highest levels of the nation. Incidents like BP's Texas City explosion on 23 March 2005, Tesoro's Anacortes fire on 02 April 2010, and BP's Macondo blowout on 20 April 2010 continue to draw the spotlight to the ugly reality of risk in the energy industry. When people die or get hurt, the risk component to the industry merits examination and mitigation. During such examinations, accurate and accessible information is paramount to defining and understanding the true problem. Without information, society (as do engineers) tends to err on the conservative end of the risk scale. Erring on the conservative end can impose unnecessary cost in the form of regulations, protocol, time, and/or money.

During times of extreme scrutiny and examination, I often observe that a massive fact-finding exercise is performed – interviews are conducted, page after page of reports are read and catalogued, and surveys are administered and analyzed. As indicated above, the point of this exercise is to build a defense (or offense depending on which side you sit on) in the form of information to be used at a congressional hearing or trial or investigative hearing. Millions and millions of dollars are collectively spent by the involved parties to ensure superior preparation for reactive assessments and resolutions of serious incidents. The end goal is to try to prevent the same event from occurring again in the future.

The current process safety fever will subside at some point. But how do we improve the reality and perception of process safety within the energy industry and keep it away from the firing squad? The answer is simple and I hate singing an old tune that we all know. Prevent all incidents.

So, how do we prevent all incidents?

2. Constant Application of Deep and Broad Knowledge by Every Responsible Entity for Every Resource is Vital to Process Safety Excellence

The above macrocosmic presentation of process safety's delicate relationship with society can be distilled down to a microcosmic parallel at the asset level (i.e. platform, gathering/distribution system, refinery, plant, and station). After a significant incident occurs at a facility, an incident investigation is typically performed. The investigation requires a massive fact-finding exercise – interviews are conducted, experts are consulted, page after page of reports and inspections are read and catalogued, and data is downloaded from the control system. The point of this exercise is to gather information that can be used to figure out what happened, how it happened, why it happened, and how to keep it from happening again. Many dollars are spent to ensure superior preparation for reactive assessment and resolution of the incident. Does this sound familiar? I assert that there would not be a process safety macrocosm to scrutinize, if the integrity of the process safety microcosm was kept uncompromised.

Preserving the integrity of process safety at the microcosmic level requires constant application of broad and deep knowledge by each entity (i.e. individuals, organizations, and cultures) responsible for the performance of a resource (i.e. individuals, cultures, physical assets). Failure to accomplish this simply-stated directive is the core problem plaguing the process safety component of the industry.

Constant application of knowledge requires continuous presence and monitoring to afford proactive solutions. Platforms, refineries, and plants are typically operated 24/7 with multiple shifts ensuring continuous supervision. While many operators and technicians have mastered the technical elements of their job, it is unclear how many have dedicated the time necessary to be considered an expert in any one field. The same can be said for many engineers. So while constant presence may exist, constant application of knowledge necessary to ensure process safety excellence is typically not realized.

The depth of knowledge required to ensure process safety excellence is overwhelming to say the least. Recognized and Generally Accepted Good Engineering Practices (RAGAGEP) have ballooned to a body of knowledge that nobody can master. Process safety knowledge includes a wide horizontal axis of subject matter including leadership, process knowledge, human factors, risk analysis, pressure relief design, mechanical integrity, process control, unit operations, process safety protocols, operating procedures, standards, codes, regulations, and the list goes on. The vertical axis is as deep or deeper with topics and sub-topics detailing each subject matter.

Rare is the scenario where one entity is solely responsible for all facets of production and process safety administration. The more common scenario involves multiple individuals and/or organizations, each with a portion of responsibility for production and process safety administration. The problem with multiple layers of responsibility is that overall responsibility and, more importantly, accountability are diluted and more susceptible to marginalization. Another problem with multiple entities interacting to deliver process safety excellence is that each entity carries and brings a unique set of experiences. Experience is good when it is available to impart wisdom and knowledge. Experience is useless when it is not around. A larger problem is not knowing what has NOT been experienced.

The last component to process safety excellence is the resource base. Remember: process safety excellence results from constant application of broad and deep knowledge by each entity responsible for the performance of a resource. A facility's resource base includes the people, the work processes, the physical assets, and the intangible assets (e.g. culture). Once again, it is unreasonable to expect anyone to claim mastery and expertise over an entire resource base. Leading an organization to achieve process safety excellence is a very different skillset from executing PHAs or MOCs. Achieving recognition as a pressure relief analysis expert takes years, as does recognition in the fields of mechanical integrity, risk analysis, or process design.

Despite all of the above, plant engineers are required to participate every day in and lead various elements of a facility's process safety management program (e.g. management of change reviews, process hazards analyses, and mechanical integrity program implementation). Although most plant engineers have an undergraduate degree in their technical field of choice, few have any formal training or recognized credentials in the field of process safety. Furthermore, from my observation it appears as though few companies have implemented a formal competency assessment and technical authorization program. Without such training and competency verification requirements, I contend that a greater likelihood exists for human error in the administration of a facility's process safety program.

Unless a change is made it is unrealistic to believe that process safety excellence can be achieved. The change must directly address and mitigate the concerns regarding constant application of broad and deep knowledge by each entity responsible for the performance of a resource.

3. The Resurgence of the Checklist

Have you ever heard of Atul Gawande? His recent book, The Checklist Manifesto: How To Get Things Right, has sparked a resurgence of sorts for the age-old tool of making a list of actions/items necessary to perform a task the right way and the same way, every time. While doing research in the medical industry to better understand why clusters of excellence exist, he observed that many high-performing teams or organizations effectively used checklists.[1]

I know some of you are saying to yourselves, "Really, a checklist? This is your plan for process safety excellence?"

No, the checklist is not my plan. *Effective use of effective checklists* is my big "Aha" for you.

Most of us already incorporate checklists into our lives. Walkarounds are typically performed during each shift to check on critical assets. These walkarounds typically require operators or technicians to document their observations on a checklist. Many work processes for technical and business functions are driven or at least accompanied by a checklist. Failure to use a checklist is not the problem. *Failure to effectively use an effective checklist is the problem.*

The remainder of this paper offers suggestions for effective use and construction of checklists for safe operations by engineers of all experience levels.

4. Effective Use and Construction of Checklists

The following guidelines for construction of effective checklists have been consolidated from the sources cited in the References section of this paper.[1,2,3] I have attempted to incorporate my additions into the referenced body of knowledge on checklist development.

1. In determining the checklist's ***purpose***, one should ensure the following:
   a. The checklist is focused on a specific task;
   b. The checklist addresses a critical task of a complex system/problem;
   c. The checklist addresses a task that cannot be verified any other way and therefore is not vulnerable to marginalization.
2. A checklist's ***development*** should address and optimize the following elements:
   a. Checklist objectives should:
      i. Be accurate, clear, and concise;
      ii. Be located at the front of the checklist for quick and easy access;
      iii. Provide the who, what, when, where, and how of the checklist;
   b. Checklist ***content*** and ***format*** should:
      i. Prominently display the creation and revision date;
      ii. Be complete and technically correct;
      iii. Use precise terms and verbs consistently;
      iv. Spell out acronyms upon first-time usage;
      v. Use common/basic words and language;
      vi. Limit each checklist item to one activity;
      vii. Fit on one page with fewer than ten items per break in workflow;
      viii. Minimize the use of color;
      ix. Use easy-to-read font in a large-enough size for the entire user pool;
   c. Checklist ***structure*** should ensure:
      i. Similar tasks are grouped together;
      ii. Order of execution is addressed;
      iii. Visual breaks and textual devices are used appropriately;
      iv. Critical words, such as "not", are underlined;
      v. Communication queues are clearly defined and evident;
   d. Visual content should:
      i. Be located on the left side of the page with corresponding text to the right;
      ii. Serve an obvious purpose and contain only essential information.
3. The ***testing*** and ***validation*** of a checklist should ensure the following:
   a. Testing should be done one person at a time by future users;
   b. No help should be provided during testing;
   c. Testing and revising should continue until future users can use the checklist as intended on their own;

d. Testing should occur under normal and abnormal conditions – day and night, routine and non-routine operations, new and experienced users;
e. Checklist should be completed in a reasonable amount of time specific to purpose and task;
f. Prior to official use a continuous monitoring and improvement plan should be established.

Additional details for some of the above guidelines are provided in the papers cited in the References section. The above guidelines are by no means exhaustive and are meant to promote additional examination and understanding of effective checklist construction principles.

5. The Obvious-to-Obscure Process Safety Checklist

In line with the guidelines from Gawande, Stufflebeam, and Bichelmeyer, I offer the following advice and information. These guidelines are also direct learnings from assignments as a plant engineer, technical consultant, and business owner with respect to process safety protocols and pitfalls. The content is divided into the following four categories:

1. The Obvious - information, knowledge, and responsibilities conveyed through schooling and company's technical onboarding process.
2. The Potentially Obvious - information, knowledge, and responsibilities conveyed through training from entry-level to mid-level engineer.
3. The Potentially Obscure - information, knowledge, and responsibilities conveyed through training from mid-level to senior-level engineer/manager.
4. The Obscure - information, knowledge, and responsibilities conveyed through advanced/expert training, application, and experience.

Some of the below guidelines are designed to serve as prompts for reflective pauses rather than physical actions. The intended audience is anyone that falls within the scope of preserving the integrity of process safety. Some sections may apply more to young engineers and others to seasoned experts. The goal is to ensure that all of the below guidelines are considered when developing and executing checklists, especially checklists that impact the process safety integrity of a system.

5.1 The Obvious

The Obvious section of the checklist should include information, knowledge, and responsibilities conveyed through schooling and an organization's technical onboarding process. Obvious elements should include:

1. Standards and guidelines;
2. Processes and protocols;
3. Process safety information;
4. Operating procedures.

The above items do not have to be memorized, but an engineer should ask, read, and become familiar with all of the above items.

In addition, an engineer is expected to enter the professional ranks with a fundamental understanding of technical principles corresponding to their field of study. More importantly, an engineer is expected to arrive with the capability to problem-solve and think independently. They should be able to independently ask and answer obvious questions, such as:

1. Do I understand the process?
2. Do I understand how this piece of equipment works?
3. What real-time information can I check myself?

5.2 The Potentially Obvious

The Potentially Obvious section of the checklist should include information, knowledge, and responsibilities conveyed through training from an entry-level to mid-level engineer. A good way to think about Potentially Obvious elements is to consider your expectations or the organization's expectations of what someone should know a handful of years after entering the professional ranks. Potentially Obvious elements should include:

1. Moral courage – Who is responsible for blowing the whistle and slamming on the brakes?
2. Experience and history – What has happened in the past and where is the historical information for review?
3. Organizational wisdom – Who are the experts and how do I reach them?

5.3 The Potentially Obscure

The Potentially Obscure section of the checklist should include information, knowledge, and responsibilities conveyed through training from mid-level engineer to senior-level engineer/manager. We are now getting into the vulnerable area of the checklist where expectations of knowledge mastery are less defined. By this time in someone's career, it is expected that they operate fairly independently with little or no direction. Because of this expectation, the openness and security to ask "dumb" questions are not available. Potentially Obscure elements should include:

1. Leadership and accountability consolidation – Who is ultimately responsible and accountable for the overall effort?
2. Experience level of individuals and group responsible for process safety functions – Do we have the right people working on this?
3. Scope – Are we trying to do too much here?
4. Self-awareness prompt – Am I the best person to be leading this effort? Am I falling prey to any cognitive biases?

5.4 The Obscure

The Obscure section of the checklist should include information, knowledge, and responsibilities conveyed through advanced/expert training, application, and experience. Obscure elements are:

1. Experience and expertise biases – Who has different experiences than me? How do I know the expert is truly an expert? Have their credentials been vetted for the specific task at hand?
2. Expert bias – Who will challenge my thought process?
3. Cultural check – What is the pulse/morale of the unit or plant?
4. Big picture pause – What current or recent activity can impact multiple systems?
5. Persuasion and influence monitoring – Are people listening to me? If not, how do I make my need important to those I need help from?
6. In the moment evaluation – Are others in the moment? Are they being competent and experienced right now?
7. Self-awareness test – Am I leading by example? Am I living process safety? Have I made it part of my everyday job by integrating it into everything I do? Process safety is not 50% of the job or time – it is 100% of the job all the time … in and out of the plant.

6. Conclusion

Our industry continues to suffer fatal incidents. Process safety is under intense domestic scrutiny. Prevention on a microcosmic level is the first step to accomplishing macrocosmic process safety excellence. Assuring process safety integrity on the microcosmic level requires constant application of broad and deep knowledge by each entity (i.e. individuals, organizations, and cultures) responsible for the performance of a resource (i.e. individuals, cultures, physical assets). Failure to accomplish this simply-stated directive is the core problem plaguing the process safety component of the industry. The effective use of effective checklists is one measure that can be taken to realize process safety excellence. Checklist effectiveness includes modifying the typical approach to checklist development and content. Obvious elements should direct; however, Obscure elements should query. While many may believe that process safety checklists are prevalent in our industry, at the time of writing this paper (December 2010) my submitted abstract showed up on the first page of Google's search results for "process safety checklist". So I conclude that we are either not sharing our checklists publicly or we are not creating them – both of these problems must be remedied to achieve process safety excellence.

7. References

[1] Gawande, A. The Checklist Manifesto: How To Get Things Right. Picador, 175 Fifth Avenue, New York, N.Y. 10010. Copyright 2009: First edition.

[2] Stufflebeam, D. L. "Guidelines for Developing Evaluation Checklists: The Checklists Development Checklist." July 2000. http://www.wmich.edu/evalctr/archive_checklists/guidelines_cdc.pdf. Accessed 24 December 2010.

[3] Bichelmeyer, B. A. "Checklist for Formatting Checklists." 04 October 2003. http://www.wmich.edu/evalctr/archive_checklists/cfc.pdf. Accessed 05 January 2011.

(60b) Incident Investigation- Addressing Common Oversights

Julie D. White, Scandpower Inc., a member of the Lloyd's Register Group, Houston, TX

Principles of incident investigation may be used in daily process troubleshooting and near miss events to identify root cause. Identifying and correcting daily process issues intervenes with the natural escalation to near miss events and incident investigations.

Fundamentals of root cause analysis and incident investigation are discussed, with examples of common oversights in routine daily troubleshooting which may lead to significant incidents or repeat incidents following incomplete incident analyses.

Common oversights of incident investigations in the areas of data gathering, root cause analysis, management, communication and leadership are discussed. Data gathering topics will address collection and preservation of process data, field evidence, process sampling, and interviewing. The employment of basic statistical analysis in root cause analysis will also be examined.

See more of this Session: Things Every Plant Engineer Should Know about Process Safety
See more of this Group/Topical: Topical 1: Global Congress on Process Safety

Management of Change (MOC) Basics for Plant Engineers

R Wayne Garland,
Eastman Chemical Company
P O Box 511 Bldg 54D
Kingsport, TN 37662
423-229-5518
wgarland@eastman

Abstract

The essentials of Management of Change (MOC) for plant engineers will be presented. The meaning of MOC will be explained and its central role in safe plant operation will be shown, including how it fits with other aspects of plant safety. The objectives of MOC will be discussed along with the general definition of "change" and the different types of changes. Examples will be used to illustrate. The essential elements of the MOC process will be covered including approvals, safety reviews, notification, training, and Process Safety Information (PSI) updates. Temporary changes will also be discussed. Special emphasis will be given to the critical role of communication in the MOC process. The importance of MOC will be further reinforced by explaining potential consequences of not using an MOC process.

Definition and Scope

The general definition of "change" from the PSM regulation 29 CFR 1910.119 is any alteration (except for replacements-in-kind) to process materials, processing conditions, equipment, maintenance materials, procedures, utilities, facilities, control systems, etc.

MOC should be utilized at each chemical handling site to cover potential hazards associated with changes to:

- Process chemicals
- Technology
- Reactions
- Materials of construction
- Process controls
- Facilities

Example 1 - According to this definition for "change", which of the following definitely IS a change?

1. Adding a new control valve
2. Relocating an electrical outlet or light fixture in an office area
3. Modifying an operating procedure to correct misspelled words
4. An emergency action in response to an accidental discharge which is discontinued immediately upon termination of the emergency
5. Replacing a worn out valve with an essentially identical valve which meets the same specifications

The answer is number 1. MOC covers alterations to manufacturing processes that are not replacement in kind. Alterations to office areas, editorial changes, or certain emergency actions may not be considered changes. Replacing a valve with essentially identical equipment is a replacement in kind.

Example 2 - According to the definition for "change", which of the following is usually NOT considered a change?

1. Temporarily by-passing an interlock
2. Going to a different schedule of pipe in a pipeline if the current piping specification for the same service does not call for that schedule
3. Adding a new nozzle to a tank
4. Changing a temperature target or alarm limit within the range defined in a standard operating procedure/control strategy.
5. Adding a new step to an operating procedure

The answer is number 4. If safe operating limits are defined in a standard operating procedure, process set points can be changed within that range without the need for MOC. However if the set point was being changed outside the pre-approved safe operating limits then MOC should be used.

Why We Need It

Why do we need MOC? Here is a list of major process safety incidents that occurred in the years just prior to promulgation of the PSM regulation.

Year	Location	Deaths	Injuries
1974	Flixborough, England	28	?
1976	Seveso, Italy	?	?
1984	Mexico City, Mexico	650	?
1984	Bhopal, India	2000+	?
1985	Institute, WV	0	135

1988	Norco, LA	5	23
1988	Henderson, NV	2	350
1989	Richmond, CA	0	9
1989	Pasadena, TX	24	132
1989	Channelview, TX	17	0
1990	Cincinnati, OH	2	41
1991	Lake Charles, LA	6	6
1991	Sterlington, LA	8	128
1991	Charleston, SC	9	33

The textbook case that created widespread awareness of the need for MOC was the 1974 incident in Flixborough, England at the Dutch State Mines Nypro plant. The plant used 6 reactor vessels in series. According to *Incidents That Define Process Safety*[1], "a crack was detected on Reactor No. 5. A Maintenance Engineer recommended complete closure for 3 weeks. The Maintenance Manager, whose role had been carried out for several months by the head of the laboratory while awaiting a reorganisation of the company, proposed instead to dismantle Reactor No. 5 and connect numbers 4 and 6 together by a 20 inch (500 mm) diameter temporary connection using piping bellows units, supported on a structure made from conventional construction industry scaffolding. During restart, the temporary piping ruptured at the bellows and 30 tons of cyclohexane escaped in 30 seconds, forming a vapor cloud 200 meters across and 100 meters high. The vapor cloud was ignited by a furnace and the resulting vapor cloud explosion (VCE) resulted in 28 deaths and many injuries. "

Learning from "Sam Shortcut"

In our training, we illustrate the value of MOC through the story of "Sam Shortcut", a staff process engineer in a manufacturing department. Sam is assigned a project to facilitate production of a new product. The only facility change required is modification of a convey system route to go to an existing product storage bin (bin 99) that is not currently in service. This requires altering some piping and a control valve. What could be simpler? Sam is already overloaded with a long list of urgent projects. He does not see the need to use the MOC process for such a simple project. Rather he sees MOC as unnecessary extra "paperwork".

Example of Not Following the Process

Sam bypasses the MOC process and gets the alteration done by going directly to the area mechanics and electricians. He is proud of his efficiency until he starts getting calls about the product bin from various people:

1) The Material Transfer operator can't get any product to go to bin 99.

2) The Area Operations Manager has a quality problem because material was transferred to wrong product bin.

3) An Operator in a unit upstream returns from his off days and uses the old targets for the process variables because he didn't know about the changed targets for the new product.

4) The Area Operations Manager is upset again. There has been an accidental discharge because the primary level device on the product bin failed and there was no backup level device.

5) The Shift Team Manager is concerned about the relief valve on bin 99 opening frequently (cycling) and possibly releasing inert gas into production area.

6) The Pressure Vessel Inspector becomes aware of the change and is concerned that state pressure vessel codes could apply. He asks Sam if the bin is rated for the new operating pressure and if the relief device is set correctly. .

7) The Environmental Coordinator becomes aware of the relief valve opening frequently and is concerned that it could be a potential deviation from the environmental permit.

8) To make matters worse, Sam comes to work Monday morning and gets a call from a Control System Mechanic (electrician). It seems that there was a problem in Material Transfer that shut down production over the weekend. Electricians troubleshooting the problem could not locate the source because the electrical elementary drawing was not up-to-date due to changes made around bin 99.

Sam now realizes that not using MOC may seem to have saved some time up front but he is spending more time in the long run due to problems that weren't addressed in the beginning. Also, Sam didn't realize that such a minor change could involve significant safety and environmental issues, or that it was so important to notify those impacted by the change prior to startup. Due to all the problems, Sam acquires the unfortunate nickname "Bin 99 Engineer".

The primary reason that we use MOC is to assure that changes to a manufacturing process are made safely. However, as the story of Sam Shortcut illustrates, it also makes good business sense.

The Basic MOC Workflow

How is MOC done? Let's take a look at a simple project workflow chart without MOC in Figure 1 (this is how Sam did it).

Figure 1 – Simple Project without MOC

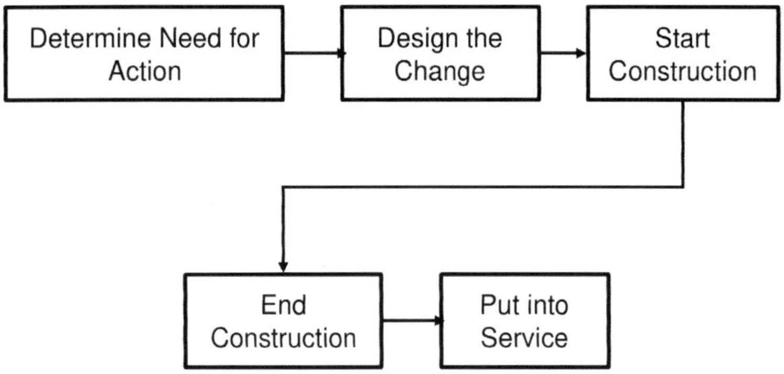

MOC is a process for changing facilities and/or documentation that includes:

- Origination
- Approvals and Reviews
- Engineering Design
- Execution
- Field Verification, Notification, Training and Startup
- Completion of Documentation including PSI updates

With essential MOC elements added the project workflow chart it looks like Figure 2. Relevant sections of the PSM regulation are cited.

Figure 2 – Simple Project with MOC (Minimum Requirements)

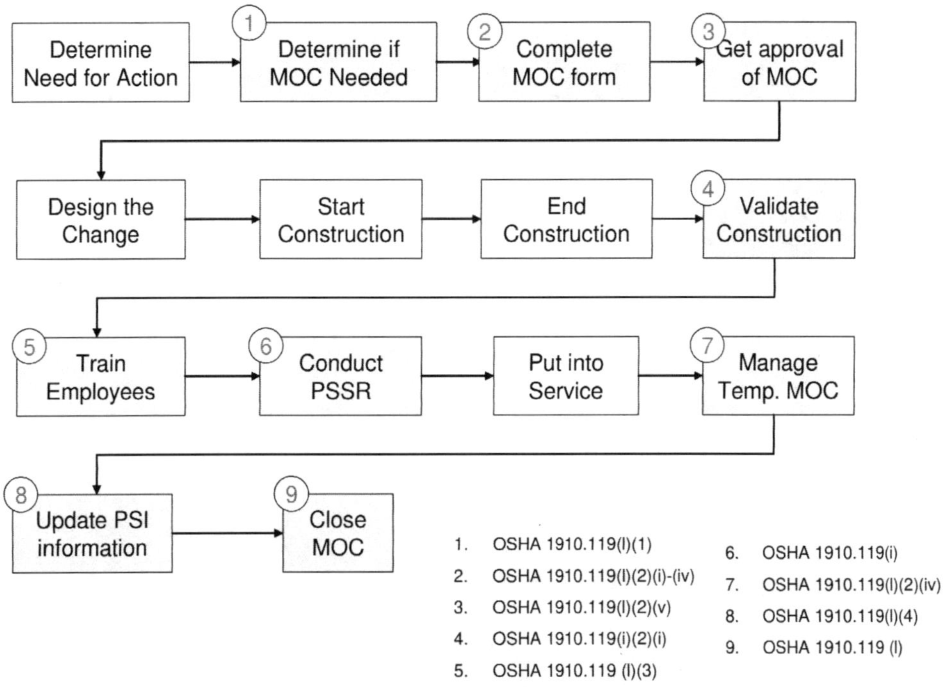

Simple Project with MOC (Minimum Requirements)

Several individuals and groups have roles within the MOC process. The performance of each of these roles is essential to providing proper checks and balances.

How MOC Fits Into an Overall Safety Program

As illustrated in Figure 3, MOC is central to success of a company's overall safety program, especially process safety.

- MOC provides updates to PSI such as P&I drawings and therefore identifies when a PreStartup Safety Review (PSSR) is needed
- MOC keeps operating procedures up-to-date
- MOC assures that people are trained on new procedures
- MOC adds new equipment mechanical integrity test and inspection schedules
- MOCs are reviewed to revalidate Process Hazard Assessments (PHAs)
- MOCs help prevent incidents

Figure 3 – The Central Role of MOC

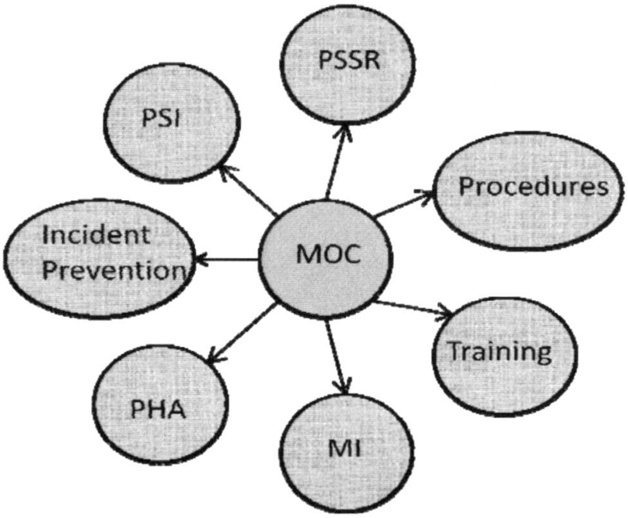

Types of Changes

Types of changes managed by the MOC Process include the following:

- Facility changes
- Control System changes
- Information Systems changes
- Procedural changes

MOC changes can also have different durations as follows:

- Permanent
- Temporary

If a temporary change is made, it must be tracked and properly reversed.

Consideration should also be given to the need for emergency changes, which are changes that are needed quickly during off-hours (nights, weekends and holidays) when all the normal approvers may not be available. Your company may have a procedure to handle these types of changes.

Keep in mind that there are certain items that must be completed before startup of the change (which is usually defined as introduction of chemicals to the modified process). These are as follows:

1. Notification of all affected persons
2. Any needed training
3. For facility changes, a field validation to verify that what has been constructed matches the engineering drawings and specifications
4. For MOCs with any change in the PSI, a PSSR

Keys for Successful MOC

The following items are keys for a successful MOC system:

- People who use the MOC process should be trained on it
- Change should be managed, not just documented. Roles in the MOC process should be clear.
- Management must set expectations that the MOC process will be followed by everyone
- Communication, communication, communication – everyone involved in the MOC process must communicate. This is critical.
- The MOC process should be audited regularly to assure it is working properly

As you can see, many of these keys are related to the company organization and culture.

For more information on MOC, a good source is the CCPS Guidelines book on MOC.[2]

References

1. Atherton, John and Gil, Frederick, *Incidents That Define Process Safety,* Center for Chemical Process Safety, New York, 2008, pp. 202-205
2. *Guidelines for Management of Change for Process Safety,* Center for Chemical Process Safety, New York, 2008

A Work Process for Revalidating LOPAs and Other Risk Analyses

Tim Wagner
The Dow Chemical Company
Dow Corporate Center, 2020 Bldg.
Midland, MI 48674
twagner@dow.com

John Champion
The Dow Chemical Company
6519 State Highway 225
Deer Park, TX 77536
johnwchampion@dow.com

Keywords: revalidation, risk analysis, LOPA, QRA

Abstract

The Chemical Process Industry (CPI) has long recognized the necessity of periodically revalidating Process Hazard Analyses (PHAs), such as Hazard and Operability Studies and What-if analyses. CCPS has published a book which describes this work process titled Revalidating Process Hazard Analyses[1]. However there is limited guidance for revalidating supplemental risk analyses, such as Layer of Protection Analyses (LOPAs), Quantitative Risk Assessments (QRAs), and Fault-Tree Analyses (FTAs). The objectives of performing periodic revalidations are to address the cumulative changes in risk that occur over time and to continuously improve our understanding of that risk. Ultimately the goal is to avoid incidents.

An objective revalidation is focused primarily on safety while a project review focuses on many other aspects beyond safety, such as productivity and efficiency. The revalidation is generally more effective at reviewing the external factors that impact the results of an assessment. By revalidating a risk analysis, the team takes a holistic approach to analyzing factors that affect risk to surrounding people or the environment imposed by the continued operation of the facility. Particular focus is placed on changes that impacted the analysis, but may not have been recognized or fully addressed in the Management of Change (MOC) review. Some of these external factors include population changes, culture and organizational changes, changes and improvements in the risk analysis methodology, and discovery of conflicting data.

This paper introduces a work process for revalidating various types of risk analyses, and discusses many of the factors that need to be considered. This can improve the quality of the revalidation while reducing the overall effort expended. The primary discussion will center on revalidating LOPAs, but other types of studies will be discussed as well. Case studies and experiences from actual revalidation efforts will be incorporated into the discussion.

Introduction

During a Layer of Protection Analysis (LOPA) or Quantitative Risk Assessment (QRA) thousands of decisions are made. *Should we pursue a specific cause/consequence pair? Could this event produce a toxic cloud capable of fatal consequences? Should I build a Safety Instrumented Function (SIF) or rely on an alarm? Would the human operator be reliable enough? Can I vent my relief device to the atmosphere? What is the chance of static ignition in a specific situation? And on and on...* Even the same team (i.e. all the same individuals) may not choose the same decision at each juncture given a second chance. These decisions can vary from one study to the next. Even the same individual will learn more in, say, five intervening years. Team interaction and experience operating the protection layers is likely to multiply the number of differences.

The Chemical Process Industry (CPI) has long recognized the necessity of periodically revalidating Process Hazard Analyses (PHAs), such as Hazard and Operability Studies (HAZOPs) and What-if analyses. CCPS has published a book which describes this work process titled Revalidating Process Hazard Analyses[1]. However there is limited guidance for revalidating supplemental risk analyses, such as Layer of Protection Analyses (LOPAs), Quantitative Risk Assessments (QRAs), and Fault-Tree Analyses (FTAs). The objectives of performing periodic revalidations are to address the cumulative changes in risk that occur over time and to continuously improve our understanding of that risk. Ultimately the goal is to avoid incidents.

According to the CCPS book to "revalidate" means to confirm that a study (e.g. LOPA) accurately reflects the process hazards given the current equipment configuration, process operations, staffing, and surroundings. One might also add "to confirm that the analysis meets the requirements of internal and external rules and guidelines".

The act of revalidating can refer to a wide spectrum of approaches, with differing effort expended relative to the original analysis. The revalidation can range from a simple update focusing only on specific changes that have been introduced and/or incidents that have occurred since the previous risk analysis, all the way to a complete redo – a complete new risk analysis that may or may not incorporate key issues from the previous study. The time and resource commitment can vary dramatically and strongly influences the value of the resulting information.

When facility or company leadership decides to revalidate a risk analysis, the level of review should be carefully considered. Some criteria that should be evaluated include

- time since the last update,
- type of scenarios involved (complexity and hazards),
- quality of the team previously involved, and
- type of study and how often the study has been reviewed over time.

For example, many companies have just started implementing LOPA in the last 5-10 years, so they are on their first round of revalidations. There will typically be a need for a more detailed revalidation this pass based on a continuing evolution of industry best practices and internal work processes for implementation of the technique; plus the relative inexperience of the first team will likely be a factor.

It is important to establish a revalidation frequency – a good practice is to tie these supplemental risk analysis revalidations to the PHA revalidation, as there are a number of synergies, including team members, impact of the PHA on the risk analysis, and possibly impact of the risk analysis on the PHA (e.g. discovery of new scenarios).

Methodology for Revalidation

Larger corporate organizations, with many analyses to maintain for multiple geographies, should establish a framework for their facilities to follow. This can be in the form of a checklist, such as that in Appendix A. Most of the items in the Appendix A checklist will be discussed in the following sections of this paper.

1. Preparation

The steps in this section of the Appendix A checklist are intended to be performed by individual team members prior to convening the larger team to conduct the revalidation. A lot of background work can be done by one person such as looking up specific company guidelines that might have changed or it can be split amongst individuals with different expertise.

1.1 Chemical Property Information

Update any chemical property information for the materials that are being handled. While things such as physical properties are not often updated (except for new processes), hazard information such as ERPG[7] (Emergency Response Planning Guideline) values, probit coefficients, and even flammability limits are sometimes updated. Check for updated toxicity data and Industrial Hygiene guidance, which could come from an internal source, or could be external from a government, supplier, or literature source. The American Industrial Hygiene Association (AIHA) establishes ERPG values and the US Environmental Protection Agency generates Acute Exposure Guideline Levels[8] (AEGL) which pair exposure time with effect. These lists are large and changes for a specific chemical occur sporadically due to updated toxicology studies so it's generally advised to go directly to the source and compare with what was used in the study and any documented internal guideline value. Probit coefficients may be chosen from a source such as the Dutch "Purple Book"[9] or a new source from the US Dept. of Homeland Security which established probit coefficients for chlorine, ammonia, and phosgene (and others) within the past several years. Whether your company chooses to use concentration-based toxic endpoints like ERPGs (established for a 60 minute exposure) or use a more complex model such as an AEGL or probit (which models the time-concentration relationship), its important for each company to establish a strategy for justifying the values used and managing updates.

1.2 Changes to Surroundings

Update for any changes in the plant's buffer with the community and industrial neighbors if these factors are used in the risk analysis. Sometimes referred to as the "greenbelt", this factor can easily change over time due to a variety of reasons such as construction of a new apartment complex, shopping mall, or chemical plant. Changes can significantly impact the consequences of scenarios based on large releases of Toxic Inhalation Hazard (TIH) chemicals, which has implications both for LOPA and QRA studies. Has the surrounding community encroached on your facility? The greenbelt can also change in the positive direction, as some companies have adopted a strategy of buying up neighboring property to enhance their greenbelt (and some have even decided to move plants to a different site altogether). This should be accounted for in the risk analysis. On the other hand, some companies have sold property that is not in use, resulting in a smaller greenbelt and a greater need to evaluate changes in surrounding population during revalidations.

Similar to changes outside the property line, there can be changes inside the fence that need to be evaluated. For example, if the company has built or significantly modified a nearby facility, or installed a new occupied building (either temporary or permanent) risk impact to the new population needs to be considered. These changes could impact the study by adding to the number of people within a flammable or explosion effect zone. Construction of buildings or equipment can also create additional congestion that increases the possibility of a vapor cloud explosion, etc. Additionally, such changes add to the number of people potentially impacted by exposure to a toxic cloud and should also be considered.

1.3 Changes in the Plant

In a perfect world, all changes are thoroughly evaluated through a Management of Change (MOC) review which identifies a need to update the risk analysis and tracks this activity to completion. While it's always beneficial to improve the company MOC process, there are fundamental differences between this and a Revalidation that may not be possible to overcome. MOC reviews address the impact of changes to the plant as they are made but chemical plants are enormously complex and humans sometimes make incomplete, or even the wrong choices. Changes in the plant represent the largest group of internal factors that can affect risk, and as a result are key drivers for periodic revalidation of risk analyses.

If the skills, fields of expertise, or experience of the MOC team members are not sufficient, it is likely that the impact of a change will not be fully integrated into a risk analysis. Other changes occur out of the purview of the individual plant engineer who is typically responsible for MOC. Some of this group are changes outside the plant altogether.

In addition, it's easy to miss the cumulative impact of multiple changes if they occur at separate times or are evaluated by separate people. A team focused on a single change, may not appreciate the cumulative or synergistic effects of changes over time. For an example of this cumulative effect, see the 2008 Loss Prevention Symposium paper from John Wincek entitled "A Series of Small Changes." Therefore one of the key activities for revalidation is to perform a review of all key changes to a plant since the last assessment and determine if they need to be incorporated into the study. Several categories of changes that are most likely to impact an assessment are discussed below. Teams should focus on these types of changes.

1.3.1 Changes in Plant Throughput

Changing the amount of storage required impacts the magnitude of the hazard. Sometimes the overall quantity of throughput doesn't change but the product mix may gradually emphasize the more hazardous products. Changing the number of times shipping containers are loaded or unloaded impacts the frequency. Changes to throughput might necessitate change to the way equipment is operated or even the equipment design itself.

1.3.2 Changes in Equipment

If more flow is needed, a larger impeller may be installed on a pump, which could impact deadhead pressure or the amount of time an operator has to respond to a high level alarm before a vessel is overfilled. The authors are also aware of at least one situation where overpressure scenarios were eliminated by shaving a pump impeller, so that the deadhead pressure could not exceed the Maximum Allowable Working Pressure of a downstream vessel. Of course, maintenance practices emphasize the need for an MOC for any impeller change.

1.3.3 Changes in Operating Conditions

Raising the operating temperature or subtly adjusting compositions might not be recognized for its impact on the time to respond in the event of a runaway reaction. Raising a flowrate might impact several scenarios. While the overpressure scenario for a downstream tank would still have to be analyzed in LOPA, the new flow rate may leave less time to respond to a potential overfill and invalidate an operator response to alarm.

1.3.4 Changes in Control Scheme

A modified control strategy can impact scenarios. Take for instance the change shown in the before and after diagrams of Figure 1. Before the change, the base level of the distillation system was set by the operator and controlled by adjusting the bottoms flow, and the feed flow was also set by the operator. After the change, the base level was still set by the operator, but controlled by the feed rate. The bottoms flow was set by the operator directly.

Before the change, a failure of the bottoms pump could cause overfill of the column with potentially hazardous results (e.g. a similar result as occurred at BP Texas City in 2005). There was no automated response because the feed flow was controlled by operator setpoint. This situation required operator intervention, which may or may not be feasible based on the time it would take to overfill the column and how closely the operator was monitoring the situation.

After the change, a failure of the bottoms pump could still cause overfill of the column with the same undesired consequence. However, because the base level is controlled by the feed rate, the automated process control system may be able to respond if the loop is tuned for a fast enough response.

The effectiveness of the protection layers both before and after the change are very dependent on the specific system design, but this example is meant to illustrate that the potential protection systems may change when a change is made to the process control strategy.

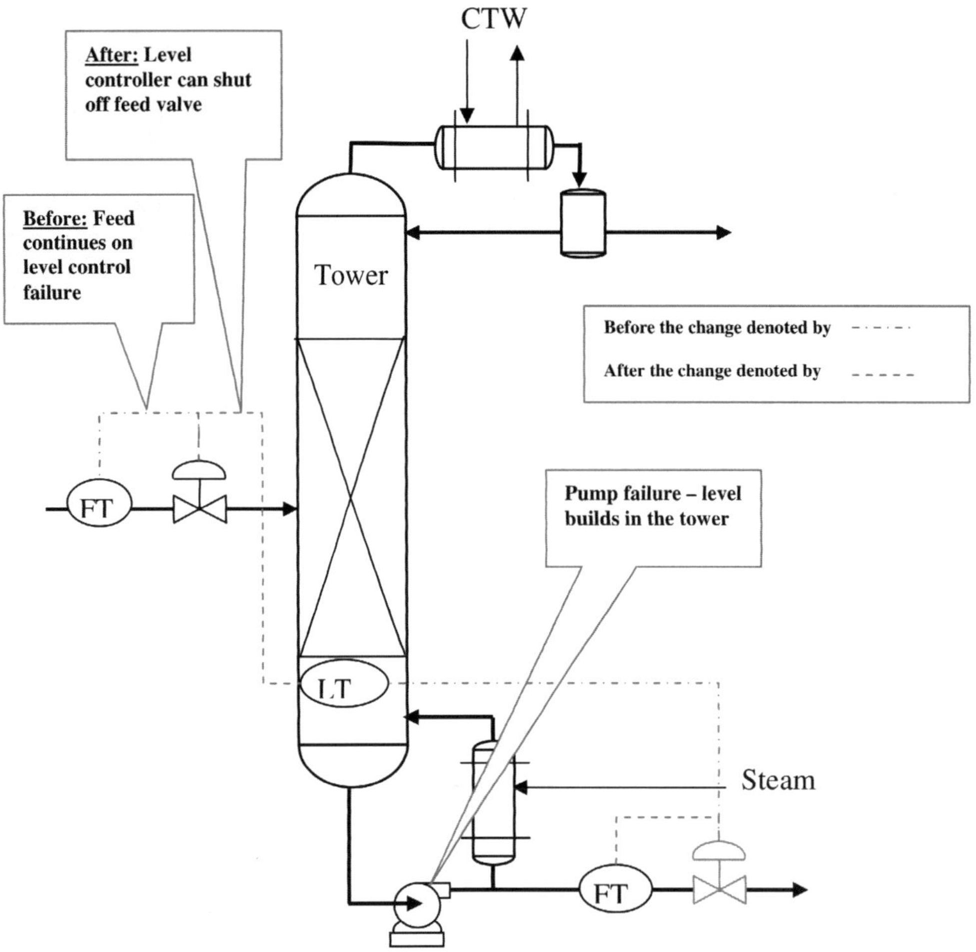

Figure 1 - Distillation System before and after Control Changes

1.3.5 Changes in Chemicals Handled

Occasionally, the chemicals used in the process (or their purity) change, impacting the results of a risk analysis in a variety of ways. For example, a change in solvent in a finished product due to a reformulation from n-Butyl Acetate to Toluene could impact the assessment. Both are flammable with similar boiling points, so that aspect hasn't changed. However, toluene is more toxic, with an ERPG-3 value of 1000 ppm versus Butyl Acetate, which has an ERPG-3 value of 3000 ppm. Also Butyl Acetate is conductive, while toluene is not, so there may very well be new scenarios as a result of static ignition hazards.

Another subtle change involves an increase in aqueous Hydrogen Chloride (HCl) concentration to save money on shipping and improve reactor cycle time. At 30% (by weight), the amount of HCl vapor generated in a spill is significantly lower than for a 37% solution. For a 2" leak of liquid to a diked area, the cloud to the ERPG-3 toxic endpoint extends less than 50 meters at 30% concentration. See Figure 2 below for generic dispersion model results. Contrast that with over 100 meters for the ERPG-3 toxic endpoint at 37% concentration (given other conditions and parameters are equal), illustrated in Figure 3.

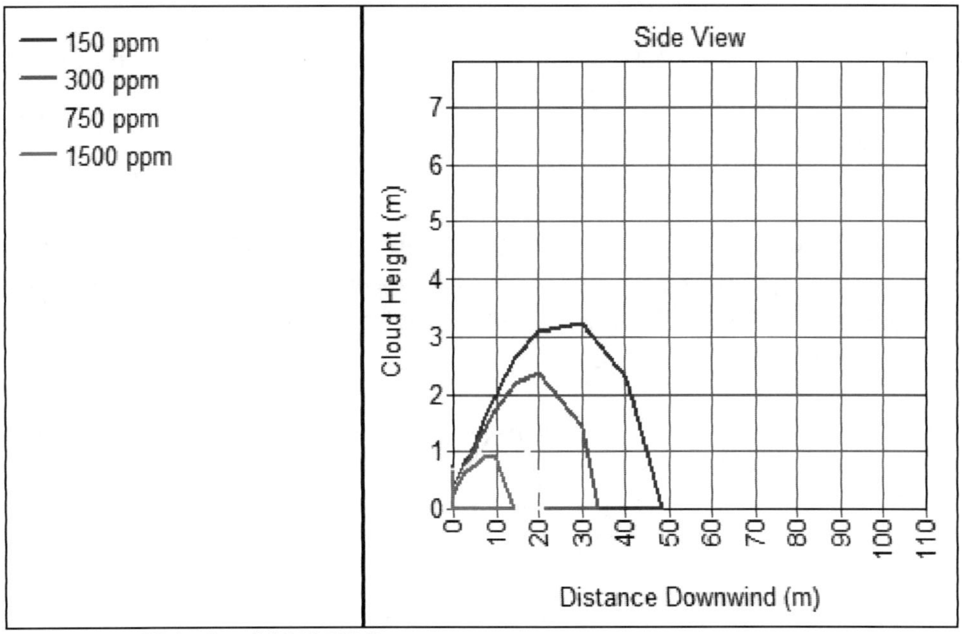

Figure 2 – 2" leak of 30% HCl

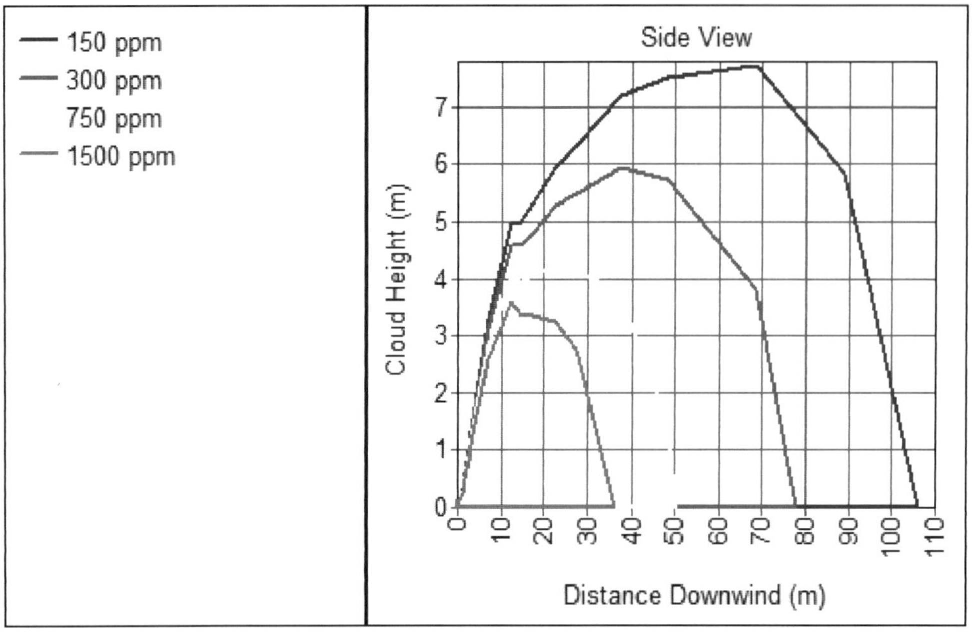

Figure 3 – 2" leak of 37% HCl

1.3.6 Changes in Operational Staffing Levels

It's easy to ignore changes in the organization when performing a revalidation since there may not be an operator involved in the study (and this is but one reason to involve them). How many operators have moved on since the initial training? This is a critical factor if there is significant dependence on human intervention to prevent or mitigate a scenario. A reduction in operating staff, swapping of duties, or increase in workload may reduce an operator's ability to respond to an alarm or monitor an activity in the field. Any change that results with less experienced people executing critical operations without the necessary training and observed practice is worth discussing in the context of initiating event factors and protection layer integrity.

Significant changes in the emergency response capabilities for the plant could reduce the ability of an Emergency Response team to prevent a situation from escalating. Whether this is credited in the risk analysis or not (e.g. whether it's a basic LOPA or QRA study) this point requires consideration and action from the plant and is less likely to be identified in a PHA.

Besides the direct operators of the plant, other peripheral roles are often involved in major accidents and sometimes included in the risk analysis. Changes in the skills of the maintenance staff can compromise equipment integrity (e.g. the temporary connection left in place that caused the Flixborough disaster). The authors are aware of an incident where a grease pencil was used to mark the inside face of a blind flange. Later the flange was put into dry chlorine service resulting in a chlorine-iron fire and significant release. High reliance is therefore placed on the maintenance staff in a chlorine plant. A reduction in laboratory sampling frequency, relaxation of practices, or reduction in the number or skills of the technologists might also affect the integrity of a layer of protection. Even the division of labor between departments might leave one group or the other in a compromised state relative to that assumed when the original risk analysis was done.

2. Revalidation

2.1 Review the Sources for Scenarios

Chemical processing is continually changing due to improvements in the technology, new or modified products, optimized operating conditions and many other factors. These changes may introduce new hazards in the plant thus continuous improvement is of utmost importance in hazard and risk management. Over time, as more and more people move through roles associated with a particular process technology, sometimes discoveries are made which surprise the technical staff. We obviously want these discoveries to be made before any serious accident happens but we're not always so fortunate. This even occurs in a mature technology, from time to time, such as when an explosion occurred in an Ethylene Oxide distillation tower in 1991 due to a previously unknown hazard[2]. One of the early steps in revalidating a risk analysis is to ensure that everyone involved is aware of the (new) hazard and this new hazard awareness is incorporated in the mitigation plan. There are several activities that can drive this new understanding, including

(1) updates to chemical reactivity data,
(2) hazard identification studies, and
(3) incidents and near-miss.

Companies institutionalize processes to assess and capture improvements when formal projects are done but at other times a potential hazard or scenario just occurs to someone based on something they see or hear or a particular train of thought. It's imperative that this knowledge is "captured" when discovered as the next time it occurs to someone may be during the accident investigation. Plants need a mechanism to capture these ideas on the spot for later consideration and possible inclusion in the risk analysis.

2.1.1 Updates to Chemical Reactivity Data

Determine if there has been an update to reactivity hazards based on new testing of raw materials, intermediate mixtures or finished products (or based on new related incidents) since the last hazard assessment. This may be internal analytical testing such as Accelerating Rate Calorimetry (ARC), or it could be based on updates to external knowledge from a source such as Bretherick's Handbook of Reactive Chemical Hazards[6]. If there is new information, its impact on consequences should be evaluated. There might also have been a new discovery of a decomposition reaction once a higher abnormal temperature has been reached. Any changes should be reviewed with reactivity subject matter experts, even if a new grade of a material is being used since the last review – either with more or different impurities, or something that's even more concentrated

Inadvertent mixture of materials that are normally separated can also represent "discovered" hazards. The authors are aware of a situation where the cooling media for a heat exchanger system came into contact with the process material when a tube catastrophically failed and caused an energetic reaction. This became a new scenario to assess and the plant chose an inherently safe solution to install an extra heat exchanger and pump with a non-reactive "barrier" heat transfer fluid.

Conversely, there are also situations where new data has justified the elimination of a scenario because the consequences are not as bad as they were once thought to be. Whatever the situation, ensure that the right expertise is involved in the gathering of data and its subsequent review.

2.1.2 Hazard Identification Studies

Periodic revalidation of qualitative studies such as HAZOPs and What-if analyses are required in most countries of the world and that is the subject of an entire CCPS book, Revalidating Process Hazard Analyses[1]. These studies are highly dependent on the team's collective knowledge, the data package provided and the skill of the facilitator to lead the team to identify all the scenarios that need to be further evaluated in a risk analysis. A revalidation can uncover new scenarios for existing equipment, along with taking a fresh look at the hazards of new equipment, above and beyond the MOC review.

2.1.3 Incidents and Near-Misses

It's important to evaluate scenarios that have actually occurred in the technology whether they happened inside or outside of your company. Auditors and government regulators (and, after a major incident, plaintiff lawyers) almost never accept "it's never happened here" as justification to avoid evaluation of the scenario. It's also important to tap into individual memories of technology and subject matter experts as they may have heard or read something suggesting a new hazard mechanism though they may not be part of the core Revalidation team. It's equally important to involve operators to identify unexpected events that may not have been recorded.

So first and foremost, any past incidents from the facility or a facility practicing similar technology should be considered in the analysis. Secondly, the team should review the significant near-miss events that have occurred that could have resulted in an incident, had the circumstances been slightly different. Third, review if the effectiveness of a protective system was compromised or proven to be ineffective in any of these events. And finally, review the incident and near-miss data as a whole to determine if there are any trends or repeat issues that warrant a higher degree of scrutiny, including failure rates of protective systems, initiating event frequencies, and even cultural issues that can impact the likelihood or outcome of a scenario.

2.2 Changes in Company Guidance for Risk Analysis

As LOPA is applied more widely a demand grows for a continually greater variety of protection layers. Our first generation LOPAs relied on guidance from the original CCPS

Layer of Protection Analysis book which outlined the methodology but included few specific examples. As questions arose, each situation was studied to provide immediate answers which evolved whenever a slightly different twist was considered. As this demand grew within industry as a whole, CCPS recognized the need to provide further guidance and will publish a new book (titled Guidelines for Independent Protection Layers and Initiating Events[5]) intent on providing answers to the "how to do this" type questions – along with guidance for many additional protection layers and initiating events. As a new methodology with much more complexity than a HAZOP, LOPA will take many years to become standardized within US industry, and after that, global industry.

Our company implemented the first version of LOPA as a fairly simple process which then drove wide application of a number of detailed consequence analysis tools and methods. One of the most impactful changes involved the transition from a subjective analysis of vapor-cloud explosion hazards in many of the first generation LOPAs to a very robust analysis of confinement, congestion, and release factors using the Baker-Strehlow-Tang Methodology. This more detailed analysis resulted in changing the integrity requirements for protection layers – in both directions. It seems the initial assessment team decisions weren't validated by sufficient analysis, to pass this reevaluation without additional work or require additional but unnecessary protection layers, in many cases.

Other consequence analyses involving toxic exposure to operators in the plant or those in neighboring buildings have evolved over the years to reduce subjectivity. The objective is to drive consistency in the analysis to avoid a situation where one group identifies a fatality potential for a case and another determines a different consequence for the exact same scenario with similar chemicals. The authors believe the most likely way to eliminate subjectivity on the decisions of fatality potential is for the company to establish standards or detailed analysis guidance.

Our earliest implementation of LOPA guided users to identify a specific initiating event and challenged them to specify the piece of equipment and failure mode. Many LOPA scenarios were written based on the failure of a particular sensor, block valve, or motor starter. However, hazard assessment teams would routinely evaluate failure of a sensor and might forget about the possible (and more likely) failure of the control valve in the same control loop. Once this was communicated, hazard assessment teams were coming up with more and more scenarios based on individual component failures. These documents and studies became unmanageable with simple spreadsheets so more complex tools were created to handle the overload. A project was chartered to determine whether there were more efficient ways to study the plant and later a new practice was adopted whereby a Basic Process Control Loop was considered the basic control unit to fail in the initiating event - rather than each instrument. The resulting protection requirements were virtually identical but with half (or less) of the analysis time.

Some more specific examples are listed below.

- In a cursory analysis, flame arresters were readily accepted as a protection layer with a probability of failure on demand of 0.1/opportunity. And, since detonation arresters require more detailed analysis, they are often thought to be worth an additional order of magnitude of risk reduction. However, further study generated more questions than answers, starting with the need to detect challenges. If an upset generates a flammable gas mixture in the protected pipeline (where it is normally non-combustible) then an arrester will protect against flashback until the device overheats. Obviously it's important to detect the event(s) and shutdown the system or correct the problem immediately. Also, while detonation arresters generally provide greater protection than flame arresters, detonations themselves are less predictable than a stable flame often leading to greater uncertainty in the design. Are they really worth more credit in all cases? And, like any other equipment, maintenance is critical to validate the condition of the device, particularly for the water-seal type arresters.
- The function of an Explosion Suppression System is also more complex than a cursory analysis might recognize. Once it deploys, it is usually important to shut off the fuel supply. These devices put out the fire but may not sufficiently reduce the temperature to avoid reigniting if fuel flow continues within the equipment.
- Single relief devices are credited with a PFD per the original CCPS LOPA book. However, is the credit still the same if you install a rupture disk upstream of a PSV or two rupture disks in series (e.g. for environmental or cost reasons)? Likely, this design is not used without additional monitoring for premature failures. Both the analysis and solution often become significantly more complex.

Many of these questions will be answered in the upcoming CCPS book but more and better data along with further study may change even these conventions, as the science in each area develops. The latest techniques to handle these more complex issues as well as changes in your corporate methodologies may well be incorporated in the revalidation.

2.3 Review List of Commonly Misapplied Rules

If your organization does not have this list of commonly misapplied rules compiled from past experience, we suggest you start one. Here are a couple of examples from the authors' list:

- One of the protection layers commonly used for overpressure scenarios is a relief device. However, the activation of this relief device can generate a secondary scenario, depending on the material released and the discharge location. It is very common for a first-generation assessment team to consider one scenario, but not the other. Both need to be considered.
- Alarms are frequently used as protection layers for scenarios. However, many times the teams do not account for the human response in that protection strategy.

> If the alarm existed, it was credited as a protection layer. In reality, once the alarm annunciates, the operator has to acknowledge it, diagnose the cause of it, and then take the proper response. These "detect and decide" actions must all take place between the time it takes for the alarm to come in and the time the undesired event can no longer be stopped.

Checking for misapplication of the study rules should be a part of the quality check of the previous assessment.

2.4 Changes in Frequency/Probability Data

Consideration should be given to the adequacy and quality of failure data used in a risk analysis. At a minimum, revalidation should include a spot check, if feasible. For example, if a generic rate of 1 failure in 10 years is being used for a pump or control loop failure, but the actual plant experience is that 5 failures have occurred in the last 10 years, the plant experience should generally be used. In an order of magnitude LOPA analysis this would not warrant a frequency credit for the initiating event and an additional protection layer would be required. The revalidation team would make sure to update every scenario using this factor.

Ideally data used are specific to the plant being studied (i.e. collected in that facility). However, when risk analyses are done for new technology facilities or projects, generic failure data is often used because there is no operating history on which to base failure predictions. Similarly, if there is no formal data collection occurring within an organization, generic data or, better, that derived from plants employing the same or similar technology may also be used, even for existing plants. The revalidation team might recommend in-plant data collection where use of generic data is judged inadequate.

It's important to carefully examine any use of vendor failure data, which may or may not reflect actual plant experience. Vendor supplied data are often based solely on factory returns and/or bench testing, which does not account for environmental factors. For instance, it's impossible for the vendor to incorporate nozzle plugging as this is based on specific process conditions. Similarly, external corrosion such as due to an acidic environment is not a condition vendors generally test under.

2.5 Adequacy of Protection Layers Used

During a revalidation effort, systems that are used to protect against the undesired consequence normally should be reviewed. This review should account for whether the protective system can effectively prevent or mitigate the consequence and whether the collection of systems represents the current best option. The review should also account for whether a protection system has fallen into a state of disrepair. Several types of protection systems are discussed below.

2.5.1 Instrumented Protection Systems (IPS)

During a revalidation, a spot check should be performed on the maintenance records to determine if proof testing is being performed and whether the failure frequencies credited for initiating events and protection layers are appropriate, if feasible. It may be worth consulting the maintenance leaders for each appropriate area (e.g. instrumentation, mechanical equipment) to identify reliability issues. Plant operators or supervisors may also be aware of these issues and can often generate a list of equipment where operational upsets regularly occur.

2.5.2 Human-driven Protection Systems

Human actions are sometimes credited in a LOPA with the optimism of a bright future profitability for the plant but later (e.g. when the product market matures) operations positions are cut back to meet economic targets or training falls off when the training coordinator takes a new job. This demand to increase individual productivity occurs for many reasons. Personnel changes or cutbacks should trigger an evaluation of all responsibilities of the remaining roles to make sure key responsibilities are being met. For plant operators, these duties include response to process alarms and correctly following procedures for scenarios with potential to cause a serious incident. In addition, they may be responsible for performing periodic preventive maintenance or checks that have been credited as a protection layer. Operators need to be trained on their newly appointed roles and should be able to describe the hazards and appropriate responses when asked. As a reminder, each operating procedure with incident potential should reference the associated LOPA or other risk analysis scenario and hazard. If this hasn't been appropriately planned or identified in an audit then the revalidation should uncover any issues.

2.5.3 Mechanical Protection Systems

Relief devices need to be effective for the specific LOPA scenario. The authors have found situations where the device was sized for the fire scenario but was later determined to not be large enough to cover a reactive overpressure due to catalytic contamination or initiated by fire and yet the original LOPA team had allowed credit based on the original assessment. Additionally, blocking devices and pressure measurement in between two devices in series need to be appropriately managed.

2.5.4 Protection Strategy

The new team needs to finish the LOPA study with confidence in the resulting protection strategy for each scenario. A complete redo of the study would involve a line by line evaluation but some revalidations may be effective with a general discussion and detailed examination of a sampling of scenarios. The team should feel that the protection strategy is the most effective given any new information. They could ask whether a review of another similar process unit has identified a new initiating event or a different protection strategy. If so, these may need to be reconciled.

If your plant could suffer from, or cause, any scenarios that depend on protection layers located outside of your plant or site, special effort will be needed to assure they remain effective. How do you know they're implemented correctly? Are they well documented? How do you know they'll be maintained? If not, will your plant be notified when these protection layers fail or are terminated or changed? At the time of revalidation it's advised to discuss the scenario with the other facility's technical staff and refresh the protection strategy.

Is all of the equipment (all operations) within the facility that can generate a process safety incident included in the scope of the LOPA (e.g. process equipment tied to non-process equipment, headers, nitrogen usage, sources of pressure)? Also consider all utilities that may impact the facility. For example, changes in steam header pressure that can affect the pressure generated during boil-up with loss of cooling.

2.5.5 Supporting Studies

Have any supporting studies such as toxic consequence or dispersion modeling, FTA, Human Reliability Analysis (HRA), or vapor cloud explosion modeling been done that have not been properly addressed and documented into the applicable scenarios? Sometimes these studies must wait in long queues for knowledgeable resources (people and tools) and occur long after the assessment team has disbanded.

3. Documentation of Revalidation Results

As with any process safety study or, for that matter, any engineering activity, quality of the documentation has the single greatest influence on the overall value of the effort. It is important to develop a thorough record of what was checked and what was changed or added. Use of a checklist such as that provided in Appendix A is an excellent way to ensure a consistent effort and improve by capturing the learning value.

Poor document control can waste time. The authors are aware of a LOPA revalidation begun with an old version of the LOPA. The team wasted a lot of time reviewing scenarios that had been recently updated by another group. Once the revalidation is complete the old version should be archived and a new version should be promptly issued with any new scenarios and changes.

A final report should be issued to plant and company leadership noting the date, participants (by name and role), scope of the equipment or unit(s) involved in the review, what was done, and findings (including an attached copy of the checklist). Much of this improvement work will need to be completed over the coming weeks and months by people not involved in the revalidation so some sort of action tracking tool is essential to assign each task to an appropriate resource or group. The report should include an electronic link to the action tracking database. Implementing some of the identified changes may require an engineering project so that all appropriate plant documents such as P&IDs, procedures, instrument specifications, and process control documents are properly updated, Whatever is learned during the effort should be considered for leveraging across other similar plants

One of the final steps is to execute a MOC to communicate any changes to the appropriate roles (e.g. training, procedures, hardware and software). Lastly, before the effort is closed out the next revalidation should be scheduled.

Conclusions

There is no "if" about it. Without revalidation, any complex analysis will become "out of date" with time and some will become irrelevant, depending on the nature of the technology, change in plant operations and business and amount of time elapsed. A comprehensive revalidation of a risk assessment should be useful in capturing and maintaining improvements in Safe Process Operations. Managing the Unexpected: Resilient Performance in an Age of Uncertainty[3] lists five points of a safe mindset. These are easily applied to revalidation of risk analyses:

1. "Do not over-simplify" - It's far too easy to shorten analysis time by omitting scenarios or assuming consequences. While this might get you through a first pass analysis, your company will often derive value from more thorough and detailed analysis and documentation. Those who oversimplify hazards are often guessing and ultimately will not understand how to interrupt the mechanisms that may result in incidents.

2. "Look out for what could go wrong" - Chemical plants are complex and understanding requires (often extensive) analysis. Engineers spend thousands of hours understanding and optimizing the one way in which their chemical plant works right. Isn't it worth a few hundred person-hours to understand the many ways in which that same chemical plant could go wrong?

3. "Keep an eye out for small change - Do not let the abnormal become normal." Subtle changes made to keep an operation running may not be properly analyzed for impact on process safety without oversight by a knowledgeable person. Even with this oversight, it's imperative to look at all the changes made since the previous review to determine cumulative impact on the assumptions and basis for the analysis.

4. "Learn from our history and teach what you learn" - History teaches us that the chance might be small but extreme consequence is worth respect and study. Thus it's important to draw learning from incidents and include this in design, operating, and maintenance discipline. Everyone who operates an automobile needs to understand how to manage certain risks (e.g. driving on ice), in order to be prepared if an unexpected event were to occur. A chemical plant with its much greater hazard potential needs an even more robust knowledge management and learning process.

5. "Defer to Expertise" - Industrial history has had periodic disasters resulting from inattention to issues that seemed subtle to those interviewed after the incident. It's essential that the revalidation team includes expert opinions where necessary but sometimes new "discoveries" are made that the experts haven't seen before and they might express doubt or confusion. This is a time to proceed cautiously while studying the problem rather than steamrolling ahead with production.

Revalidations coupled with an attitude of periodic, continuous improvement are essential and complementary parts of any safety culture. How to manage this periodic, continuous improvement is a prerogative of each company. We spend so much time improving our products, rightsizing our workforce and tweaking the equipment and recipe to find the single best way to make on-spec product that it would be utterly inconsistent to ignore the impact of this tweaking on the safety of the plant.

References

1. Frank, Walter L. and Whittle, David K., Revalidating Process Hazard Analyses, Center for Chemical Process Safety, 2001.

2. G.A. Viera, L.L. Simpson and B.C. Ream, Lessons Learned From the Ethylene Oxide Explosion at Seadrift, Texas. Chem. Eng. Progress 89 8 (1993), pp. 66–75.

3. Weick, Karl and Sutcliffe, Kathleen, Managing the Unexpected: Resilient Performance in an Age of Uncertainty. (Revised edition) Wiley and Sons, 2007.

4. Layer of Protection Analysis, Simplified Process Risk Assessment, Center for Chemical Process Safety, 2001.

5. Guidelines for Independent Protection Layers and Initiating Events. Wiley and Sons, 2011 (pending).

6. P. Urben and L. Bretherick, Bretherick's Handbook of Reactive Chemical Hazards, Butterworth-Heinemann (Elsevier), 1999.

7. American Industrial Hygiene Association (AIHA) Current ERPG™ Values (2010), http://www.aiha.org/foundations/GuidelineDevelopment/ERPG/Documents/ERPG_Values2010.pdf, 2010 ERPG/WEEL Handbook, AIHA Guideline Foundation.

8. U.S. Environmental Protection Agency Acute Exposure Guideline Levels (AEGLs) web site, http://www.epa.gov/oppt/aegl/pubs/chemlist.htm.

9. RIVM, Reference Manual Bevi Risk Assessments, version 3.2, Dutch "Purple Book", http://www.rivm.nl/milieuportaal/images/Reference-Manual-Bevi-Risk-Assessments-version-3-2.pdf.

APPENDIX A – Example Risk Analysis Revalidation Checklist

Step	Source	Description
		Prepare for the Revalidation
1	Tracking tool	Risk Analysis tracking tool indicates that revalidation is due.
2	Facility document management system	Obtain previous version of assessment, accompanying reports, action resolution documents.
3	MSDS, physical property database, etc.	Check for any changes in chemical properties, such as ERPG or AEGL values, probit coefficients, or flammability limits.
4	Site plans, Landview, census data	Determine if the facility greenbelt has changed, due to changes in fenceline location or changes in the community.
5	Site plans	Determine if distances to neighboring facilities (both internal and external) has changed.
6	Facility MOC database	Check for any changes in the facility that could impact the results of the assessment.
		Revalidate the Risk Analysis
7	Reactivity database and public domain	Review chemical reactivity data and determine if there are any new scenarios from new data. Also, ensure there are no gaps in the data.
8	PHA	Review the facility PHA and incorporate any new scenarios or changes to scenarios.
9	Incident and near-miss database	Review any incidents and near-misses, both on this facility or any similar facilities since the last assessment and incorporate any new scenarios or changes to scenarios.
10		Create and evaluate any new scenarios generated in the 3 steps above.
11	Guidance document change log	Review changes in the company guidance for risk analysis and incorporate into revalidation effort.
12	Compiled list	Review list of commonly misapplied rules and check previous assessment for any of these misapplications.
13	Reliability database	Review the frequency/probability data for initiating events and protection layers and ensure the most current data is in use.
14		Review the instrumented protection systems in place and ensure their adequacy.
15		Revalidate the effectiveness of any human actions that directly provide protection.
16		Review any mechanical protection systems and ensure that they are still effective.

17		Identify any gaps against risk tolerance, and make recommendations to address those gaps.
		Document the Revalidation
18		Capture recommendations in tracking system, provide due dates and assign responsible parties
19		Issue final revalidation report and archive in facility document management system.
20		Schedule next revalidation in Risk Analysis tracking tool.

How to Efficiently Perform the Hazard Evaluation (PHA) Required for Non-Routine Modes of Operation (Startup, Shutdown, Online Maintenance)

By: William Bridges
Process Improvement Institute, Inc. (PII)
1938 Riversound Drive, Knoxville, TN, 37922, USA
wbridges@piii.com

Dr. Tony Clark
Process Improvement Institute, Inc. (PII)

Keywords: Human Factors, Hazard Evaluation, Process Hazard Analysis, Human Error, PHA, Layer of Protection Analysis, LOPA, Process Safety, PSM, Risk-Based Process Safety, RBPS

Abstract

Process safety is about controlling risk of failures and errors; controlling risk is primarily about reducing human error. All elements of Risk-Based Process Safety (RBPS) and alternative standards for process safety (such as US OSHA's standard for Process Safety Management [PSM] or ACC's Process Safety Code™ [PSC]) have many elements, and each of these in turn help to reduce the chance of human error or else limit its impact. One core element is the process hazard analysis (PHA), also called a hazard identification and risk assessment (HIRA). PHAs have been performed formally in gradually improving fashion for more than five decades. Methods such as HAZOP and What-If Analysis have been developed and honed during this time. But, one weakness identified twenty years ago still exists in the majority of PHAs performed around the world; most PHAs do not thoroughly analyze the errors that can occur during startup mode, shutdown modes, and other non-routine (non-normal) modes of operations. This is despite the fact that most major accidents occur during non-routine operations (about 70%), even though the process/plant may only be in that mode of operation for 5% or less a year. Instead of focusing on the most hazardous modes of operation, most PHAs focus on normal operations. In a majority of both older operations and new plants/projects, the non-routine modes of operations are not analyzed at all. This means that perhaps 60 to 80% of the accident scenarios during non-routine operations are being missed by the PHAs. **If the hazard evaluation does not find the scenarios that can likely occur during these non-routine operations, the organization will not know what safeguards are needed against these scenarios.**

This paper shows practical ways to efficiently and thoroughly analyze the step-by-step procedures that are used to control non-routine operating modes, as well as those for batch and between batch operations. This paper builds upon the methods and rules provided in papers beginning in 1993 and brings them up-to-date. Experienced PHA leaders should be able to use the rules and approaches provided in this paper to improve their PHAs. And others will be able to use the results of this paper to estimate the number of accident scenarios they may be missing and to estimate the time it would take to complete an efficient and thorough PHA of the non-routine modes of operation.

1. Introduction

During the period 1970 to 1989, 60% to 75% of major accidents in continuous processes occurred during non-routine modes of operation (principally startup and online maintenance modes).[1] This trend has continued unabated for most the process industry to the present day. A compilation of 47 major process safety accidents in the past 23 years is given in Appendix A; of these, 66% occurred during non-routine operations. In addition, a poll of over 50 clients indicates that 70% of their moderate and major accidents occurred during non-routine modes of operation. This data is particularly disturbing when factoring in the time at risk, since most continuous processes are typically shut down 5% or less per year. Therefore, for many continuous processes the workers and other stakeholders are 30 to 50 times more likely to have a major accident during the time frame of startup, shutdown, or on-line maintenance modes of operation. One reason for processes being at higher risk during these operating modes is many of the safeguards (independent protection layers) are bypassed or may not be fully capable. A hazard evaluation is necessary to help a company identify the layers of protection necessary to lower the risk to acceptable levels. To fulfill this need, a company operating a continuous process should **Fully** evaluate the hazards during **All** modes of operation. Unfortunately, in the first four decades of hazard evaluation use (beginning after the Flixborough disaster in the UK in 1974 – another accident that occurred during startup in a temporary configuration), many companies have done a poor job of identifying and evaluating accident scenarios during startup, shutdown, and online maintenance modes of operation, while usually doing a good job of evaluating hazards of normal modes (continuous or normal batch modes) of operation.

Recent US OSHA PSM National Emphasis Programs underscore the need for companies to identify potential accident scenarios during non-routine modes, and to reduce the frequency and consequences of such errors as part of an overall process safety management (PSM) program. Paragraph (e) of the US OSHA regulation on PSM, 29 CFR 1910.119,[2] and similar requirements in US EPA's rule for risk management programs (RMP), 40 CFR 68.24,[3] specifically require that PHAs consider human factors and address all hazards (stated as "during all modes of operation, both routine and non-routine," in Appendix C to the OSHA PSM regulation).

How does someone responsible for coordinating or performing hazard evaluations (including PHAs) uncover potentially important accident scenarios during all modes of operation without consuming too many resources? To correctly answer this question, we must (1) understand the root causes of human error and (2) develop a strategy for systematically finding the scenarios that are caused by human error, during all modes of operation. The strategy must be thorough, yet provide for a practical allocation of resources. This paper provides a strategy that uses widely accepted hazard evaluation techniques (such as those referenced by OSHA and EPA for PHAs, which include what-if analysis and hazard and operability [HAZOP] analysis). This strategy has proven effective for hundreds of facilities over the past two decades since it was first published.[4] In addition to identifying accident scenarios during non-routine modes; this approach helps to more fully address human factors, which is a specific requirement of OSHA's PSM regulations and EPA's RMP rule.

Human factor deficiencies can make operations during non-routine modes extremely hazardous – since operators have less operating experience for non-routine modes, and these times of operations rely heavily on operator decision-making and tasks. In addition, there are less layers of protection in effect during non-routine operations. Analyzing procedure steps can identify steps where the operator is most likely to make mistakes and suggest ways, ranging from adding hardware to improving management systems, to reduce risk of an accident scenario.

The approach outlined in this work applies equally to any hazard evaluation where the steps for a non-routine mode of operation are well defined (i.e., written), including PHAs of existing units, hazard evaluations during preliminary and detailed design phases of projects (for new/revised processes), and large or small management-of-change hazard reviews.

2. Background: What are Human Factors?

Human error in research, design, construction, installation, operation, maintenance, manufacturing, inspection, management, etc., can be considered the cause of almost all industrial accidents. (Experts typically quote that about 85% of accidents are caused by human error, though some say that except for natural disasters this figure is 100%.) However, simply attributing these accidents to "human error" without evaluating the root cause implies that the errors are inevitable, unforeseeable, and uncontrollable. Nothing could be further from the truth.

Human errors are sometimes mistakenly called procedural errors. This is not true anymore than saying all equipment errors are due to design errors. People make mistakes for many reasons, but experts estimate that only about 10% of accidents due to human errors in the workplace occur because of personal influences, such as emotional state, health, or carelessness. Over the past five decades of research and observation in the workplace on human error, we have come to know that human error probability depends on many factors. These factors (described in more detail in *Human Factors Missing from PSM* [5]), include:

- Procedure accuracy and clarity (the number one, most sited root cause of accidents):

  - A procedure typically needs to be 95% or better accuracy to help reduce human error; humans tend to compensate for the remaining 5% of errors in a written procedure.
  - A procedure must clearly convey the information (there are about 25 rules for structuring procedures to accomplish this) and the procedure must be convenient to use.
  - Checklist features – These should be used and enforced either in the procedure or in a supplemental document.

- Training, knowledge, and skills
  - Employees must be selected with the necessary skills before being hired or assigned to a department.
  - Initial Training – There must be effective, demonstration-based training on each proactive task and each reactive (e.g., response to alarm) task.
  - Ongoing validation of human action is required and usually must be repeated (in either actual performance or in drills/practice) at least once per year (as discussed later in this paper). For human IPLs, the action must be demonstrated to be "fast enough" as well.
  - Documentation – the human performance must be documented and retained to demonstrate the error rates chosen are valid.

- Fitness for Duty – Includes control of many sub-factors such as fatigue (a factor in a great many accidents), stress, illness and medications, and substance abuse.

- Workload management – Too little workload and the mind becomes bored and looks for distraction; too many tasks per hour can increase human error as well.

- Communication – Miscommunication (of an instruction or set of instructions or of the status of a process) is the second or third most common cause of human error in the workplace. There are proven management systems for controlling communication errors.

- Work environment – Factors to optimize include lighting, noise, temperature, humidity, ventilation, and distractions.

- Human System Interface – Factors to control include layout of equipment, displays, controls and their integration to displays, alarm nature and control of alarm overload, labeling, color-coding, fool-proofing measures, etc.

- Task complexity – Complexity of a task or job is proportional to the (1) number of choices available for making a wrong selection of similar items (such as number of similar switches, number of similar valves, number of similar size and shaped cans), (2) number of parallel tasks that may distract the worker from the task at hand (leading to either an initiating event or failure of a protection layer), (3) number of individuals involved in the task, and (4) judgment or calculation/interpolation, if required. For most chemical process environments, the complexity of the task is relatively low (one action per step), but for response actions (human IPLs) there are almost always other tasks underway when the out-of-bounds reading occurs or the alarm is activated.

In addition to the human factors listed, other considerations for use of a human as an IPL include (1) time available to perform the action and (2) physical capability to perform the action. Table 1 on the next page lists the main human factors as well as the multiplying effect of deficiencies in control of these human factors.

Table 1. Summary Human Factor Categories[a]

Human Factor Category	Human Factor Issue/Level	Multiplier for Cognitive & Diagnosis Errors
Available Time (includes staffing Issues) – for responses only	Inadequate time Barely adequate time (≈2/3 x nominal) Nominal time 1 Extra time (between 1 and 2 x nominal and >than 20 min) Expansive time (> 2 x nominal and > 20 min	P(failure)=100% 10 1 0.1 0.01
Stress/ Stressors (includes staffing issues)	Extreme High Nominal	5 2 1
Complexity & Task Design	Highly complex Moderately complex Nominal Obvious diagnosis	5 2 1 0.1
Experience/ Training	Low Nominal High	10 1 0.5
Procedures	Not available Incomplete Available, but poor Nominal Diagnostic/symptom oriented	20-50 20 5 1 2
Human-Machine Interface (includes tools)	Missing/Misleading Poor Nominal Good	20-50 10 1 0.5
Fitness for Duty	Unfit (high fatigue level, illness, strong medication, not physically capable of job today) Degraded Fitness Nominal	10-50 5 1
Work Processes & Supervision	Poor Nominal Good	2 1 0.8
Work Environment	Extreme Good	5 1
Communication	No communication or system interference/damage No standard for verbal communication rules Well implemented and practiced standard	10 5 1

[a] Based in part on *The SPAR-H Human Reliability Analysis Method,* NUREG/CR-6883.[6] Another US NRC initiative, NUREG/CR-6903, *Human Event Repository and Analysis (HERA)*, 2007[7], builds on the human factors categories described in SPAR-H in order to develop a taxonomy for collection of human error data from events at nuclear power plants. PII has modified the list to account for general industry data and terminology and to incorporate PII internal data.

These human-error causes (human factors), which in turn result from other human errors, are all directly within management's control. When using human error data for controlling initiating events (IEs) and independent protection layers (IPLs), the site must ensure that the factors above are consistently controlled over the long-term and that they are controlled to the same degree during the mode of operation that the PHA, HAZOP, What-if, FMEA, or LOPA covers. For instance, if the workers are fatigued following many extra hours of work in a two week period leading up to restart of a process, then the human error rates can increase by a factor of 10 times or 20 times during startup.

3. Overview of Methodology for Hazard Evaluation of Non-routine Modes of Operation

The hazard evaluation of non-routine modes of operation involves reviewing procedures using a HAZOP, simplified HAZOP, or What-if analysis to uncover potential accident scenarios associated with non-routine operations, for continuous or batch operations. As mentioned earlier, human error is more likely and more critical during non-routine operations. By analyzing procedural steps where human error is more likely, and where human error or component failure could lead to a consequence of interest, risk can be reduced. The hazard evaluation team objective is to evaluate the risk associated with skipping steps and performing steps wrong.

FMEA cannot be applied to procedure-based deviations, unless you create a "human" component, in which case you have simply merged HAZOP deviations for "steps" into FMEA. Pre-Hazard Analysis (P_rHA) and other hazard evaluation methods are not applicable for accomplishing a detailed hazard evaluation of non-routine modes of operations.

Checklist of human factors issues (see an earlier paper[4] and also in the Guidelines for Hazard Evaluation Procedures[8]) can be very useful after the detailed hazard evaluation of deviations of steps. Such analysis can indicate where generic weaknesses exist that can make errors during any mode of operation more likely, or that can make errors during maintenance more likely. Such human factors checklists are normally used at the end of the analysis, they can be done piecemeal during an analysis (on breaks from the meetings) by individuals on the team, and then the results of each individual review can be discussed as a team at the end.

As with scenarios uncovered during continuous modes of operation, the company may need to perform analysis (including quantitative analysis such as LOPA or HRA) to more fully address any unresolved or complex issues raised in the hazard evaluation of non-routine modes of operation.

Case studies illustrate the analysis approach and the usefulness of this strategy.

4. Purpose of Hazard Evaluation of Procedures-Based Modes of Operation

Although incorporating human factors considerations into hazard evaluation studies of continuous operation is straightforward by asking why the human might make a mistake that leads to a parametric deviation, this approach only addresses a small fraction of the potential human errors that can affect process safety. Many analysts have tried to find accident scenarios in non-routine modes of operations by adding generic guide words such as "deviations during startup" and "deviations during maintenance/sampling" to the hazard evaluation of equipment nodes/sections. Unfortunately, this only catches a fraction of the accidents that can occur in non-routine modes since a hazard evaluation team is focused on "continuous" mode of operation during HAZOP or What-if of equipment sections/nodes.

From an informal survey of more than 100 companies, most do not currently perform process hazard evaluations of procedures, although many do perform some type of job safety analysis (JSA). The JSA is an excellent starting point for an evaluation of procedures because a JSA identifies the tasks that workers must perform and the equipment required to protect workers from typical industrial hazards (slips, falls, cuts, burns, fumes, etc.). Unfortunately, a typical JSA will not usually identify process safety issues or related human factors concerns. For example, from a JSA perspective, it may be perfectly safe for an operator to open a steam valve before opening a feed valve; however, from a process safety perspective, the operator may need to open the feed valve before the steam valve to avoid the potential for overheating the reactor and initiating an exothermic decomposition. The primary purpose of a JSA and other traditional methods for reviewing procedures has been to ensure that the procedures are accurate and complete (which is required of employers in 29 CFR 1910.119(f)(3)[2]).

By contrast, the purpose of a hazard evaluation is ***not*** to ensure the procedures are accurate and acceptable, but instead, to ***evaluate the accident scenarios if the procedures are not followed***. Even the best procedure may not be followed for any number of reasons, and these failures to follow the prescribed instructions can and do result in accidents. In fact, in the chemical industry and most other process industries the chance of an operator or other worker making a mistake in following a procedure is greater than 1/100, and in some cases much greater. When taking into account common human factor deficiencies that accompany non-routine operations, such as fatigue, lack of practice, the rush to restart and return to full production, etc., the probability of errors can climb to 1/10 chances per task (a task being about 1 to 10 detailed steps). Table 1 presented earlier lists the factors that can increase or decrease human error rates.

The purpose of a hazard evaluation of non-routine modes of operation (governed by written procedures) is to make sure an organization has enough safeguards for the inevitable instance when a step is either performed wrong or skipped (inadvertently or due to shortcutting or other reasons)

Regulators have repeatedly recognized the need for full evaluation of hazards during all modes of operations. US OSHA obviously recognized the importance of this category of hazard evaluation when emphasizing in CPL 2-2.45 (Systems Safety Evaluation of Operations with Catastrophic Potential)[9] that a human error analysis should address:

1. *Consequences of failure to perform a task.*
2. *Consequences of incorrect performance of a task.*

3. *Procedures and controls to minimize errors.*[9]

In the first major PSM inspection in 1992 using 29 CFR 1910.119[2], OSHA assessed a serious violation when the PHAs did not address "human factors such as board operator error, line breaking mistakes, and improper lockout and isolation of process equipment,"[10] all of which are errors originating from failure to either perform tasks or perform them correctly.

In another citation[11] OSHA alleged a serious violation because the company did not address all of the hazards of a process. In particular, the company was cited for **not** evaluating the hazards (during the unit-wide PHA) associated with non-routine procedures such as "startup, shutdown, emergency shutdown, and emergency operations." There were several other violations assessed in this citation because these non-routine procedures did not (allegedly) address the **consequences** associated with operators failing to follow the prescribed procedures. The OSHA inspector was convinced that a hazard evaluation of the non-routine operating procedures should have been part of the PHA scope.

In an article by Woodcock (an OSHA staff member) entitled, "Program Quality Verification of Process Hazard Analysis"[12] (for use in OSHA's training program for PSM inspectors), the author states that a PHA should include analysis of the "procedures for the *operation* and *support* functions" and goes on to define a "procedure analysis" as evaluating the risk of "skipping steps and performing steps wrong."

In the Risk Management Program rule (40 CFR 68.24)[3] EPA also recognizes the importance of procedural analysis, by defining the purpose of a PHA to be to "examine, in a systematic, step-by-step way, the equipment, systems, and *procedures* (emphasis added) for handling regulated substances."

Industry has found that a HAZOP or what-if analysis, structured to address procedures, can be used effectively for finding the great majority of accident scenarios that can occur during non-routine modes of operation.[4, 8, 13, 14] Experience shows that reviews of non-routine procedures have revealed many more hazards than merely trying to address these modes of operation during the P&ID driven hazard evaluations.

> ***Example:*** *For the BP Texas City, Texas refinery, pre-2005, if the isomerization unit had a hazard analysis of the startup mode (using What-If and 2-Guide Word analysis [explained later in this work]), the team would have likely identified that the high-high level switch in the column was a critical safety device. They also may have recommended moving the switch to a location higher in the column and then interlocking the high-high level switch to shutdown feed to the column. However, the site performed a parametric deviation HAZOP of the equipment nodes which focusing on continuous mode of operation, and so the team decided that the high-high level switch was not critical since devices in upstream and downstream process units (during continuous operation) would indicate possible level problems in the column – and besides, the operator would certainly notice the high-level condition on the sight glass during the rounds twice per shift. Unfortunately, these are not necessarily safeguards during startup of the column (1) since the routine practice was to overfill the bottoms (raise the level above the upper tap of the level controllers transmitter*

and above the nozzle for the high-high level switch and (2) since swings in upstream and downstream units are expected (and so likely such swings would not have led to intervention by the operators of the other units).

To reinforce the need for and to explain the method for analysis of deviations of steps in a procedure, Section 9.1 was included in the 3rd Edition of *Guidelines for Hazard Evaluation Procedures,* 2008[8]; this was one of the major changes to the hazard evaluation procedures.

5. HAZOP Method for Analyzing Deviations of Procedural Steps

The Hazard and Operability (HAZOP) method has two major variations; one for the continuous mode of operations (where the team brainstorms what would happen if there were deviations of parameters) and procedure-based (where the team brainstorms what would happen when the steps of a procedure are not followed correctly). The procedure-based variation of HAZOP is the oldest form of HAZOP (from ICI in 1960s). It was an expansion of a Hazard Evaluation method based strictly on asking:

- **What happens if the step is skipped?**
- **What happens if the step is performed wrong?**

In turn, the "pre-HAZOP" method for brainstorming accident scenarios from not following procedures (including because the procedure is itself wrong) is based on the understanding that human errors occur by someone not doing a step (errors of omission) or by doing a step incorrectly (errors of commission). So, simply asking what would happen if the operator omitted a step or performed a step wrong is one way to structure a hazard evaluation of a step-by-step procedure. ***(We will discuss the usefulness of this simple approach to hazard evaluation of steps later.***)

5.1 Seven (7) Guide Word method

In an effort to be more thorough, the inventors of HAZOP (at ICI) broke these two types of errors into subparts and agreed on using the following 7 Guide Words:

Omission:	**Skip (or Step Missing)**
	Part Of
Commission:	**More**
	Less
	Out of Sequence
	As Well As
	Other Than
	Reverse

In the early 1990s, the guide word Skip was augmented by adding the option of discussing "are there any steps **missing** from the procedure."[4]

Table 2: Definitions of 7-8 Guide Words for HAZOP of Procedure-Based (non-continuous mode) Operation

Guide Word	Meaning When Applied to a Step
Missing (optional guide word)	A step or precaution is missing from the written procedure prior to this step (similar to "Out of Sequence", except the missing step is not written)
Skip (No, Not, Don't)	The specified intent of this step is not performed
Part-of	A portion of the full intent is not performed. Usually only applies to a task that involves two or more nearly simultaneous actions ("Open valves A, B, and C".)
More	Too much of the specified intent is done (does not apply to simple on/off; open/close functions); or it is performed too fast
Less	Too little of the intent is done, or it is performed too slowly
Out of sequence	This step is performed too early in the sequence
As well as	Something happens, or the user does another action, in addition to the specified step being done correctly (could be a short cut)
Other than (or Reverse)	The wrong device is operated, selected, read, etc., or operated in a way other than intended. Or the wrong material is selected or added. "Other than" errors always imply a "Skip" as well.

To apply HAZOP to procedural steps for startup, shutdown, online maintenance, and other modes of operation, the facilitator (or team) must first divide the procedure into individual actions. This is already done if there is only one action per step. Then, the set of guide words or questions is systematically applied to each action of the procedure resulting in procedural deviations or what-if questions. The guide words (or procedural deviation phrases) shown in Table 2 above were derived from HAZOP guide words commonly used for analysis of batch processes. The definition of each guide word is carefully chosen to allow universal and thorough application to both routine batch and non-routine continuous and batch procedures. The actual review team structure and meeting progression are nearly identical to that of a process equipment HAZOP or what-if analysis, except that active participation of one or more operators is even more important and usually requires two operators for a thorough review; a senior operator and a junior operator. For each deviation from the intention of the process step (denoted by these guide words applied to the process step or action), the team must dig beyond the obvious cause, "operator error," to identify root causes associated with human error such as "inadequate emphasis on this step during training," "responsible for performing two tasks simultaneously," "inadequate labeling of valves," or "instrument display confusing or not readable." The guide word *missing* elicits causes such as "no written procedural step or formal training to obtain a hot work permit before this step," or "no written procedural step or formal training to open the discharge valve before starting the pump."

5.2 Two (2) Guide Word Method for Analyzing Deviations of Procedural Steps

A more streamlined guide word approach has also proven very useful for (1) procedures related to less hazardous operations and tasks and/or (2) when the leader has extensive experience in the use of the guide words mentioned previously and can therefore compensate for the weaknesses of a more streamlined approach. The two guide words for this approach (as defined in Table 3 below) encompass the basic human error categories: errors of omission and commission. These guide words are used in an identical way to the guide words introduced earlier. Essentially "omit" includes the errors of omission related to the guide words "skip," "part of," and "missing" mentioned earlier. The guide word "incorrect" incorporates the errors of commission related to the guide words "more," "less," "out of sequence," "as well as," and "other than" mentioned earlier. Note that these two guide words (Table 3) fill the basic requirements for a human error analysis as outlined in OSHA's CPL 2-2.45.[9]

Table 3: 2 Guide Word (Guide Phrases) for Modified-HAZOP of Procedure-based Operation

Guide Phrase	Meaning When Applied to a Step
Step not performed	The step is not done or part of the step is not done. Some possible reasons include the employee forgot to do the step, did not understand the importance of the step, or the procedures did not include this vital step
Step performed wrong	The employee's intent was to perform the step (not omit the step), however, the step is not performed as intended. Some possible reasons include the employee does too much or too little of stated task, the employee manipulates the wrong process component, or the employee reverses the order of the steps.

Table 4: Example of 2 Guide Word HAZOP of a Critical Step in a Procedure

Drawing or Procedure: SOP-03-002; ***Cooling Water Failure***	**Unit:** HF Alkylation	**Method:** 2 Guide Word Analysis	**Documentation Type:** Cause-by-Cause

Node: 23 — **Description:** STEP 2: Block in olefin feed to each of the 2 reactors by blocking in feed at flow control valves

Item	Deviation	Causes	Consequences	Safeguards	Recommendation
23.1	Step not performed	Operator failing to block in one of the reactors, such as due to miscommunication between control room operator and field operator; or control valve sticking open or leaking through	High pressure due to possible runaway reaction (because cooling is already lost), because of continued feeding of olefin (link to 11.7 - High Rxn Rate; HF Alky Reactor #1/#2) High pressure due to high level in the reactor, because of continued feeding olefin (link to 11.1 - High Level; HF Alky Reactor #1/#2)	High temperature alarm on reactor High pressure alarm on reactor Field operator may notice sound of fluid flow across valve Flow indication (in olefin charge line to reactor that is inadvertently NOT shutdown) Level indicator, high level alarm, and independent high-high level switch/alarm	
		Operator failing to make sure bypass valve is also closed, since this precaution is not listed in the written procedure; or the bypass valve leaks through	High pressure due to possible runaway reaction (because cooling is already lost), because of continued feeding of olefin (link to 11.7 - High Rxn Rate; HF Alky Reactor #1/#2) High pressure due to high level in the reactor, because of continued feeding olefin (link to 11.1 - High Level; HF Alky Reactor #1/#2)	High temperature alarm on reactor High pressure alarm on reactor Field operator skill training requires always checking bypasses are closed), when blocking in control valves Field operator may notice sound of fluid flow across valve Flow indication in olefin charge line (but likely not sensitive enough for small flows) Level indicator, high level alarm, and independent high-high level switch/alarm	
		Operator failing to close flow control valve manually from the DCS because the phrase "block in" is used instead of the word "close"	Valve possibly opens full at restart, allowing too much flow to reactor at restart, resulting in poor quality at startup and/or possibly resulting in runaway reaction and high pressure	Control room skill training requires always manually commanding automatic valves closed before telling field operator to block in control valve	37. Implement best-practice rules for procedure writing, which includes using common terminology.
23.2	Step performed wrong	Operator closing the olefin charge flow control valves before shutting down the charge pump, primarily because the steps are written out of the proper sequence	Deadheading of charge pump, leading to possible pump seal damage/failure and/or other leak, resulting in a fire hazard affecting a small area (link to 5.12 - Loss of Containment; Olefin Charge Line/Pump)	Step 3 of procedure that says to shutdown charge pump The step to shut down the charge pump (Step 3) is typically accomplished before Step 2 (in practice)	41. Move Step 3 ahead of Step 2.
		Field operator closing both upstream and downstream block valves	Possible trapping of liquid between block valve and control valve, leading to possible valve damage (due to thermal expansion)	Field operator skill training stresses that only one block valve should be closed	

6. What-if Method for Analyzing Deviations of Procedural Steps

The What-if method for analyzing procedure-based modes of operations is free brainstorming without the aid (or constraints) of guide words. This method is described in detail the *Guideline for Hazard Evaluation Procedures (CCPS)*[8]. The hazard evaluation team using this method would read the procedure and then answer the question: "What mistakes will lead to our consequences of interest?" The team would list these mistakes and then brainstorm the full consequences, causes, and existing safeguards – the same analysis approach described for the guide word approaches mentioned earlier in this section. What-if brainstorming **is not** applied to each step of the procedure, but rather covers the entire task (procedure) at one time.

7. Choosing the Right Method for Analysis of Non-Routine Modes of Operation

Obviously the What-if approach takes far less time than the 2-Guide Word method, and the 2 Guide Word method takes much less time than the 7-8 Guide Word method of HAZOP of procedures. Experience has shown that hazard evaluation facilitators, newly trained in the three techniques above, tend to overwork an analysis of non-routine procedures, so a tiered approach is best. In this tiered approach, the first step in choosing the right method of analysis in the hazard evaluation of procedures is to screen the procedures and select only those procedures with extreme hazards. These procedures should be subjected to a detailed HAZOP analysis (7-8 guide word set) presented above. The 2-Guide Word set is efficiently used for less complex tasks or where the consequences are lower. The What-if method is applicable to low hazard, low complexity, or very well understood tasks/hazards.

Experience of the leader or the team plays a major part in selecting the procedures to be analyzed, and then in deciding when to use each guide word set.

Figure 1 shows the typical usage of the three methods described above for a typical set of operations procedures within a complex chemical plant or refinery or other process/ operation. Most of the procedures are simple enough, or have low enough hazards to warrant using the What-if method. Currently, the 7-8 Guide Word approach is used infrequently, since most tasks do not require that level of scrutiny to find the accident scenarios during non-routine modes of operations.

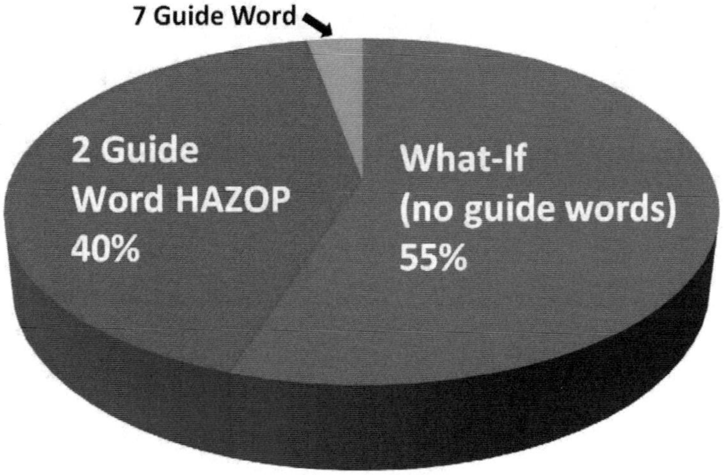

Figure 1. Relative Usage of Techniques for Analysis of Procedure-Based Modes of Operation

The experience of the leader or the team plays a major part in selecting the method to use for each task/procedures to be analyzed. However the first decision will always be "Are these procedures ready to be risk reviewed?" If the procedures are up-to-date, complete, clear, and used by operators, then the best approach for completing a complete hazard evaluation of All modes of operation, including routine modes of operation, is shown in Table 5 below:

Table 5. Example Choice of Methods for Hazard Evaluation of all Modes of Operations if the Operating Procedures are More Than 80% Accurate[15]

For Continuous Mode	For Discontinuous Mode (batch, startup, shutdown, on-line maintenance, emergency shutdowns)

• HAZOP of Parameters of Continuous Mode
• FMEA of Continuous Mode
• What-if of Simple Sub-systems
	• HAZOP of Steps
	  o 8 Guide Word
	  o 2-3 Guide Word
	• What-if of Simple Tasks

Notes: Do this first for a continuous process. Do not do this at all for a Batch process.	Note: Do this second for a continuous process. For a batch process, use Only this analysis approach.

If you do **not** have accurate procedures then the best approach is to develop accurate and up-to-date procedures as quickly as possible and in the meantime follow the approach shown in Table 6 below:

Table 6. Example Choice of Methods for Hazard Evaluation of *Partially All* Modes of Operations if the Operating Procedures are Less Than 80% Accurate[15]

For Continuous Mode	For Discontinuous Mode (batch, startup, shutdown, major maintenance, emergency shutdowns)
• HAZOP of parameters of Continuous Mode • FMEA of Continuous Mode • What-if of Simple Sub-systems	You may find 10% of the scenarios that can occur during non-routine operations by adding guides such as "Deviations During On-line Maintenance" to your set of Parametric Deviations.
Note: After this "incomplete" PHA, you must develop accurate operating procedures (better than 80% accuracy) and the you can complete the PHA by either What-If or HAZOP methods applied to these accurate procedures	

Any procedure (even a computer program) can be analyzed using these techniques. Reviews of routine procedures are important, but reviews of non-routine procedures are even more important. As mentioned earlier, the nature of non-routine procedures means that operators have much less experience performing them, and many organizations do not regularly update these procedures [though this should change as companies comply with 29 CFR 1910.119(f)][2]. Also, during non-routine operations, many of the standard equipment safeguards or interlocks are off or bypassed.

Using the approaches above, a company doing a complete hazard evaluation of an existing unit will invest about 65% of their time to evaluate normal (e.g,, continuous mode) operation and 35% of their time for evaluating the risks of non-routine modes of operation.

Many companies do **not** perform a thorough analysis of the risk for startup, shutdown, and on-line maintenance modes of operation; the reason normally given is that the analysis of these modes of operation takes "too long." Yet, actually, the hazard evaluation of the normal mode is taking too long and so the organization feels it has no time left for the analysis of procedures for startup and shutdown modes of operation. But, if these hazard evaluations for the normal mode of operation are **optimized** (such as using rules presented elsewhere[16]), the organization will have time for thoroughly analyzing the non-routine modes (typically discontinuous modes) of operation and the organization will still have a net savings overall! This point is critical since 60-75% of catastrophic accidents occur during non-routine modes of operation. Figure 2 illustrates (for a continuous process unit) the typical split of meeting time for analysis of routine mode of operation versus non-routine modes of operation.

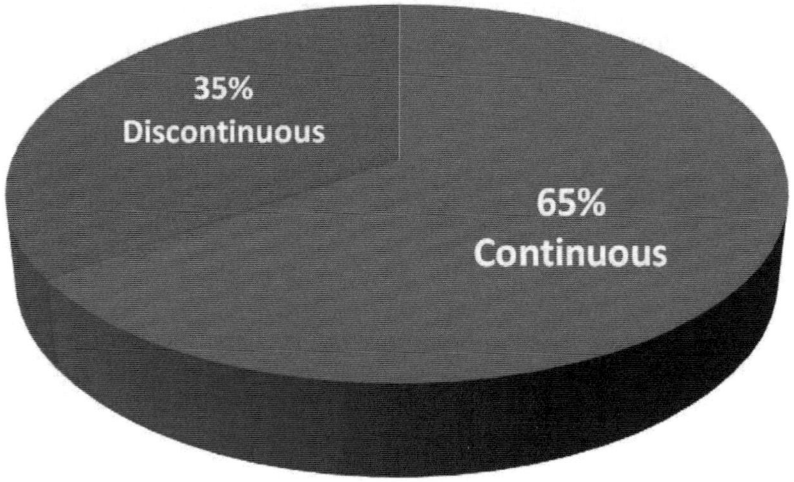

Figure 2. Relative Amount of Meeting Time Spent for Analysis of Routine and Non-routine Modes of Operation for a Continuous Process

8. General Guidelines for Analyzing Non-routine Modes of Operation or Batch (Step-by-step) Processes

- Define the assumptions about the system's initial status. "What is assumed to be the starting conditions when the user of the procedure begins with Step 1?"
- Define the complete design intention for each step. "Is the step actually 3 or 5 actions instead of one action? If so, what are the individual actions to accomplish this task?"
- Don't analyze safeguard steps that start with ensure, check, verify, inspect, etc., or where the consequence of skip is "loss of one level of safeguard/protection against ….." There is no reason to analyze these steps since they will show up as safeguards of deviations of other steps. This approach is similar to not analyzing a PSV during a HAZOP of continuous mode (parametric deviation analysis); instead the PSV is shown as a safeguard against loss of containment.
- Together with an operator before the meeting, identify the sections of the procedures that warrant use of:
  - 7-8 Guide Words (extremely large consequences can happen if deviations occur)
  - 2 Guide Words (the system is complex, mistakes are costly, or several consequences could occur)
  - On others, use What-If (no guide words or guide phrases; for use on simpler or lower hazard systems)
- Decompose each written step into a sequence of actions (verbs)
- Apply guide words directly to the intentions of each action

The Following Preparation Steps May Also Be Needed:

- Walk through procedure in the plant with one or more operators to see the work situation and verify the accuracy of the written procedure. This is optional and should have also

been performed as part of validation of the procedure after it was originally drafted.
- Determine if the procedure follows the best practices for "presentation" of the content; the best practices will limit the probability of human error.
- Discuss generic issues related to operating procedures, such as:
    - staffing (normal and temporary)
    - human-machine interface
    - worker training, certification, etc.
    - management of change
    - policy enforcement
- Review other related procedures such as lock out/tag out and hot work.

9. Case Studies

The following case studies illustrate the usefulness of the process outlined in this paper.

9.1 Case Study 1: Phillips Polyethylene Plant 6, Pasadena, Texas

In 1991-1992, a PHA was performed for the first of the rebuilt polyethylene plants at the Pasadena, TX, Phllips 66 plant. The accident there two years prior claimed 24 lives, injured hundreds of others, destroyed all three polyethylene plants, and cost Phillips an estimated $1.4 billion. Following the investigation of the accident, one of the requirements of the settlement agreement between Phillips and the US government was to ensure the PHA of the rebuilt units addressed hazards during ***All modes of operation***.

The PHA team varied in size, but always included at least two operators. The team leader was a process engineer with 15 years of experience, who was also trained in human factors. The PHA first covered the continuous mode of operation for the approximately 200 nodes of equipment (from feed stock through pellet handling) using the "parametric deviation" form of HAZOP (and some What-If). Then, to complete the analysis of all modes of operation, the PHA team performed a step-by-step analysis of all steps of all startup and shutdown and online maintenance procedures (about 700 steps changed the state of the system and each of these steps were analyzed) using the 7 Guide Word HAZOP method (2 Guide Word analysis was not known to the team at this time). For deviations such as "operator skips a step," the causes identified by the team included "the operator doing this step miscommunicates with the operator who performed steps earlier in the day and went to the wrong reset panel/switch in the field". In this example, an "other-than" error led to the "skip" error; so two errors occurred at once: the wrong switch was flipped and the correct switch was not flipped. Other causes included: "label not distinct enough" or "thinking/believing the previous operator completed this step." The additional safeguards suggested by the PHA team sometimes lower the likelihood of the error by addressing a human factors weakness. But in many cases, the solution was a change to the hardware or instrumentation, including adding new interlocks (these would be called Safety Instrumented Functions today) and adding mechanical interlocks and installing larger relief valves. In a couple of cases, isolated sections of the process were redesigned to lower the inherent risk, such as adding error-proofing (Poke Yoka) features.

The 7 Guide Word HAZOP of non-routine modes of operation took 2.5 weeks of meetings, 40 hours a week. This was in addition to the 3.5 weeks of meetings to complete the parametric deviation analysis HAZOP of the continuous (normal) mode of operation (as mentioned before, 200 nodes of equipment). *Note that if the team had known of and been trained in 2 Guide Word HAZOP for procedure steps, they likely would have chosen that for many of the tasks and it is estimated that the meeting time for analysis of non-routine procedures would have been reduced to less than 2 weeks, with little or no loss of thoroughness.* The completed PHA report was submitted to US OSHA for review and was approved almost immediately; OSHA particularly reviewed the analysis of all modes of operation and coverage of human factors.

After the PHA was completed, the settlement agreement also required a quantified human reliability analysis (HRA) of the online maintenance task of clearing a plugged settling leg – mistakes during this task led to the accident in 1989. The HRA was performed similar to as required for nuclear reactor risk assessments, using a human reliability event tree to model the error probability for the task. The HRA results indicated that if the three IPLs in the new design are maintained, the probability of a similar event occurring when using this procedure was less than the risk of fatality when driving to work. This report was also approved by OSHA.

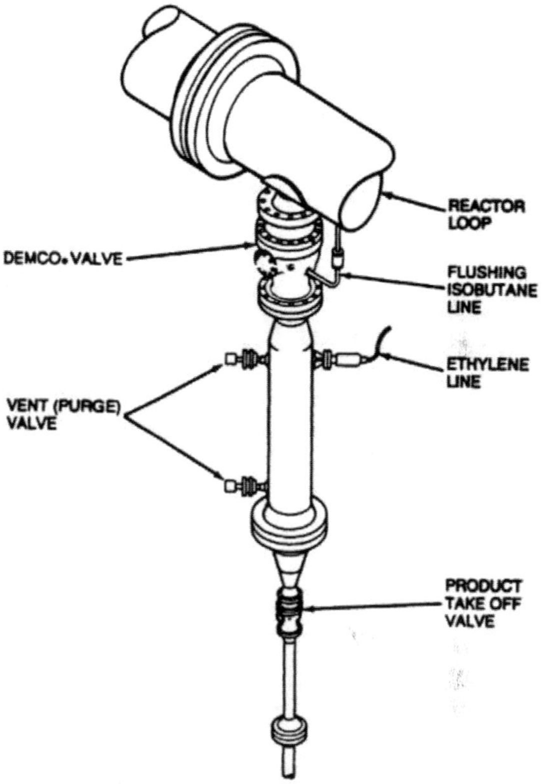

Figure 3: Typical Settling Leg Assembly for Phillips Polyethylene Reactor

The HRA shed light on new aspects of making errors and recovering from the errors during this task, but the HRA results did not result in changes to the process steps (at least not much), or the training program, or the human factors engineering, or the hardware/IPLs. In this instance, the HRA validated the results of the 7 Guide Word HAZOP, but did not make new recommendations. The HAZOP of the procedures had already found the major accident scenarios and had already identified well enough the changes needed to reach tolerable risk.

9.2 Case Study 2: Catalyst Addition System

After several near misses and accidents, a company completely upgraded its PSM program. Detailed operating and maintenance procedures were written for all non-routine tasks. Training experts were called in to ensure that workers understood these new procedures. During process design, particular emphasis was placed on reducing human error. The company also decided to perform PHAs (using the HAZOP analysis technique) to uncover potential accident scenarios in its process design and procedures, with the PHA teams instructed to focus on human error sources. Detailed human reliability analysis was reserved for critical, complex tasks identified by the PHA teams as having potentially severe consequences.

The following results were taken from the analysis of a continuous catalyst addition system for a reactor, shown in Figure 4. Though the normal mode of operation was continuous, the addition system was frequently isolated, depressurized, refilled, and then restarted while a standby addition system maintained feed flow to the reactor. It was interesting to note how the team's perception of "likelihood" for a given human error scenario changed after the procedures were reviewed.

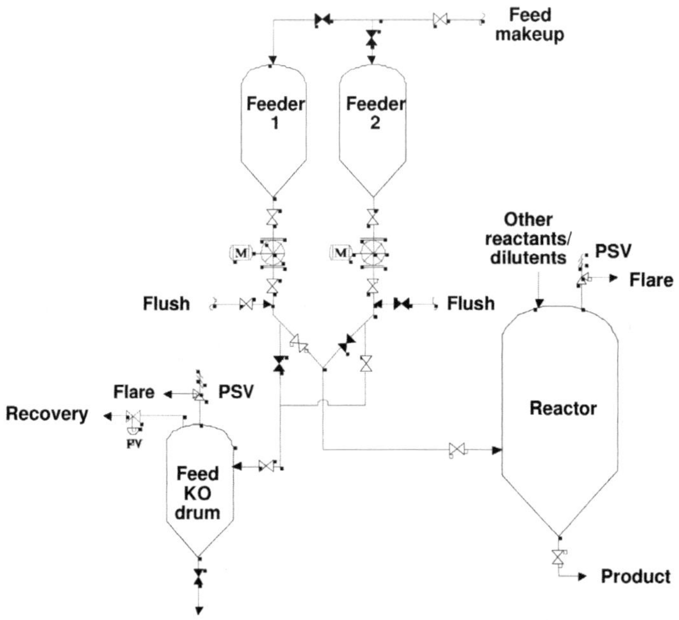

Figure 4: Catalyst Addition System for Reactor

The HAZOP review of the procedures for switching between feeders indicated that the steps telling the operator to close the drain line valve, and later to verify the drain valve was closed, were both "missing." Also, the operators recalled that formal (classroom) training did not mention the operation of this manual valve and field training did not always cover operation of this valve. The senior operators on the review team began to realize that an inexperienced operator might not understand that the sound made by the rushing backflow of fluid is abnormal and might react too slowly to prevent over-pressurization. Therefore, the team concluded that if hands-on training failed to correct the procedure deficiencies, backflow from the reactor to the KO drum was a likely accident scenario, especially with new operators. The team recommended the procedures be modified to (1) reflect the proper sequence of steps and (2) emphasize the consequences of leaving the drain valve open and then later opening the feed line to the reactor. A checklist of the proper sequence of steps was recommended for this procedure. In addition, a recommendation from the continuous mode HAZOP involving the relief valve on the KO drum was modified to ensure the relief valve was also capable of handling reverse flow from the reactor (the relief size was ultimately increased).

Note about 3% of the relief valves across the entire chemical complex (which contained more than 3000 relief valves) were resized to account for accident scenarios that were only found during analysis of non-routine modes of operation.

9.3 Case Study 3: Urea and Ammonia Plants in the Arabian Gulf Region

Several ammonia and urea plants had hazard evaluations performed using HAZOP of nodes (parametric deviations for continuous mode of operation) and then non-routine modes of operation were analyzed using What-if and 2-Guide Word analysis for selected, critical procedures. The HAZOP of nodes (which focuses necessarily on the continuous mode of operation) missed many accident scenarios that were later found during review of non-routine modes of operation. The analysis of critical procedures resulted in a number of new recommendations to reduce the risk of accident scenarios (these accident scenarios were not identified during the HAZOP of the nodes a few days before). One example was the recommendation to install a second, independent level transmitter on the urea reactor to prevent overflow to the high pressure scrubber which will lead to overpressure of the entire synthesis loop and poor reaction kinetics during startup and normal operation.

Another example related to the start-up of high pressure ammonia pumps in the synthesis section of the ammonia plant follows. A recommendation was made during the analysis to include more detail on the required rate of temperature increase together with notes explaining the potential consequences for heating too quickly. In this case, a high heating rate may damage the 316L stainless steel liner in vessels and would have caused corrosion, deformation and ultimately a vessel failure.

It was also identified during the analysis that the development of trouble-shooting guides should be considered for responding to severe system upsets, including control valve failure, sticking of a hand valves, and to address contamination issues. These can be developed with the help of the PHA tables, which contain 40% of the scenarios and information related to trouble-shooting guides.

10. Conclusion

Qualitative analyses of non-routine operating procedures is an extremely powerful of tools for uncovering deficiencies that can lead to human errors and analyzing accident scenarios during all modes of operation. This approach of step-by-step HAZOP and/or What-If analysis is not new to industry and regulators have required similar approaches for decades. And regulators continue to note lack of analysis of the risk of non-routine operations and lack of risk review of changes to procedures.

> ***From the Wall Street Journal[17] referencing the presidential commission investigating the Deepwater Horizon accident of April 2010:*** *BP had rules in place governing procedural changes, but its workers didn't consistently follow them,*

according to BP's September [2010] internal report on the disaster and the report released earlier this month [January 2011] by the presidential commission on the accident. "Such decisions appear to have been made by the BP Macondo team in ad hoc fashion without any formal risk analysis or internal expert review," the commission's report said. "This appears to have been a key causal factor of the blowout."[17]

From CSB Report on August 2008 Bayer CropScience Explosion:[17] *"The accident occurred during the startup of the methomyl unit, following a lengthy period of maintenance ... CSB investigators also found the company failed to perform a thorough Process Hazard Analysis, or PHA, as required by regulation...In particular, for operational tasks that depend heavily on task performance and operator decisions, the team should analyze the procedures step-by-step to identify potential incident scenarios and their consequences, and to determine if the protections in place are sufficient."*

Further, such analyses make it easier to provide a thorough consideration of human factors. Regardless of what hazard evaluation technique is employed, it is imperative for PHA teams to ask, "Why would someone make this mistake?" whenever a human error is identified as a cause of a potential accident. "To err is human" may be a true statement, but the frequency and consequences of such errors can be effectively reduced with a well-designed strategy for analyzing risk of non-routine operating modes.

11. References

1. Rasmussen, B. "Chemical Process Hazard Identification," *Reliability Engineering and System Safety*, Vol. 24, Elsevier Science Publishers Ltd., Great Britain, 1989.

2. "Process Safety Management of Highly Hazardous Chemicals; Explosives and Blasting Agents," US OSHA Final Rule, 29 CFR 1910.119, February 24, 1992.

3. "Risk Management Programs for Chemical Accidental Release Prevention," US EPA Final Rule, 40 CFR 68, 1994-2000.

4. Bridges, W.G., et. al., "Addressing Human Error During Process Hazard Analyses," *Chemical Engineering Progress*, May 1994.

5. Tew, R. and Bridges, W., "Human Factors Missing from PSM," *Loss Prevention Symposium (part of the Global Congress on Process Safety [GCPS])*, AIChE, March, 2010.

6. "The SPAR-H Human Reliability Analysis Method," NUREG/CR-6883, U.S. Nuclear Regulatory Commission, prepared by Gertman, D.; Blackman, H.; Marble, J.; Byers, J. and Smith, C., of the Office of Nuclear Regulatory Research, Washington, DC, August 2005.

7. "Human Event Repository and Analysis (HERA)," NUREG/CR-6903, U.S. Nuclear Regulatory Commission, prepared by B. Hallbert, A. Whaley, R. Boring, P. McCabe, and Y. Chang, 2007.

8. "Guidelines for Hazard Evaluation Procedures, 3rd Edition, with Worked Examples," Center for Chemical Process Safety, AIChE, New York, 2008.

9. "Systems Safety Evaluation of Operations with Catastrophic Potential," page C-11, US OSHA Instruction CPL 2-2.45, Directorate of Compliance Programs, September 6, 1988.

10. OSHA Inspection Number 103490306, page 7 of 77, Issued November 2, 1992.

11. OSHA Inspection Number 123807828; pages 2, 3, and 7 through 22, of 25; Issued November 18, 1993.

12. Woodcock, Henry C., "Program Quality Verification of Process Hazard Analyses (for instructional purposes only)," US OSHA, 1993.

13. Hammer, W., "Occupational Safety Management and Engineering, 3rd Ed.," Prentice Hall, 1985.

14. Kongso, H.E., "Application of a Guide to Analysis of Occupational Hazards in the Danish Iron and Chemical Industry," *International Conference on Hazard Identification and Risk Analysis, Human Factors and Human Reliability in Process Safety,* Center for Chemical Process Safety, AIChE, New York, 1992.

15. "Hazard Evaluation (PHA) Leadership Course," Process Improvement Institute, Inc., 2003.

16. Tew, R., et.al., "Optimizing Qualitative Hazard Evaluations (or How to Complete A Qualitative Hazard Evaluation Meeting in One-Third the Time Currently Required)," *5th Global Congress on Process Safety,* AIChE, April 2009.

17. *Wall Street Journal*, January 29, 2011.

18. "Investigation Report: Pesticide Chemical Runaway Reaction Pressure Vessel Explosion, at Bayer CropScience, LP, Institute, WV, on August 28, 2008", US Chemical Safety Board, Report No. 2008-08-I-WV, January 2011.

Appendix A. Random Selection of Major Accidents versus Mode of Operation; limited to Processing Facilities[a,b]

Yr	Event	Company	Location	Deaths	Injuries	Losses (USD)	Mode of Operation
1987	Overpressure of and release of distillate from LP Separator, resulting in explosion	BP Oil	Grangemouth, Scotland	2	2	+50 million	Online maintenance
1987	External impact and release of HF	Marathon Oil	Texas City, TX	0	100+	4000 evacuated	Maintenance turnaround
1988	Natural gas explosion and fire	Occidental Petroleum	North Sea, Near UK	167	61	+2 billion	Startup after maintenance
1989	Release and ignition of isobutene and ethylene	Phillips	Pasadena, TX	24	300+	1.4 billion	Online Maintenance
1990	Internal explosion of waste water tank	ARCO Chem.	Channelview, TX	17	5	100+ million	Online Maintenance
1990	Explosion and fire	BASF	Cincinnati, OH	3	40	1.1million in fines Site closed	Non-routine batch
1992	Release and explosion	Texaco	Wilmington, CA	0	107	73 million	Normal Operation
1992	Ignited sludge sending a jet of flame over 50m-long through the plant control building into the site's main office block	Hickson & Welch	Castleford, UK	5	18	Unreported	Non-routine procedures - Removal of sludge
1993	Internal explosion of fired heater	Ashland Oil	Catlettsburg, KY	2	NA	Unreported	Startup
1994	Release of ammonia and nitric acid	Terra Ind	Port Neal, IA	4	18	Unknown	Start-up
1994	Release of hydrogen cyanide	BASF	Seal Sands, UK	0	11(	Not significant	Startup
1994	Series of fires and then explosion	Texaco	Milford Haven, South Wales, UK	0	26	>700 million	Startup after emergency shutdown
1995	Explosion and fire	NAPP Tech	Lodi, NJ	5	4	Unknown	Non-routine batch steps
1995	Explosion and fire with flammable hydrocarbons	Pennzoil	Rouseville, PA	5	1	40 million	Online maintenance
1997	Phenol/Formaldehyde explosion	Georgia Pacific	Columbus, OH	1	13		Normal operation, wrong sequence

Yr	Event	Company	Location	Deaths	Injuries	Losses (USD)	Mode of Operation
1997	VCE from spill of LPG	Hindustan Petroleum Corporation (HPCL)	Visakhapatam, India	50	Unknown	64 million	Normal Operation
1997	Release and ignition of distillate	Tosco	Martinez, CA	1	46	6 million	Online maintenance
1997	Spill of HCL	Surpass	Albany, NY	0	43	Unknown	Startup
1997	ETO fire and explosion	Accra Pac	Elkhart, IN	1	65	Unknown	Shutdown
1998	Release and ignition of natural gas	Esso	Longford, Australia	2	8	1.1 billion	Startup, after maintenance
1998	Release of flammable and ignition	Morton	Patterson, NJ	0	9	Unknown	Normal Operation
1998	Catastrophic vessel failure, release and ignition	Sonat	Pitikin, LA	4	0	Unknown	Normal Operation
1998	Release of more than 4 tons of sulfur dioxide and hydrogen chloride	Nipa Laboratories	Lancashire, UK	0	3	> 100,000	Normal Batch operation
1998	Explosion and fire in refinery coker	Equilon	Anacortes, WA	6	Unknown	Unknown	Startup
1999	Release of distillate and ignition	Tosco	Martinez, CA	4	1	Unknown	Online maintenance
1999	Release, fire and explosion	Thai Oil-Royal/Dutch Shell	Sri Racha, Thailand	8	13	37 million	Normal Operation
1999	VCE resulting due to leak from valve failure	Chevron	Richmond, CA	0	0	79 million	Normal Operation
1999	Hydroxylamine explosion	Concept Sciences	Allentown , PA	5	14	4 million	Normal Operation (New Process)
2000	Release and explosion	Kuwait	Mina Al-Ahmadi, Kuwait	5	50	433 million	Normal Operation
2001	Release and ignition	BP Amoco	Augusta, GA	3	0	Unknown	Online maintenance
2001	Sulfuric acid tank explosion	Motiva	Delaware City, DE	1	8	Unknown	Maintenance
2002	Decomposition reaction in distillation column	First Chemical Corp	Pascagoula, MS	0	3	Unknown	Normal Operation

Yr	Event	Company	Location	Deaths	Injuries	Losses (USD)	Mode of Operation
2003	Benzoyl peroxide explosion and fire	Catalyst Systems	Gnadenhutten, OH	0	1	Unknown	Normal Operations
2004	Fire at explosion from LNG release	Sonatrach	Skikda, Algeria	27	74	Unknown	Startup
2004	Release of toxic allyl alcohol and allyl chloride	MFG Chemical	Dalton, GA	0	169	Unknown	Initial Startup
2004	Release and explosion of vinyl chloride	Formosa Plastics	Illiopolis, IL	5	3	Permanent closure	Cleaning during reactor shutdown
2004	Release of alkylate and explosion	Giant Industries	Cinizia (Gallup), NM	0	6	Unknown	Normal operation (switching pumps)
2005	Release and ignition of distillate	BP	Texas City. TX	15	170+	2+ billion	Startup
2006	Release and ignition of propylene	Formosa Plastics	Point Comfort, TX	0	16	Unknown	Normal operation (external impact)
2006	Release and ignition of flammable vapors	CAI/Arnel	Danvers, MA	0	10	Unknown	Normal operation
2007	Runaway reaction and explosion	T2 Labs	Jacksonville, FL	4	32	Unknown	Routine batch
2007	Release and ignition of natural gas	ARAMCO	Hawaiyah, Saudi Arabia	40	9	Unknown	Online maintenance
2008	Oleum release and evacuation of community	Occidental (INDSPEC Chemical Corporation)	Petrolia, PA	0	0	Unknown	Non-routine batch
2008	Runaway reaction and overpressure and explosion/fire	Bayer CropScience	Institute, W VA, USA	2	8	Unknown	Startup
2009	Solvent vapor release and explosion	Veolia	W. Carrollton, OH	0	2	Unknown	Shutdown
2010	Fire and explosion in crude oil distillation unit at Lindsey Oil Refinery	Total	Humberside, UK	1	2	Unknown	Maintenance turnaround
2010	Explosion and fire at oil refinery	Tesoro Corp	Anacortes, Washington State	5	2	Unknown	Startup after maintenance

[a] Data in the table excludes storage and terminals and also excludes exploration (drilling, etc.) and transportation
[b] Forty-seven major accidents are listed above, of which thirty-one are during non-routine operation

Using the HAZID Study, HAZOP Study and LOPA to Generate Ideas for Inherently Safer Designs

Steven T. Maher, PE CSP
Risk Management Professionals
300 Goddard, Suite 200, Irvine, California 92618
Steve.Maher@RMPCorp.com

Kristin D. Norton
Senem Surmeli
Risk Management Professionals

Keywords: Inherently Safer Design, HAZID, HAZOP, LOPA, Design Cycle, Hazard Analysis

Abstract

Fundamental changes in a chemical process are most cost-effective at early stages of a design, and specific inherently safer design evaluations are typically performed at these early stages to identify fundamental improvements that can remove hazards or significantly decrease the magnitude of the consequences associated with plant accidents. As the detailed design comes into focus, thoughts of inherently safer designs are rarely revisited because the designer's paradigm has shifted. Although fundamental changes are more difficult, there may still be opportunities during the detailed design phase to entertain new ideas of inherently safer design, especially due to the rapidly changing environment of process plant technologies. Thus, it is the duty of the process designers to look for new mechanisms for improving plant safety, and throughout the design process. As teams assemble for various activities, good opportunities for revisiting inherently safer design concepts do materialize. The focus of this paper is to present these opportunities to the reader, provide examples of inherently safer design concepts that can be applied at different phases of design, and provide guidance as to how these concepts can be applied specifically during the HAZID, HAZOP study, and LOPA processes that are routinely carried out during most major plant process designs.

1. Introduction

The past decades have seen an increased consideration of risk associated with processing of chemicals that are considered highly hazardous from a toxicity, flammability, or reactivity perspective. Much of this interest, and the first priority, has been focused on application of various Safety Management System (e.g., Process Safety Management) elements to lower the likelihood of events or to reduce the consequences. Evolution in consideration of risk issues has placed increased emphasis on infusing inherently safer design concepts early in the design process, e.g., during the application of HAZID (Hazard Identification) study in the early portion of the Front-End Engineering and Design (FEED) process.

The following figure provides an overview of the different tools that can be utilized as the design evolves from a conceptual idea to an operating process. Acronyms identified in the figure below will be defined in the following sections.

Figure 1. Process Design Evolution

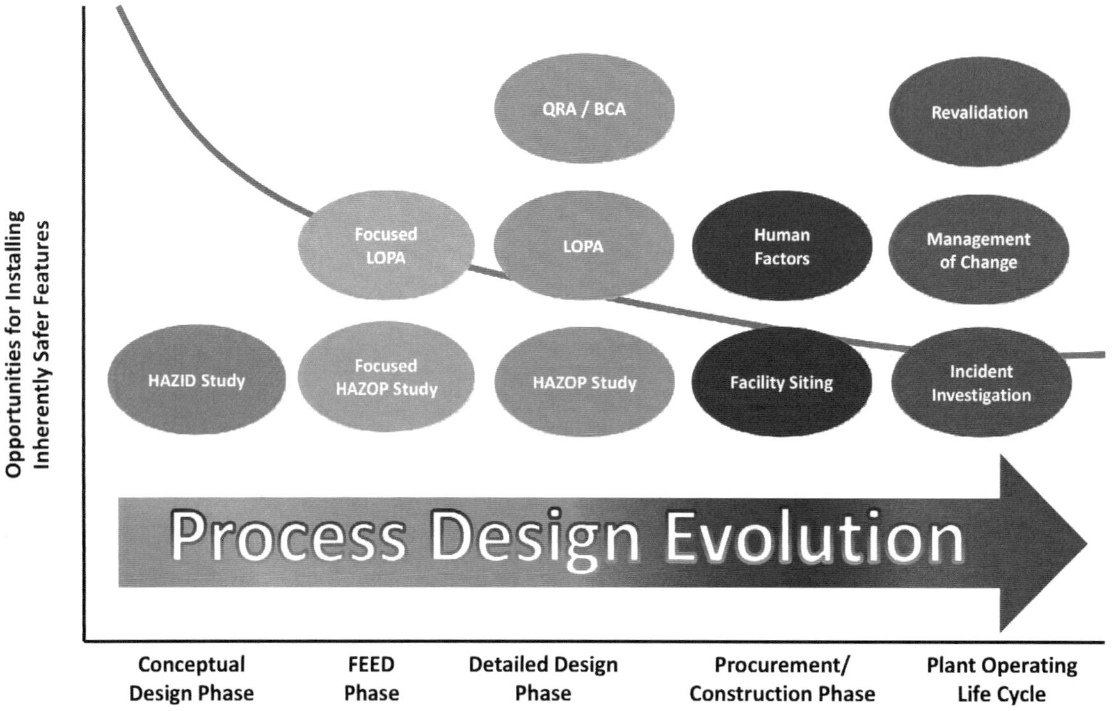

As depicted above in Figure 1, opportunities to install inherently safer features reduce due to associated costs, and other impediments, as the process design evolves; however, it is the process designer's responsibility to infuse whatever tools may be available throughout the design progression, as each one of these tools can provide different, yet valuable insights.

The following sections describe each project phase and available tools to help identify opportunities for implementation of traditional inherently safer design methods:

- Minimization
- Substitution
- Moderation
- Simplification

2. Conceptual Design Phase

During conceptual design more opportunities exist for the identification and incorporation of inherently safer technologies into the design. Additionally, the identification of these opportunities prior to detailed design can yield measurable savings in cost and schedule that would precipitate from future design re-work. Thus, it is extremely important to emphasize the benefits of identifying potential inherently safer design (ISD) considerations during the conceptual design phase in order to maximize these benefits.

Since detailed Piping and Instrumentation Diagrams (P&IDs) are not available during the conceptual design phase, the typical tool utilized for identifying potential process hazards and the associated inherently safer design mitigation opportunities is the Hazard Identification (HAZID) study performed using the Process Flow Diagrams.

2.1 HAZID Study

Application of HAZID at early stages of the design cycle allows for identification of high level ISD opportunities, specifically including the following categories:

- Minimization – minimizing the hazardous material inventory.
- Substitution – substituting a hazardous material for a less hazardous material.

Table 1 provides an overview of the ISD considerations and mitigation examples that are typically considered as part of the HAZID Study. During the HAZID study, in order to maximize the potential for identifying opportunities, ISD considerations are applied prior to recommending an engineered system or administrative controls.

Table 1. Potential Inherently Safer Design Mitigation Examples – Conceptual Design Phase

Consequence / Condition	Potential Inherently Safer Design Mitigation
Loss of Containment / Excess Inventory	**Minimize** • Minimize hazardous feed stock, intermediate, and product inventories. • Reduce equipment size • Reduce piping lengths
Loss of Containment / Hazardous Material/Form	**Substitute** • Non/less hazardous reaction chemistry, • Use lower volatility solvent • Non/less hazardous heat transfer media • Use of liquid form vs. vapor (if risk is reduced) • Use less concentrated form

3. Front-End Engineering and Design (FEED) Phase

The Front-End Engineering and Design (FEED) phase consists of activities to determine the project feasibility and to develop the initial project cost estimates. During the FEED phase, the scope of work as well as responsibilities of all involved parties are identified and agreed upon. Although more detailed than the Conceptual Design Phase, the FEED phase process variables are still considered estimates, and detailed design information is not yet available.

Identification of ISD options plays an important part in determining project feasibility and initial cost estimates. As the decisions are made to estimate project cost, the process designer gains new opportunities to include ISD alternatives into the design and minimizes the potential for re-work, thereby increasing project efficiency.

The tools available to the process designer to help in identifying ISD alternatives during the FEED phase include:

- Focused Hazard and Operability (HAZOP) Study
- Focused Layer of Protection Analysis (LOPA)
- Simplified Quantitative Risk Assessment (QRA)

3.1 Focused Hazard and Operability (HAZOP) Study

As the first draft of the P&IDs is completed to match the project specifications, a HAZOP Study is often conducted to determine the potential worst-case consequences of failures in the design. During this study, a team consisting of individuals knowledgeable in the design and controls of the process discuss the potential hazards associated with loss of containment and equipment failures as it relates to personnel and/or community exposure as well as environmental and economic impacts. Using a risk-ranking matrix that matches the operating company's risk tolerance criteria, the team determines the severity (based on the worst consequences of that scenario) and the likelihood of the consequence. Although the likelihood of the scenario consequence decreases with increased layers of protection (i.e., safeguards), the goal of the process design team should be to first consider potential ISD alternatives (i.e., reducing the severity) by determining the applicability of the following traditional method for all hazardous scenarios:

- Moderation – designing less hazardous conditions.

Table 2 provides an overview of the ISD considerations and mitigation examples that are typically considered as part of the FEED HAZOP study.

Table 2. Potential Inherently Safer Design Mitigation – FEED Phase

Consequence / Condition	Potential Inherently Safer Design Mitigation
Loss of Containment / Hazardous Condition	**Moderate - Less hazardous conditions** • Reduce pressure • Reduce temperature • Refrigerated storage (except ammonia) • Limit escalation – equipment spacing to reduce congestion; layout considering wind direction; buffer zones between units; secondary containment; barriers; separate hazardous inventory • Limit sources of ignition – identify, remove, separate • Design building to withstand blast

3.2 Focused LOPA

Layer of Protection Analysis is another risk assessment tool that can be used during the FEED phase to determine opportunities for re-design. Although LOPAs are frequently conducted in conjunction with the HAZOP Study (discussed in Section 3.1), some facilities and design companies perform the LOPA separately to provide objectivity. LOPA uses data compiled from industry on equipment reliability to calculate the initiating event frequency and combined effectiveness of existing independent layers of protection (e.g., safeguards), and compares the LOPA scenario frequency to the risk tolerance criteria (i.e., target frequency for an event of that consequence) of the subject company. Large gaps between the probability of reaching the scenario consequences in question and the associated target event frequency would indicate that the process designer either needs to install additional or more effective safeguards to further lower the probability or change the design such that the consequence identified is no longer credible. Conducting LOPA early in the FEED phase makes it feasible to do the latter, thus encouraging inherently safer design alternatives.

4. Detailed Design Phase

Although during the detailed design phase, the design is more mature, there are still opportunities to incorporate inherently safer design mitigation features for high severity hazards or to correct design flaws. These opportunities are typically present in the following category of inherently safer design:

- Simplification – designing equipment to be robust instead of adding equipment or controls to mitigate hazards.

While these opportunities may be more limited due to costs associated with engineering re-work, it is still vitally important to systematically evaluate these opportunities to maximize the benefits of inherently safer design. Additionally, simplification may also reduce costs associated with additional equipment and controls. The following techniques allow for the evaluation of inherently safer design mitigation in the detailed design phase:

- HAZOP study
- LOPA
- Benefit-Cost Analysis (BCA)

4.1 HAZOP Study

A HAZOP study in the detailed design phase differs from a HAZOP study in the FEED phase in that the P&IDs available are more detailed, and the specific design parameters are available for the team to evaluate. Thus, the team is able to characterize the hazards and risks more accurately and find hazards that were not necessarily identified during the conceptual design or FEED phase. When presented, these hazards are typically mitigated with engineered solutions (e.g., additional equipment redundancy, safety instrumented systems); however, the addition of equipment intrinsically introduces additional risks into the system. Instead, the team should consider mitigating these hazards through simplification. Some examples are presented in Table 3 below.

Table 3. Potential Inherently Safer Design Mitigation – Detailed Design Phase

Consequence / Condition	Potential Inherently Safer Design Mitigation
Loss of Containment / Complex Design	**Simplify - More Robust** • Increase equipment pressure rating to maximum achievable • Use spiral-wound and flexible graphite gaskets • Use pumps with double mechanical seals, diaphragm pumps, eductors, seal-less pumps (note seal-less pumps may overheat and leak internally) • Design vessels for full vacuum • Design equipment containing liquid to withstand maximum hydrostatic load (if full) • Select materials that are impervious to corrosion, erosion • Design heat exchanger shell to withstand maximum pressure (tube or shell side) • Use gravity or pressure/vacuum differential to transfer materials

4.2 LOPA

The application of LOPA in the detailed design phase is typically focused on determining the required Safety Integrity Level (SIL) targets for Safety Instrumented Systems (SIS). As discussed previously, for high consequence scenarios resulting from the HAZOP, LOPA evaluates the gap between the target event frequency (i.e., the frequency that it is acceptable that the event occurs) and the probability of the event (taking into account the initiating cause likelihood and the probability of failure on demand of the independent safeguards). If the probability of the event is not a tolerable risk, then either additional independent safeguards need to be incorporated into the design, or the SIS features must have a high enough SIL pedigree (e.g., SIL-1, 2, 3 or 4) to decrease the probability of the event to at least the tolerable risk category. Alternatively, the design team can evaluate an inherently safer design that would reduce the consequence of the event.

While LOPA does not directly evaluate the potential for incorporating inherently safer design features, the results of the study may produce cases where inherently safer design may be more cost effective to achieve a tolerable risk. LOPA can also be useful for quantifying the relative improvements offered by several design options. For example, if the results of the LOPA yield a required high safety integrity level (e.g., SIL-3 or 4) to close the gap, the cost associated with purchasing the robust SIS equipment to meet the reliability requirements is extremely high and it may be more cost-effective to rework the design to be inherently safer (e.g., increasing equipment and piping pressure ratings to maximum achievable, designing for full vacuum, etc.). Additionally, SIL-4 is typically only utilized in the nuclear industry due to the required complexity and reliability. Similarly, although achievable in the chemical industry, the implementation of SIL-3 systems is considered difficult due to the required SIS proof test effectiveness and frequency, especially when considering human factors. To close a SIL-3 or SIL-4 LOPA gap, multiple SIS functions can be applied to meet the overall risk reduction, though caution is advised to evaluate the configurations for common cause failures and quality of selected components. Simplifying the design using inherently safer techniques not only provides safety and environmental benefits, but also has the potential to reduce costs.

4.3 Benefit-Cost Analysis

A BCA is a quantitative technique that measures the cost-effectiveness of various design alternatives, which can be applied to evaluate the benefits/costs of inherently safer design modifications vs. robust control system modifications (as identified through the LOPA). The performance of the analysis allows decision-makers to systematically evaluate design alternatives taking into account the benefits, as well as the potential costs. In order to perform the analysis, the following numerical factors are identified:

Potential Benefits

- Direct project savings
- Increased efficiency (shortened project timeline, automation, operations, maintenance, start-up/shut-down, etc.)

Potential Costs

- Direct project costs (engineering re-work, equipment costs, etc.)
- Hazard severity (potential worst-case scenario risks – monetary, reputation, personnel, business)
- Hazard frequency (annual probability of reaching consequence with associated economic impacts)

Using these values, the annualized hazard probability is determined to calculate the Benefit-Cost Ratio for each project alternative. When comparing the project alternatives side-by-side, the project with the highest BCA Ratio is the most cost-effective solution.

5. Procurement / Construction Phase

During the Procurement / Construction Phase, design change is one of the last (and least desirable) thoughts crossing a process designer's mind. As such, inherently safer design is often left out of the thought process since it is mostly branded as changing the design. This section of the paper is intended to provide simple checklist tools that can be used to ensure that ISD is not left out during the procurement/construction phase.

5.1 Human Factors Checklists

Human factors are unarguably one of the most significant contributors to industry incidents. The need for increased attention to human factors has been highlighted by the recent industry incidents, including the events that led up to the tragedy at the BP Texas City Refinery in 2005. Whenever possible, the process should be simplified to minimize the potential for human error.

The following table provides an overview of the human factors issues that are typically considered to address ISD concepts as part of the design and construction of new processes.

It should be noted that for all facilities subject to OSHA's Process Safety Management of Highly Hazardous Chemicals (29 CFR 1910.119) standard, it is a requirement to address human factors as part of the Process Hazard Analysis (PHA).

Table 4. Potential Inherently Safer Design Mitigation – Procurement/Construction Phase

Consequence / Condition	Potential Inherently Safer Design Mitigation
Loss of Containment / Complex Design	**Simplify - Human Factors** • Reduce instrumentation complexity, avoid information overload • Design field conditions that allow operators to easily determine equipment condition; lighting and ease of access; • Valve position easily identified; manually controlled equipment in line-of-sight of measuring element • Signage is easy to read, clear and unambiguous • Housekeeping and general work environment is conducive to efficient performance • Board and screen displays match the actual equipment configuration • Layout of the console is logical, consistent and effective Design components with unique shapes to prevent mis-connection/assembly.

5.2 Facility Siting Checklists

Although it is the intention of all involved parties to design a process that is safe, it is impossible to avoid risk when dealing with hazardous materials. By utilizing facility siting checklists during design, process designers can ensure that an incident can be prevented from spreading throughout the facility thus avoiding domino effects. Commonly used facility siting checklists bring attention to the need for adequate spacing between units or sections of a large unit, orientation of cylindrical tanks such that the heads are not facing other equipment or communities, installation of adequate containment areas and locating commonly occupied portable buildings away from the process units. Facility siting must be addressed in the PHA for PSM-covered facilities.

6. Plant Operating Life Cycle

The plant operating lifecycle is dynamic and provides numerous opportunities for incorporating inherently safer design into expansion projects and facility upgrades, as well as through safety evaluations associated with the US Environmental Protection Agency (EPA) Risk Management Plan (RMP) and OSHA PSM. Specifically, the following elements of these programs provide mechanisms for evaluating potential ways to incorporate inherently safer design elements:

- Process Hazard Analysis (PHA) revalidation
- Management of Change (MOC)
- Incident investigation

Figure 2 illustrates the cyclical nature of performing plant operations and maintenance, evaluating the implementation of the programs, and developing improvements to lessen the impact of hazards or eliminate them through inherently safer design.

Figure 2. Inherently Safer Design in the Plant Operation – Evaluation – Improvement Cycle

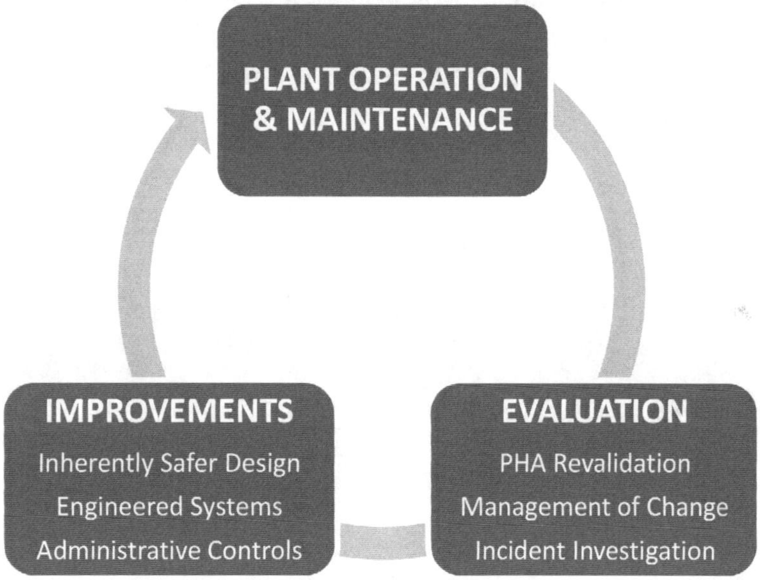

6.1 HAZOP/LOPA Revalidation

The RMP and PSM Programs require that facilities revalidate the PHA on a five-year basis, or as major changes are implemented for the unit. As discussed in the previous sections, HAZOP evaluates potential hazards and allows the team to discuss mitigation measures, including engineered systems, administrative controls, and most importantly – inherently safer design. Revalidating these studies periodically also allows the facility to evaluate the hazards and associated mitigation measures utilizing contemporary strategies and design elements. One mechanism to ensure that ISD options are discussed is to have a HAZOP Study "Deviation" that focuses the team to at least spend a few minutes considering the possibility of introducing ISD concepts into the design.

6.2 Management of Change (MOC)

The MOC Program is a specific procedure to manage changes (except for "replacement in kind") to process chemicals, technology, equipment, procedures, and changes to facilities that affect the hazardous process. As changes are proposed, they are specifically evaluated to determine whether new hazards are introduced and that the changes are designed utilizing good engineering practices. As part of this evaluation, the team may be required to revisit the HAZOP study, providing an additional opportunity to incorporate inherently safer design. Even if a HAZOP study is not required, the MOC process should pose the question of whether ISD options exist.

6.3 Incident Investigation

When a facility has an incident or a near miss, there is a requirement to conduct a detailed incident investigation to determine the root cause and to develop recommendations to prevent the incident from occurring in the future. As part of the team brainstorming of recommendations to mitigate the hazard or the contributing events for the incident, the team should specifically evaluate whether modifying the design to be inherently safer will prevent the incident from occurring in the future. Specifically, the Incident Investigation Procedure can specifically pose the question of whether ISD options exist.

7. Conclusion

While inherently safer design technologies can be applied at any stage of the design to minimize hazards, the table below illustrates the primary and most effective methods for each design phase:

Table 5. Inherently Safer Design Techniques by Design Phase

Most Effective Inherently Safer Design Techniques Per Design Phase				
	Minimization	**Substitution**	**Moderation**	**Simplification**
Conceptual Design	X	X		
Front-End Engineering and Design			X	
Detailed Design				X
Procurement / Construction				X
Plant Operating Life Cycle	X	X	X	X

Thus, although the application of inherently safer design techniques earlier in the design/construction/operations cycle provides the most flexibility, design options, and cost-benefit; there can be many tangible benefits to their consideration throughout the life cycle of the facility. Use of the approaches identified in this paper can facilitate the sensible application of inherently safer design concepts (even after the original designers are longer involved with the plant) and also introduce operational experience.

8. References

Center for Chemical Process Safety, "Inherently Safer Chemical Processes – A Life Cycle Approach", 2nd Edition, American Institute of Chemical Engineers, New York, NY (2009).

Ian Sutton, "Inherent Safety in Front End Engineering", paper accepted for presentation at the Process Plant Safety Symposium and Spring National AIChE Meeting, Chicago, IL (March 2011).

Trevor Kletz and Paul Amyotte, "Process Plants – A Handbook for inherently safer design", 2nd Edition, CRC Press, Taylor & Francis Group, Boca Raton, FL (2010).

(80a) Strong Reporting Culture as Stepping Stone to Continuously Drive Safety Performance Improvement

Joe Stough, Syntex Management Systems, Inc., Houston, TX

This session will review the leading indicator research findings drawn from analyzing a large Energy and Chemicals industry data set. The source of research includes data sets and Safety business practices spanning operating sites in over 100 countries involving some global industry leading companies such as ExxonMobil, ConocoPhillips, CITGO, Valero, Total, and multiple others. This session will review the research findings and specifically targets Reporting Culture as an area that was shown to have a strong affect on Safety performance.

Most companies still use lagging incident-based metrics to drive line management behavior. Often times this goes so far to include lagging metrics as part of pay incentives for line management and supervision. Such an approach typically perpetuates a focus on lagging metrics in monthly management performance reports and thus offers top leaders little insight into the effort taken in the field to proactively improve HSE performance.

By balancing these lagging metrics with a carefully chosen short list of "actionable" metrics targeting key leading areas such as Reporting Culture, line managers are afforded guidance to lead their organizational focus in a manner that positively affects HSE outcomes. And top leaders are afforded the routine monthly measurements to hold line managers accountable to continuously improve.

Through researching a vast data set of routine HSE events (such as near misses, incident investigations, observations, audit findings, hazard reporting, action item management, etc.) and employee engagement in the processes underlying those events, it became apparent that field supervision / line management was the organizational role that most directly influenced HSE outcome performance. And after careful review with Safety executives and Subject Matter Experts, it was also clear that attaining a strong Reporting Culture is a critical milestone necessary to begin the process of proactively driving Safety performance.

The areas that best predicted the Safety outcome performance of an organization included measurements of (a) proactive event reporting (Reporting Culture), (b) the discipline and consistency in execution of HSE processes, (c) the timeliness in closure of important corrective / preventive actions, and finally (d) the responsiveness and accountability of supervisors and line managers. In each of these key areas, solid performance can be attributed to the tone set by leadership and the influential day-to-day behavior of field supervision and line management.

This session will review the measurable leading factors that have been found to be uniquely strong in the organizations with top-level Safety lagging performance. Specifically, we will focus on the importance of Reporting Culture as a stepping stone to continuously improving the key factors that are shown to result in top-tier Safety performance.

We will review some examples of creative technology solutions to reach out to front-line workers and engage them directly in the reporting process. In addition to basic web-based reporting tools, hand-held technology is being used in some heavy manufacturing environments to include inspection and observation-based processes in the overall program of event reporting.

Lastly we will discuss the affects of risk-aware, accountable leadership in each of the key leading areas. And finally, we will demonstrate a process for implementing practical, "actionable" leading metrics in each key area to drive line manager / leadership behavior and ultimately improve HSE performance outcomes.

See more of this Session: Process Safety Culture – Impact of the Baker Panel Report to BP
See more of this Group/Topical: Topical 1: Global Congress on Process Safety

Improving process safety culture: An audit checklist for effective first-line supervision based on common operations failure modes

Peter T. Bullemer
Human Centered Solutions, LLC
Independence, MN USA
pbullemer@applyHCS.com

Jason C. Laberge
Honeywell Advanced Technology
Minneapolis, MN USA
jason.laberge@honeywell.com

Keywords: Process safety culture, operations practice failures, root cause incident analysis, continuous improvement, effective supervision

Abstract

The Abnormal Situation Management® (ASM®) Consortium funded a study to investigate common failure modes and root causes associated with operations practices. The study team analyzed 20 public and 12 private incidents across the refining, chemical and oil & gas industries for common operations practice failure modes including the BP Texas City Incident. A key finding from the analysis was that ineffective first-line supervision was one of the most frequent operations practice failures. Given the significant influence of first-line supervision on plant safety culture, this study provides the results of a detailed examination of the root causes of this particular failure mode. Specifically, a number of potential proactive indicators of weaknesses in operations practices were observed in the incident sample. An operations audit checklist is presented to enable plant personnel to assess their potential vulnerability to this common failure mode.

1. Introduction

Process industry plants involve operations of complex human-machine systems. The processes are large, complex, distributed, and dynamic. The sub-systems and equipment are often coupled, much is automated, data has varying levels of reliability, and a significant portion of the human-machine interaction is mediated by computers [1]. These systems are also social in that many plant operations function with a teamwork culture such that activities are managed by crews, shifts, and heterogeneous functional groups [2]. Team members have to cope with multiple information sources, conflicting information, rapidly changing scenarios, performance pressure and high workload [3].

Historically, the reporting of failures has tended to emphasize root causes associated with equipment reliability and less so on human reliability root causes [4]. Consequently, there is limited information available on the frequency and nature of operations failures pertaining to human reliability. This tendency has limited the ability of process industry operations organizations to identify improvement opportunities associated with their management systems and operations practices.

In an effort to improve on the understanding of the impact of ineffective operations practices and management systems on safe plant operations, the ASM Consortium decided to conduct root cause analysis of existing major incident reports [5] across the refining, chemical and oil & gas industries. The study team analyzed 20 public and 12 private incidents for common operations practice failure modes. A key finding from the analysis was that ineffective first-line supervision was the 2nd most frequent operations practice failure (65 out of 539 observed operations practice failures), representing 12% of all failures.

Safety culture is a part of the overall culture of an organization that influences the members' attitudes and behaviors with respect to health and safety performance [6]. The findings from the BP Texas City incident highlight the importance of addressing process safety practices as well as personal safety [7]. Moreover, a review of the literature has shown that management is a key influence on an organization's safety culture as revealed in findings that employees' perceptions of management's attitudes and behaviors regarding safety practices are the most useful metric on safety climate [6]. In addition, a survey of offshore facilities has demonstrated that first-line supervision has a direct impact on safety performance [8].

The ASM Consortium root cause analysis study also revealed that ineffective first-line leadership is a significant contributor to process safety incidents [5]. By its very nature, the first-line supervisory role is management's primary interaction with operations personnel in communicating and enforcing their policies and practices for effective process safety performance. Moreover, we assert that an organization will be challenged to establish an effective process safety culture without effective first-line supervision.

To that end, this paper examines the root cause manifestations associated with the first-line supervision common failure mode to highlight potential indicators of weaknesses in this influence on process safety culture. An operations audit checklist is presented to enable organizations to assess their potential vulnerability to this common failure mode.

2. Incident Analysis

In general, the purpose of an incident analysis is to generate information to enable an understanding of why an incident occurred and identify corrective actions to address weaknesses in operations practices or management systems that contributed to its occurrence. In the ASM Consortium project, several incidents were analyzed to identify common failure operations practice failure modes to help member companies understand where they may have unacceptable risk to human reliability failures. The project team developed a new approach that goes beyond the typical root cause analysis methodology to identify systemic operations practice failures that are not indicated when looking at root causes alone. A detailed description of the methodology is available [5]. In this section, a high level description of the incident analysis methodology is provided with the findings related to the first-line supervision operations failure mode.

Figure 1~~Figure 1~~ shows the ~~eight step~~eight-step incident analysis approach. In the second step of this methodology, all of the operations failures are clustered into common failure modes using the operational practice definitions from the ASM Consortium *Effective Operations Practices* guidelines document [9]. Since there are 52 operation practice guidelines, there were a total of 52 potential failure modes. A common failure mode represents a common problem across industry sites and characterizes 'What went wrong' across the incident sample. While this ASM guidelines document is not available to the general public, a process industry organization could develop a similar set of operational definitions based on their operations practice standards, policies and management systems.

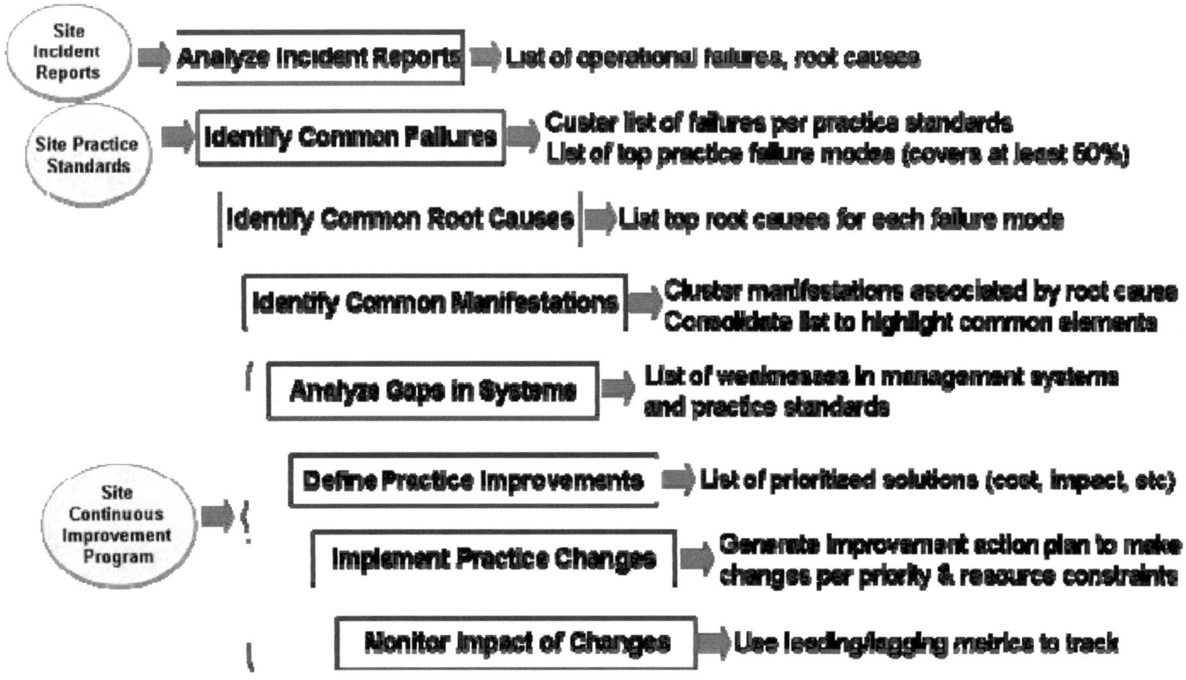

Figure 1. Summary of work process for incident analysis and continuous improvement.

The outcome of this second step in the analysis is the identification of the top ten failure modes (See **Error! Reference source not found.**~~Table 1~~).

The 10 most common failure modes accounted for 70% of the total number of failures across incidents. However, the top three failures accounted for 38% of the total and there was a clear drop in coverage between the top three failures and the rest (from 11% to 7%). Therefore, sites that are looking for critical areas to focus on should consider the following top three failures as most critical:

- Implement a comprehensive hazard analysis and communication program—15%
- Establish effective first line supervisor roles to direct personnel, enforce organizational policies, and achieve business objectives—12%
- Establish an effective and comprehensive program to continuously improve the impact of people, equipment, and materials on plant productivity and reliability—11%

To check for consistency of the findings, the sample was randomly split into two sub–samples to compare the distribution of failure modes. The comparison showed that the rank order of the failure modes was relatively consistent between the sub-samples, which suggested the overall sample of 32 incidents is relatively stable and the addition of more incidents would likely not dramatically alter the results. Therefore, the findings can be generalized across the process industry sectors represented by the incident reports.

Table 1. Top 10 common failure modes across all the incidents.

Common Failure Modes	#	%
Implement a comprehensive hazard analysis and communication program	79	15%
Establish effective first-line supervision roles to direct personnel, enforce organizational policies, and achieve business objectives	**65**	**12%**
Establish an effective and comprehensive program to continuously improve the impact of people, equipment, and materials on plant productivity and reliability	60	11%
Develop a strong safety culture[1]	36	7%
Establish initial and refresher training based on competency models that address roles and responsibilities for normal, abnormal, and emergency situations	30	6%
Establish effective protocol for task-oriented collaborative communications within operations	29	5%
Implement a comprehensive Management of Change (MOC) program that specifically includes changes in staffing levels, organizational structures, and job roles and responsibilities	28	5%
Establish good, periodic communication across plant functional responsibilities	23	4%
Ensure compliance with an explicit policy on the use of procedures in plant operations	15	3%
Use design guidelines and standards for consistent, appropriate implementation of process monitoring, control, and support applications	14	3%
Other failure modes (n=33)	160	30%
Total	539	100%

The focus of this paper is on the second most common failure mode, *Ineffective first-line supervision,* representing 12% (65 of 539) of the identified failures in the ASM approach. Focusing on the top failure modes is justified because they represent a significant portion of the total operations failures that were observed. To illustrate, if each potential failure mode had an equal probability of occurring, then one would expect on the average ~10 instances per failure

[1] A poor safety culture is reflected in a failure to follow many of the practices recommended by OSHA PSM. In general, we considered a failure to reflect a poor safety culture if many root causes were identified that would otherwise span multiple common failure modes. Based on the results in Table 8, these multiple failure modes tended to include use of procedures, management of change, and not having site standards, work processes, and standards or a failure to follow them.

mode for the 539 identified failures across the 52 potential operations practices defined in the ASM guidelines document. Hence, the observation of 65 instances of the ineffective first line leadership failure mode is much greater than one might expect by chance.

The third step in the incident analysis methodology examines the root causes associated with the common operations failure modes (See Figure 1~~Figure 1~~). A root cause describes 'Why a failure occurred.' Table 2~~Table 2~~ shows the top root causes associated with the ineffective first-line supervision failure mode.

In the incident analysis approach, the value of the analysis of operations practice failure modes is that it establishes the context for understanding the root cause information. Most importantly, understanding the causes of failures establishes the opportunity to make improvements to mitigate the risk of plant incidents. Neither the aggregation across operations failures or root causes on their own provides sufficient detail to identify potential improvement opportunities.

Hence during the fourth step in the methodology (Figure 1~~Figure 1~~), the team identified 'How' operational failure modes are expressed in real operations settings by examining the manner in which each root cause was manifested in the incident sample. The individual manifestations were clustered together around common themes indicating how the operations practice failed.

Table 3~~Table 3~~ below shows the result of the manifestation analysis for the two most frequent common root causes associated with *the Ineffective first-line supervision* failure mode.

Table 2. The top root cause profile associated with the Ineffective First-line Supervision operations practice failure mode.

Root Cause	#	%
No supervision	22	20%
Crew teamwork needs improvement (NI)	18	17%
SPAC[2] not followed	10	9%
Management of Change needs improvement (NI)	6	6%
No communication	6	6%
Pre-job briefing needs improvement (NI)	5	5%
Other	41	38%
Total	108	100%

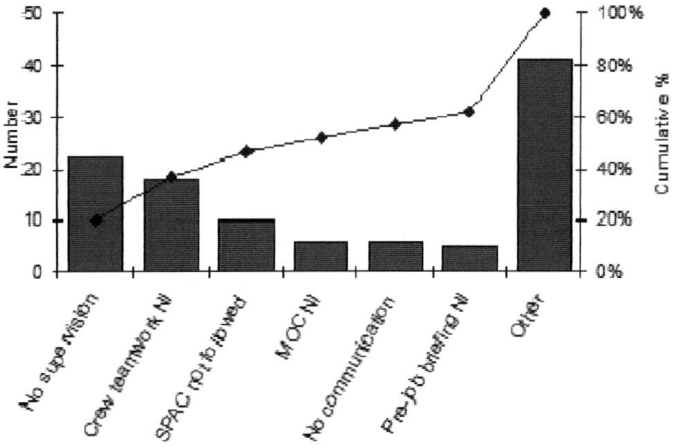

[2] SPAC (Standards, policies, administrative controls)—standardized work processes, rules, and procedures.

Table 3. Common root cause manifestations for two common root causes associated with the *Ineffective first-line supervision* operations failure mode.

Root Causes	Common Manifestations
No Supervision	Not checking procedure progress for area of responsibility
	Not at job site and maintaining situation awareness
	Fail to identify and address risk to personnel
	Fail to monitor high risk activities for problems/issues
Crew Teamwork NI	Not enforcing violations of practices/procedures
	Not ensuring team members stays coordinated
	Not correcting or communicating known problems
	Team members not questioning when evidence of problems
	Not keeping track of big picture; losing sight of hazards

The examination of all common root cause manifestations for the first-line supervision failures revealed supervision problems, teamwork, and enforcing policy and work practices (i.e., MOC and pre-job briefings). A critical component to effective supervision is being at the work site to provide work direction. Teamwork issues were often due to team members failing to question improper readings, indications, or directions by the person-in-charge. Another critical role that leaders have is to communicate and enforce policy and work practices. Several failures occurred because leaders failed to ensure standards, policies, and administrative controls were used, adhered to, and followed correctly.

These findings illustrate that the detail in the common manifestations for each root cause profile provides:

- Specific reasons the failures occurred across incidents
- Manifestations are "indicators" of failures
- Potential candidates for leading indicators of incidents

Consequently, improvement opportunities are identified by extracting the root cause manifestations for each root cause profile for the top common failure modes. After collecting this information, the continuous improvement program is in a better position to analyze gaps in their management systems and operations practices and identify specific solutions to reduce vulnerability to systemic and repeating root causes. The ASM project team used the information from common root cause manifestations to develop an audit ~~check list~~checklist. In fact, the audit checklist can serve as a gap analysis tool in step 5 of the incident analysis and continuous improvement work process (Figure 1~~Figure 1~~).

3. Audit Checklist

The purpose of the checklist is to help plant and operations managers conduct audits to identify the presence of a common failure mode for significant process industry incidents. Also, the audit checklist can serve as a gap analysis tool as part of the plant continuous improvement work process. The manifestations or expression of the root causes associated with the *Ineffective first-line supervision* failure mode provide the basis for the audit items in this ~~check list~~checklist. The audit ~~check list~~checklist items are expressed in positive terms of the practice elements that the auditor is looking for in his/her observations of current site operations practice for first-line supervision.

ASM Operation Practice Guideline: Establish effective first-line supervision roles to direct personnel, enforce organizational policies, and achieve business objectives.

- ☐ The supervisor maintains a presence in the control room and field areas with face-to-face contact periodically throughout a shift to ensure good situation awareness of operations and maintenance activities.
- ☐ The supervisor is easily accessible via radio contact by any team member to answer questions and respond to problems.
- ☐ The supervisor assigns a stand-in responsibility when leaving the job site.
- ☐ The supervisor ensures that the behaviors of personnel are compliant with site policy and work practices, and does not allow individuals to operate in the presence of known hazards without taking adequate precautions.
- ☐ The supervisor ensures that team members stay coordinated with appropriate plant disciplines, each other and him or herself.
- ☐ The supervisor establishes an open communications culture, where there is two-way dialog and team members feel free to question each other when there is evidence of problems.
- ☐ The supervisor establishes effective mechanisms to communicate known problems to all shift team members and shift teams, and ensure that corrective actions are assigned and tracked for timely completion.
- ☐ The supervisor frequently monitors progress of procedural activities and ensures compliance with site policy on use of procedures; especially as a safety observer for high risk activities.
- ☐ The supervisor enforces clear guidelines on when and how to conduct pre-job briefings.
- ☐ The supervisor is involved in the review of safety issues and hazards during pre-job briefings; specifically in the comprehensive identification and mitigation of risk to personnel.

4. Conclusion

Ineffective first-line supervision was the 2nd most frequent operations practice failure identified in an analysis of 32 major process industry incidents [5]. Not surprising, first-line supervision had previously been identified as having a critical role in the management of safety [8]. Moreover, this previous HSE research found that the aspects of supervisor behavior that impacted on subordinate safety performance included:

- valuing subordinates
- visiting the worksite frequently
- work group participation in decision making
- effective safety communication

In the present study, an examination of the common manifestations of the supervision common failure mode identified a list of potential proactive indicators of weaknesses in operations supervision practices. The operations audit checklist has a similar theme in terms of the specific aspects of first-line supervisor behaviors that contributed to major process safety incidents.

By its very nature, the first-line supervisory role is management's primary interaction with operations personnel in communicating and enforcing their policies and practices for effective process safety performance. Any organization seeking to establish a strong safety culture needs to ensure that there is an effective first-line supervision practice. The operations audit checklist provides a set of leading indicators for assessing potential vulnerabilities associated with first-line supervision.

5. Acknowledgements

This study was funded by the ASM® Consortium, a Honeywell-led research and development consortium. ASM is a registered trademark of Honeywell International, Inc. The Abnormal Situation Management® (ASM®) Consortium (www.asmconsortium.org) is a long-running and active research and development consortium of 16 companies and universities concerned about the negative effects of industrial plant incidents. The consortium identifies problems facing plant operations during abnormal conditions, and develops solutions. Deliverables from the collaboration among member companies include products and services, guideline and other documents, and information-sharing workshops; all incorporating ASM knowledge.

6. References

[1] Soken, N., Bullemer, P.T., Ramanathan, P., and Reinhart, B. (1995). Human-Computer Interaction Requirements for Managing Abnormal Situations in Chemical Process Industries. *Proceedings of the ASME Symposium on Computers in Engineering*, Houston, TX.

[2] Vicente, K. (1999). *Cognitive work analysis*. Mahwah, NJ: Lawrence Erlbaum Associates.

[3] Laberge, J.C., and Goknur, S.C. (2006). Communication and coordination problems in the hydrocarbon process industries. *Proceedings of the 50th Annual Meeting of the Human Factors and Ergonomics Society*, San Francisco, CA, USA.

[4] Bullemer, P. (2009). *Better metrics for improving human reliability in process safety.* Paper presented in the 11th Process Safety Symposium at the 5th Global Congress on Process Safety, Tampa, FL, USA.

[5] Bullemer, P.T. and Laberge, J.C. (2010). Common operations failure modes in the process industries. *Journal of Loss Prevention in the Process Industries*, 23(6), 928-935.

[6] Gadd, S. (2002). *Safety culture: A review of the literature.* Health and Safety Laboratory Report 2002/25. Sheffield, UK: HSE.

[7] CSB (2007). Refinery explosion and fire at BP Texas City. CSB Final Investigation Report 2005-04-I-TX. U.S. Chemical Safety and Hazard Investigation Board.

[8] Fleming, M. (2001). *Effective supervisory safety leadership behaviors in the offshore oil and gas industry.* Health and Safety Executive, Offshore Technology Report 1999/065. Suffolk, UK: HSE Books.

[9] Bullemer, P., Barreth, R., Laberge, J., and Nimmo, I. (2008). *Effective Operations Practices.* ASM Consortium Guideline Document. Minneapolis, MN: ASM Consortium.

Near Miss Reporting Behavior and the Safety Climate—A Scenario Based Study

Sunil D. Lakhiani, Ph.D., P.E.
Exponent, Failure Analysis Associates
185 Hansen Court, Suite 100, Wood Dale, IL 60191
slakhiani@exponent.com

Keywords: Safety climate, culture, near miss reporting, employee perception, behavior

Abstract

In the aftermath of the March 2005 incident at BP's Texas City Refinery, the U.S. Chemical Safety Board cited concerns about "the effectiveness of the safety management system" and "the corporate safety culture" at that Refinery. Included in the Baker Panel's findings were discussions about process safety leadership, the distinct process safety culture at BP's U.S. refineries, and the BP process safety management system that likely results in underreporting of incidents and near-misses.

An independent study was conducted at the University of Wisconsin-Madison that looked at the relationship between employees' perceptions of various factors that are part of the safety climate of the plants and their near-miss reporting behavior. Subjects included employees from three chemical process plants in the U.S. Results showed that subjects who perceived a more positive management culture towards safety were more willing to report near misses. This paper discusses the safety climate literature and the results of the near miss reporting behavior study.

1. Introduction

In their preliminary report on BP's Texas City refinery's March 2005 incident, the U.S. Chemical Safety Board (CSB) raised serious concerns about BP's safety management system and their corporate safety culture. As a result of CSB's recommendations, an independent safety review panel was formed, chaired by the former Secretary of State, James A. Baker, III. Based on their review and assessment of "the effectiveness of BP's corporate oversight of safety management systems at its five U.S. refineries and its corporate safety culture," (Baker Panel's report, 2007, p. ix) the Panel published their findings and recommendations in 2007. As part of their review, the Panel conducted a process safety culture survey of the workforce at the five BP refineries and found differences in the safety culture between the five refineries. An excerpt from the Baker Panel's report summarizing the Panel's finding is shown below.

> ***Process safety cultures at BP's U.S. refineries.*** BP has not instilled a common, unifying process safety culture among its U.S. refineries. Each refinery has its own separate and distinct process safety culture. While some refineries are far

> more effective than others in promoting process safety, significant process safety culture issues exist at all five U.S. refineries, not just Texas City. Although the five refineries do not share a unified process safety culture, each exhibits some similar weaknesses. The Panel found instances of a lack of operating discipline, toleration of serious deviations from safe operating practices, and apparent complacency toward serious process safety risks at each refinery.

With respect to near miss investigations, the Panel believed that BP had "an incomplete picture of process safety performance at its U.S. refineries because BP's process safety management system likely results in underreporting of incidents and near misses." The safety culture survey conducted by the panel suggested that the percentage of employees indicating an impression that workers reported minor process-related events differed between the five refineries. Table 1 lists the percent of employees in each refinery surveyed who disagree or tend to disagree to the survey item: "In general, workers don't bother to report minor process-related incidents, accidents, or near misses."

Table 1. Percentage disagreement that workers don't bother to report minor process-related incidents, accidents or near-misses

Texas City	Carson	Toledo	Whiting	Cherry Point	Average
47	54	56	62	77	57

During the timeframe of the Baker Panel review, a scenario-based study was being conducted at the University of Wisconsin-Madison to understand the link between employee perception of different human factors and employee behavior with respect to near miss reporting. Finding of this study is presented in this paper to show a link between safety climate in the chemical process industry and employees' willingness to report near misses.

2. Near Miss Reporting

The use of near-miss reports to aid in the decision making process is based on the common cause hypothesis which states that serious accidents and near misses have similar underlying causal patterns. Heinrich (1931) claimed that this hypothesis is valid. He estimated that for every 330 potential accidents, one will result in serious or disabling injury, 29 in minor injuries and 300 will lead to 'no injury'. Another such study examined 1,753,498 accident reports from 297 companies and estimated that for every serious or disabling injury, there were 10 minor injuries, 30 incidents of property damage, and 600 incidents with no visible damage (Bird, 1966, Bird & Germain, 1990).

According to Davies et al., (2003), the causes of minor events can provide hints for the causes of major events, if the hypothesis holds true and thus, "detailed scrutiny of minor events is warranted as a method of removing the causes of major incidents" (p.43).

2.1 Under reporting of near misses

While near miss incidents can provide significant information about the deficiencies in the safety system of an organization, under reporting has been discussed as being one of the most common biases (Lakhiani, 2010). In the case of BP, the Baker Panel concluded that "BP's systems were not detecting or reporting all minor process-related incidents, accidents, and near misses…Responses tended to be more negative at Texas City and Toledo, but responses at Whiting, Carson, and to a lesser extent, Cherry Point, also reveal potential reporting deficiencies." (p. 206)

The reporting of a near miss is initiated through identification of an abnormality in the system, followed by a report that alerts authorities to investigate the event and develop feasible solutions to prevent the recurrence of a similar event. Studies have shown that under some circumstances, employees involved in, or witnessing near miss events may not report events for a variety of reasons (van der Schaaf et al., 1991; van der Schaaf, 1995; Clarke, 1998). In the workplace, employees who have an obligation to disclose a near-miss event may feel reluctant to do so if they believe that their involvement was a major contributor for the event or that their reporting may reflect negatively on the organization's safety climate. The task of disclosure of a near-miss event carries certain challenges, which if overcome, would likely increase the reporting efficiency. These challenges include the fear of embarrassment, being ridiculed by coworkers, or the fear of facing retribution, especially if the event is the result of the reporter's own personal errors. As seen in the social psychology literature, one of the highly recognized reasons for not reporting is the 'blame culture' in the organization (Adams and Hartwell, 1977; Webb et al., 1989).

Reason (1997) listed some "disincentives" that affect employees' ability to participate in reporting schemes. These include: extra work, skepticism, natural desire to disregard the occurrence of the incident, lack of trust, and fear of reprisals. Management commitment has also been associated with reporting behavior of employees. Bridges (2000) found that lack of management commitment and lack of follow-through when a near miss is reported were among the reasons for lack of reporting in one of the chemical plants. This was also observed in another study (Powell et al., 1971) where near miss reporting was perceived as useless when no actions were taken based on the reports.

3. Safety Culture/Climate

Safety Culture has been defined as the organizational attribute that reflects the shared attitudes, beliefs, perceptions, and values of the employees in relation to safety (Cox and Cox, 1991) According to Zohar (1980), safety climate is the analytic set of expectations and perceptions that the employees have regarding safety in their organization. While there is still debate about whether safety climate and safety culture are synonymous, safety climate might be considered a safety performance indicator (Guldenmund, 2000).

Although the precise factors determining the safety climate remains unclear, management commitment and worker involvement appear to be the major dimensions (DeJoy, 1994). Other dimensions that have been identified are supervisor support and competence (Seo et al., 2004;

Flin et al., 2000), coworker support and employee participation (Seo et al., 2004), and perception of risk, safety system, and work pressure (Flin et al., 2000).

Actions taken by the top-management relating to the incidents is a reflection of how management views the causes of poor safety performance (DeJoy, 1985). The amount of safety training or meetings, extensive rules and regulations, rewards or punitive enforcements, and other characteristics of the safety system in place can all influence employees perceptions about their workplace safety. Clark (1998) found that the safety attitude of employees affected their reporting behavior. This suggested a link between safety climate and near miss reporting behavior.

4. Near Miss Reporting Behavior Study (2007)

A quasi-experimental study was conducted in 2006-2007 that aimed at understanding the effect of confidentiality of reporting system and perceived severity of potential consequences of near miss events on causal attribution and employees' willingness to report. This study aimed at understanding how employees in the process industry perceive near misses and whether certain attributes associated with near miss events are associated with the near miss reporting behavior, namely, employees' willingness to report. Figure 1 shows the research model that was studied.

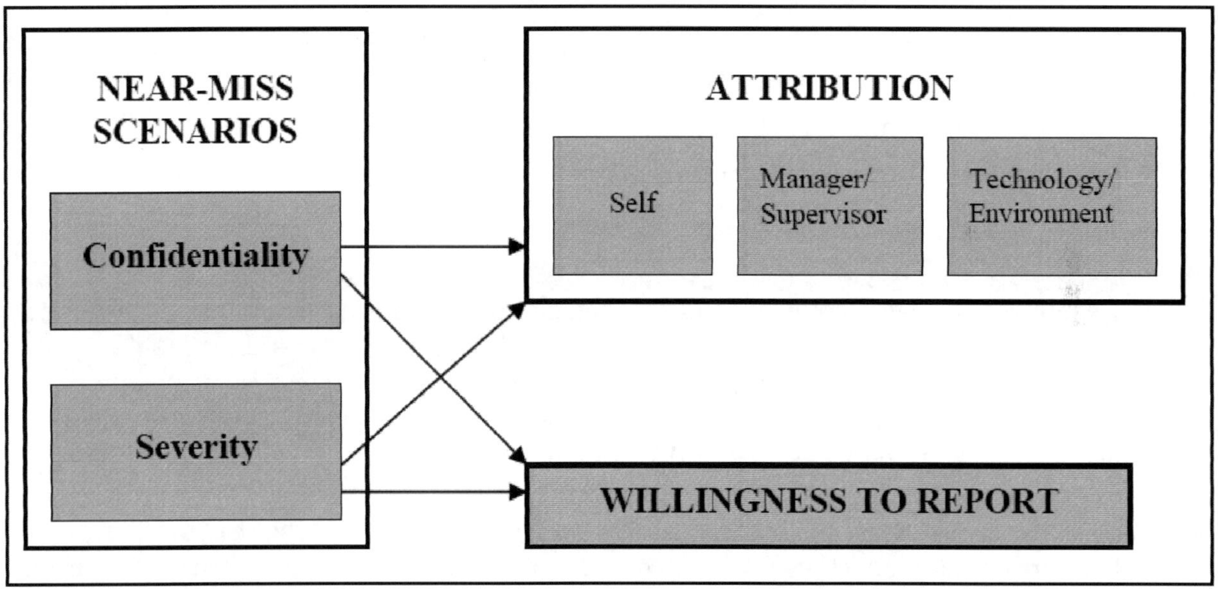

Figure 1. Near miss reporting study – Research model

Literature had reflected potential links between the safety climate and the near miss reporting behavior. Hence, it was added in as a separate factor to test any associations with employees' near miss reporting behavior. This paper only aims at discussing the link between safety climate and employees willingness to report. Results pertaining to the effect of confidentiality on employee willingness to report were published in the proceedings of the American Institute of Chemical Engineers, 2010 Spring National Meeting, 44rd Annual Loss Prevention Symposium (poster session).

4.1 Hypothetical Scenarios

Eight hypothetical scenarios of near miss events were developed, which were realistic to the chemical industry. The potential consequences of the near-miss events were manipulated in the scenarios to reflect differing levels of severity. The subjects were divided into two groups at each plant: one group was informed that the reporting system was confidential; and other group was informed that the reporting system was non-confidential. For each of the scenarios, subjects were asked what they perceived as the level of severity of the potential consequences of the near miss event reflected in the scenario as well as their causal attribution for each scenario.

Overall, 108 employees (93% males) from three chemical plants participated in this study yielding a 94% response rate. The distribution of participants is shown in Table 2.

Table 2. Participant distribution and response rates

Plant	Questionnaires Distributed	Questionnaires Returned	Response Rate
1	45	45	100%
2	34	34	100%
3	36	29	81%
TOTAL	**115**	**108**	**94%**

All three chemical plants were being managed by one company but produced different chemicals. Due to the confidentiality agreement with the participating company, other details about plant identity, location, or chemicals produced will not be discussed.

4.2 Safety climate instrument

Although safety climate was not included as the integral part of the research model, a 14-item safety climate instrument was integrated into the questionnaire that was distributed to the participating employees. The 14-item safety climate questionnaire used was developed by Mendez-Cruz (2005). Employee responses to these items or statements were collected on a 7-point Likert-scale (1-Strongly Disagree and 7-Strongly Agree). Since the sample population for the Mendez-Cruz study was vastly different from the target population for the near miss reporting study, it was suspected that this study would not support the Mendez-Cruz's three-factor structure of the safety climate survey. A principal component factor analysis with Varimax rotation was conducted that divided 13 of the items into three grouped factors: Organizational climate, Supervisor commitment, and Job influence. The grouping is illustrated in Table 3.

Table 3. Safety climate items and groups

Factors	Item Description
Organizational Climate	Top management believes workplace safety and health are very important. (Item #1) Employees are encouraged to become involved in safety matters. (Item #2)

	I feel comfortable reporting safety and health hazards to my supervisor. (Item #4) I feel comfortable reporting work injuries and health effects to my supervisor. (Item #5) The members in my work area support one another. (Item #10)
Supervisor Commitment	My supervisor frequently tells us about hazards in our work. (Item #6) My supervisor emphasizes safety procedures when we are working under pressure. (Item #7) My supervisor discusses how to improve safety with us. (Item #8) My supervisor frequently checks to see if we are obeying the safety rules. (Item #9)
Job Influence	My job duties often interfere with my being able to follow safety rules and procedures. (Item #11) I feel I have control over what happens to my safety at the job. (Item #12) Sometimes it is necessary to ignore safety regulations to get the job done. (Item #13) I have to be more interested in production than safety. (Item #14)
Uncategorized	It is likely that I might be injured in my job in the next 12-month period (Item #3)

Scores for items 3, 11, 13 and 14 were reversed since agreement to the statements reflected negative perception of safety by the participating employees.

4.3 Dependent variable - Willingness to report

Subjects' willingness to report each near-miss scenario was measured using a single item: How willing would you be to report this event? This variable was measured on a 7-point Likert scale.

5. Results

The scores were aggregated for the high and low severity scenarios in order to check for trends in responses between the plants. Results show that the average willingness to report near-miss scenarios was significantly lower in Plant 2 as compared to the other two plants for both high-severity scenarios ($p<0.01$) and low-severity scenarios ($p<0.1$). Figure 2 shows the average scores for employees' willingness to report high and low severity scenarios.

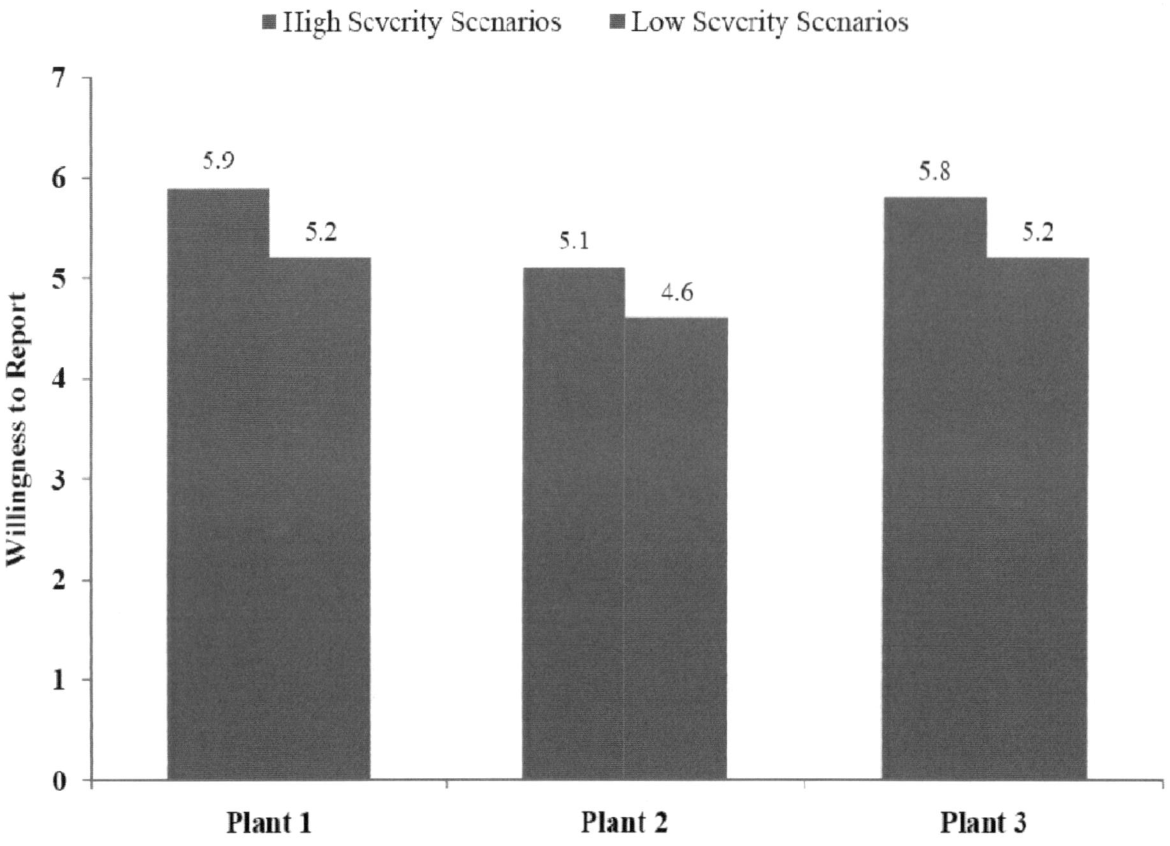

Figure 2. Mean 'willingness to report' score for high and low severity scenarios

In addition, the plants were also significantly different in the overall scores for two of the three safety climate groups: organizational climate and supervisor commitment. It was observed that the employees in plant #2 gave lower scores to items relating to these two groups.

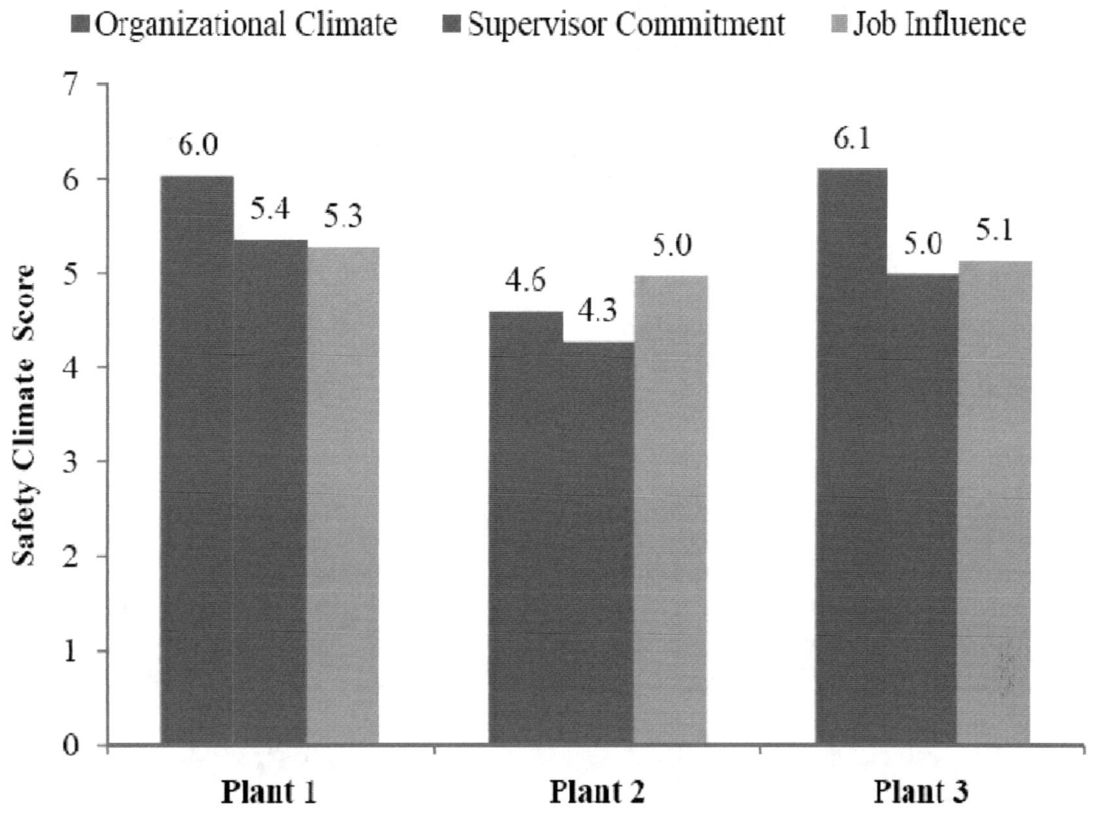

Figure 3. Mean scores for safety climate factors

Due to these patterns, a regression analysis was conducted using the safety climate groups as independent variables and overall mean willingness to report score as the dependent variable. Results showed that only organizational culture was significantly and positively related to willingness to report, $p<0.005$. This meant that subjects who perceived management climate towards safety as positive tended to be more willing to report near miss events.

6. Conclusion

The Baker Panel report concluded that "BP's process safety management system likely results in under reporting of incidents and near misses at BP's five U.S. refineries" (p. 208). Explaining reasons why incidents and near misses were being unreported or undetected, the report states "lack of process safety awareness" and "absence of a trusting environment among employees, managers, and contractors" as being the factors inhibiting the reporting process (p. 207).

A trusting environment between employees and management can be reflected in a safety climate survey aimed towards knowing employees' perception about their management, such as the one included in this study. In this study it was observed that participants from one of the three plants rated their organizational safety climate to be low as compared to the other two plants and showed similar trends in their willingness to report near miss events. Although this study was done independent of the Baker Panel's investigation, the results of this study show a correlation between employees' willingness to report and their perception of organizational factors, such as their perception of top management commitment, support of peers, and their comfort in reporting issues to their supervisors.

7. References

[1] Adams, N. L., & Hartwell, N. M. (1977). Accident reporting systems: a basic problem area in industrial society. Journal of Occupational Psychology, 50, 285-298.

[2] Bird, F. E. (1966). Damage Control. Philadelphia: Insurance Company of North America.

[3] Bird, F. E., & Germain, G. L. (1990). Practical Loss Control Leadership. Loganville, Georgia: Institute Publishing (A division of International Loss Control Institute).

[4] Bridges, W. G. (2000). Get near misses reported- process industry incidents: investigation protocols, case histories, lessons learned. In: Proceedings of the Center for Chemical Process Safety International Conference and Workshop, American Institute of Chemical Engineers, New York, 379-400.

[5] Clarke, S. (1998). Organizational factors affecting incident reporting of train drivers. Work and Stress, 12(1), 6-16.

[6] Cox, S., Cox, T. (1991). The structure of employee attitudes to safety: a European example. Work and Stress, 5(2), 93-106.

[7] Davies, J., Ross, A., Wallace, B., & Wright, L. (2003). Safety Management: A Qualitative Systems Approach: Taylor & Francis.

[8] DeJoy, D. M. (1985). Attributional process and hazard control management in industry. Journal of Safety Research, 16, 61-71.

[9] DeJoy, D. M. (1994). Managing safety in the workplace: An attribution theory analysis and model. Journal of Safety Research, 25(1), 3.

[10] Flin, R., K. Mearns, et al. (2000). Measuring safety climate: identifying the common features. Safety Science 34: 177-192.

[11] Guldenmund, F. W. (2000). Nature of safety culture: a review of theory and research. Safety Science, 34, 215-257.

[12] Lakhiani, S. D. (2010). Effect of confidentiality of reporting system on employee perception and their willingness to report near-miss. American Institute of Chemical Engineers, 2010 Spring National Meeting, 44rd Annual Loss Prevention Symposium, San Antonio, TX, March 22–24, 2010.

[13] Powell, P. I., Hale, M., Martin, J., & Simon, M. (1971). 2000 Accidents: A Shop Floor Study of their Causes (No. 21). London: National Institute of Industrial Psychology.

[14] Reason, J. T. (1997). Managing the Risks of Organizational Accidents: Ashgate Publishing Limited.

[15] Seo, D. C., M. R. Torabi, et al. (2004). A cross-validation of safety climate scale using confirmatory factor analytic approach. Journal of Safety Research 35: 427-445.

[16] van der Schaaf, T. W., Lucas, D. A., & Hale, A. R. (Eds.). (1991). Near Miss Reporting as a Safety Tool. Oxford U.K.: Butterworth-Heineman.

[17] van der Schaaf, T. W. (1995). Near Miss Reporting in the Chemical Process Industry - an Overview. Microelectronics and Reliability, 35(9-10), 1233-1243.

[18] Webb, G. R., Redman, S., Wilkinson, C., & Sanson-Fisher, R. W. (1989). Filtering effects in reporting work injuries. Accident Analysis and Prevention, 21, 115-123.

[19] Zohar, D. (1980). Safety climate in industrial organizations: Theoretical and applied implications. Journal of Applied Psychology, 65, 96-102.

Optimizing a MI program using Risk Based Inspection

Tony Poulassichidis
Hess Corporation
500 Dallas street, Houston, TX, 77002
tpoulassichidis@hess.com

Keywords: risk based inspection (RBI), mechanical integrity, asset management, risk

Abstract

The Risk Based Inspection (RBI) process delivers a calculated relative risk value for each fixed equipment item in a facility. Fixed equipment risk values are the product of the Likelihood-of-Failure (LOF) and the Consequence-of-Failure (COF) factors. The calculated equipment risk values are used to optimize the facility inspection program in accordance with industry best practices (API-580/2009, API-581/2008).

Using the RBI calculated risk values, a fixed equipment relative risk ranking is established that strengthens the facility mechanical integrity program by allowing the operator to:

–set priorities and focus their attention on the critical areas
–justify capital investment for lifetime extension projects
–proactively address loss of primary containment and process safety issues

This paper will present the actual mechanical integrity gains realized from successfully implementing the RBI analysis at a gas plant facility. Examples of inspection findings due to the RBI plan recommendations are provided. Emphasis is placed on the accurate RBI prediction of hydrogen-induced damage for the product storage tanks, as the follow-up inspection confirmed.

Specific examples of the gas plant case study are offered as "lessons learned", illustrating potential RBI analysis pitfalls when required process steps are not followed.

Risk target selection is an important part of the RBI analysis. Based on the gas plant case study, and in particular the storage tanks findings, this paper will:

- present RBI as a quantitative data analysis tool, compliant with industry best practices for fixed equipment risk management
- attest for the RBI's ability to strengthen a facility's mechanical integrity program
- identify potential pitfalls and what might cause RBI process breakdown
- review the risk target selection criteria and their significance.

1. Introduction

RBI is a well established methodology within the oil & gas industry. It allows inspection activities to be focused on higher-risk areas, usually resulting in an overall decrease in inspection program costs, while maintaining an acceptable level of risk. API-RP580 and API-RB581 RBI analyses focus only on fixed process equipment (vessels, piping, and tanks). RBI can successfully replace the time-based criteria for inspection activities that have been used across industry facilities in the past.

The RBI methodology consists of a detailed inspection, corrosion, materials and process analysis that aims to identify equipment risk and propose a risk-prioritized inspection plan for optimized equipment maintenance.

The recommended amount and type of inspection activity is not the same for all equipment items that exceed acceptable risk levels. Recommended inspection activity is a reflection of the effort that a piece of equipment requires as a result of the risk it poses to safe and efficient facility operation (process safety, environmental issues, production interruption, repair costs).

The present paper provides updated information for the Gas Processing Plant (GPP) RBI study that was presented at a previous conference [(1)]. Examples for the GPP gains from the RBI implementation are presented. In addition, three main categories of potential RBI implementation pitfalls are presented along with their respective mitigating actions.

2. RBI technology

2.1 Risk calculations

In RBI analysis the relative equipment risk value determines the inspection plan required to mitigate risks. High risk equipment becomes the focus of finite inspection resources. Equipment risk is defined as the product of Likelihood of Failure (LOF) and Consequence of Failure (COF).

$$\textbf{Risk} = \textbf{LOF} \times \textbf{COF} \qquad \textbf{[Eq. 1]}$$

The Likelihood of Failure (LOF) is a function of a Generic Failure Frequency, a Management System Factor (MSF), and a Total Damage Factor (TDF). The Generic Failure Frequency (GFF) is defined as the industry average failure rate for each equipment type (e.g. vessel, heat exchanger, tank). Management Systems related to the mechanical integrity of equipment are also evaluated. The Management System Factor (MSF) adjusts for the influence of the facility's management system on the mechanical integrity of the plant. The Total Damage Factor (TDF) is an expression of how much more or less likely a particular plant equipment item is to fail than the industry average. The product of GFF, MSF, and TDF is the equipment LOF.

$$\textbf{LOF} = \textbf{GFF} \times \textbf{MSF} \times \textbf{TDF} \qquad \textbf{[Eq. 2]}$$

The TDF factor is defined as the sum of a) external damage factor (EDF), b) internal damage factor (IDF), c) cracking damage factor (CDF) and d) mechanical damage factor (MDF). The Total Damage Factor (TDF) is an expression of active corrosion damage mechanisms for the equipment (equation 3). The Internal Damage Factor and the External Damage Factors reflect the relative level of concern about the equipment susceptibility to internal thinning corrosion and external corrosion.

$$\mathbf{TDF = EDF + IDF + CDF + MDF} \qquad \textbf{[Eq. 3]}$$

In equation (1), the LOF can be expressed as a probability (events /year) while the COF can be expressed as the area impacted by an event (ft^2/event). The resulting risk, from equation (1), is then expressed in ft^2/year.

Software programs are commonly used for analyzing data inputs and calculating equipment damage factors and risk values. The API-RBI software (version 8.03) was used for this work. Equipment that has higher than acceptable risk value or damage factor (defined by the user and agreed by management) will be an inspection candidate.

In a risk-based approach, inspections provide updated information as to whether actual corrosion degradation is as predicted, and the extent of equipment damage that is expected in the future. Recognizing the potential for failure, and specifying the right inspection method at the proper equipment location, results in a reduction of the TDF. The underlying implicit assumption is that in a competent organization inspection findings will be followed by proper action that will actually reduce the probability of equipment failure. An action plan may include one or a combination of the following activities:

- follow-up inspection,
- equipment monitoring,
- use of upgraded materials,
- equipment replacement,
- operational procedure changes,
- instrumentation upgrades.

Consequence of failure is not impacted from inspection since loss of containment will have the same consequences regardless of inspection activity. However, improvements in detection, isolation and mitigation systems may reduce potential consequence.

RBI analysis optimizes a fixed equipment inspection program by:

- "right sizing" the amount of inspection - usually the amount of inspection needed based on risk criteria is more specific to the predicted damage mechanism and is often less than the amount of inspections required using time based criteria
- establishing optimal economic levels of inspection as weighed against risk reduction.

The recommended amount and type of inspection activity isn't the same for all equipment items that exceed acceptable risk levels. Specified inspection activity is a reflection of the respect that a piece of equipment deserves as a result of the threat it poses for facility operation (process safety, environmental issues, production interruption, repair costs etc.).

Since the mid-90's several companies within the oil and chemical industry have reported significant gains associated with RBI implementation and subsequent fixed equipment inspection program optimization as described above. Maximum benefits can be extracted from an RBI analysis when it serves as the foundation for the plant's mechanical integrity program.

A case study based on the RBI implementation at a gas processing plant was presented two years ago and is used as the foundation for this paper. While revisiting the main points of the GPP RBI analysis, this paper focuses on the experience and learnings from the inspection plan implementation.

3. Gas Processing Plant – case study

The Gas Processing Plant (GPP) inlet gas consists mainly of light hydrocarbons with CO_2, and about 0.5% H_2S. The GPP utilizes a four column Ryan Holmes process to recover and sweeten CO_2. Natural gas, NGL, and sulfur are also recovered. The recovered CO_2 is returned to the field for enhanced oil recovery.

An overview of the GPP main process steps is provided below:

3.1 Process description

3.1.1 Inlet compression and dehydration

The produced gas is dehydrated in counter-flow with glycol solution. The glycol contactor is the second process vessel for the produced gas entering the GPP. The dry gas is then compressed using a combination of centrifugal and reciprocating compressors and further dehydrated to assure the water dewpoint meets the Ryan-Holmes Process requirements.

3.1.2.The Ryan Holmes process

The GPP is a four column Ryan Holmes Process. The four columns are the Ethane Recovery Column (ERC), the Additive Recovery Column (ARC), the CO_2 Recovery Column (CRC) and the Demethanizer. Industry references describing the Ryan-Holmes process in detail are available [(2)].

3.1.3. Liquids and sulfur treating

The overhead gases of the ARC are processed with a combination of Methyl Diethanolamine (MDEA) and Diethanol amine (DEA) (two separate amine units) in order to remove the H_2S and CO_2 from the NGLs that flow from the ARC. MDEA is used to capture H_2S to send it to the sulfur recovery units. The Sulfur Recovery System is composed of two independent Claus Sulfur Recovery Units (SRU's). Each Claus unit is provided with Cold Bed Absorbers to further minimize undesirable emissions. The NGLs are then treated with DEA and caustic in order to react with remaining carbonyl sulfide (COS) and to remove the last H_2S and CO_2 which is recycled to the GPP inlet.

4. GPP Corrosion review

The GPP was commissioned during the mid 80's and has had a successful operating record for the last twenty years. A critical step in an RBI project is the analysis of applicable corrosion mechanisms and assignment of appropriate inputs in the RBI model. The following is a summary of the GPP corrosion review.

4.1. Internal and external corrosion

Measured corrosion rates for all pressure vessels were available and retrieved from the extensive GPP inspection history files. During the last twenty-five years of operation there were several plant-wide turnarounds with inspection activities covering all fixed-equipment. The measured corrosion rates were aligned with industry knowledge and experience for this type of process.

The majority of GPP process areas experience low internal thinning corrosion rates. The two amine units (MDEA and DEA) along with specific equipment of the sulfur recovery units are the only areas with noted internal thinning corrosion issues.

In the GPP process the glycol contactor is positioned immediately downstream of the inlet gas separators. Therefore the inlet gas stream is dehydrated to a -50°F (-45°C) water dewpoint immediately after the plant inlet. Dehydration of the inlet gas stream acts as an effective corrosion control measure resulting in a non-corrosive stream that is processed further in the plant's compression and Ryan-Holmes sections. The inlet separators, glycol contactors and associated piping are either stainless steel or stainless clad equipment.

Corrosion in glycol units is primarily caused by the formation of formic and other organic acids resulting from glycol degradation[(3)]. In GPP's case the two glycol regeneration units are not experiencing high corrosion rates due to a) design choices for material of construction and b) operating discipline. The rich glycol stream is processed in a series of stainless steel equipment. The overhead of the glycol stripper including the condensers and overhead accumulator are stainless steel equipment. The lean glycol process stream is usually considered non-corrosive. An exception would be lean glycol with low pH. Additional GPP glycol unit corrosion control elements include the amine injection (maintain neutral pH solution) and use of triethylene glycol at operating temperature lower than 375°F (190°C) (avoiding glycol degradation products). The lean glycol equipment is carbon steel construction and has operated without issues since plant start-up.

The sulfur recovery units (SRU) have carbon steel equipment that operates at elevated temperatures above 500°F (260°C). The sulfur content of the process stream combined with the elevated temperatures and carbon steel material of construction results in increased corrosion rates for segments of the SRU's. In order to eliminate the sulfidation corrosion issues, two of the exchanger channel heads have been upgraded to stainless steel material.

The amine corrosion is a well documented mechanism that is affecting carbon steel equipment in amine service. Corrosion isn't caused by the amine itself, but results from dissolved acid gases (CO_2 and H_2S) amine degradation products, heat stable amine salts and other contaminants[(4)].

The GPP process includes a MDEA and a DEA unit. The amine stripper column, the amine stripper reboiler, the overhead amine stripper area, the lean/rich amine exchanger, the piping between the lean/rich amine exchanger and the stripper were among the areas of interest identified during the unit corrosion review. Inspection history, including recent work, confirms our predictions for corrosion problems occurring in the above unit equipment. Proper mitigation actions are currently implemented for the amine units.

External corrosion susceptibility was calculated by the RBI software and these results were in agreement with plant experience. The GPP is in a dry inland environment that isn't conducive to external corrosion. In addition, atmospheric barriers (insulation and coatings), are well maintained.

4.2. Environmental assisted cracking

Hydrogen degradation mechanisms, chloride stress cracking and alkaline amine stress cracking were the main environmental-assisted cracking mechanisms predicted for GPP process. The following is a summary of that review.

The inlet gas stream from the production field contains chlorides. Since the inlet separators and glycol contactors are stainless equipment, the possibility of chloride stress corrosion cracking (ClSCC) was evaluated for the plant inlet equipment. Based on their operating temperature of approximately 75°F these pressure vessels were not considered susceptible to ClSCC. Industry references list a minimum temperature of 100°F (38°C) for activating ClSCC mechanism. Actually API-581[(5)] uses the 100°F (38°C) threshold while API-571[(6)] states that "cracking usually occurs at metal temperatures above about 140°F (60°C) although exceptions can be found at lower temperatures". In addition to operating below minimum threshold temperature, during past internal inspections of these stainless vessels there was no evidence of ClSCC initiation.

Stainless steel equipment is used for processing rich glycol streams within GPP's two identical glycol units. Laboratory analysis of the glycol solution recorded chloride concentrations of 500mg/l to 1500mg/l respectively for the lean and rich glycol. The rich glycol stream operating temperature is between a minimum of 100°F (38°C) at unit inlet filters and a maximum of 339°F (170°C) at rich/lean glycol exchanger outlet. Therefore ClSCC is considered an applicable failure mechanism and susceptibility is assigned per API-581 guidelines.

During GPP service life, internal inspections utilizing dye penetrant techniques have been recorded for one of the two glycol units without any indications having been found. In addition to susceptible material and the presence of chlorides, sufficient stress levels must be present for ClSCC to occur. The absence of ClSCC inspection findings at GPP's glycol units could be explained by the low operating pressures within the glycol units and the resulting low operating stress levels for the vessels in question. The RBI analysis recommended an additional round of ClSCC inspection for the glycol unit's stainless vessel.

Amine Stress Corrosion cracking (ASCC) affects carbon steel equipment in aqueous alkaloamine systems used to remove/absorb H_2S and/or CO_2 and their mixtures from various gas and liquid hydrocarbon streams. In a NACE 1995 paper(7), it is mentioned as being "the most common cracking mechanism in gas treating plants". ASCC typically occurs in lean amine streams, while it is not likely in rich amine streams and does not occur in fresh amine solutions. GPP has a DEA and a MDEA amine unit, which are not considered the type of amines that aggressively promote ASCC. "Cracking is more likely to occur in lean MEA and DEA services but is also found in most amines including MDEA and DIPA (ADIP)."(8) Operating temperature is a parameter considered in assigning susceptibility to ASCC for carbon steel. Following API-945 and API-581 recommendations, equipment operating a) in lean DEA service below 140°F (60°C) and b) in lean MDEA service below 180°F (82°C) was not considered susceptible to ASCC. In addition, post weld heat treated equipment was not considered susceptible to ASCC regardless of operating temperature or amine type. A list of ASCC susceptible vessels within the GPP amine units was generated using the above inputs in the RBI model and appropriate inspection plans were developed.

There are three distinct hydrogen related cracking mechanisms applicable to carbon steel equipment in gas plant process and amine treating units within gas plants. Sulfide Stress Cracking (SSC), Hydrogen Induced Cracking (HIC) and Stress Oriented Hydrogen Induced Cracking (SOHIC) are identified and well documented applicable corrosion cracking mechanisms for these processes.

The GPP inlet gas stream is dehydrated as it flows through a series of three stainless steel vessels with the glycol contactor being the third vessel. Stainless equipment in the operating envelope of this GPP is not susceptible to hydrogen related cracking mechanisms. The dehydrated gas stream is then processed by carbon steel equipment and, due to lack of water content, it is not able to promote any of the hydrogen related corrosion cracking mechanisms. Therefore, until the stream is processed in the amine units, hydrogen environmental cracking isn't an applicable degradation mechanism. As a precautionary measure accounting for possible glycol process upsets over the plant life, we did assign HIC/SOHIC potential for the first vessel immediately downstream of the glycol contactor even though previous inspection history did not detect it.

Hydrogen assisted cracking mechanism were considered for all carbon steel pressure vessels in the rich amine streams of both GPP amine units. Because all of these vessels are post-weld heat treated, SSC was not considered as an applicable cracking mechanism.

According to industry references, the amine absorber (particularly the bottom side), amine regenerator (particularly the top side), the amine regenerator overhead system (condensers, accumulator), the exchanger rich amine side and flash separators are prone to degradation by HIC/SOHIC within the amine units. Susceptibility to HIC/SOHIC for each of the above-mentioned amine unit equipment was assigned in accordance with API-581 guidelines. In addition to relevant equipment design and operating conditions, parameters that are considered for assigning HIC/SOHIC susceptibility in API-581 include: H_2S content in water, solution pH, carbon steel grade, sulfur content of steel plate, heat treatment, and prior inspection findings.

5. RBI implementation

Software programs are commonly used for analyzing data inputs and calculating equipment damage factors and risk values. The API-RBI software (version 8.03) was used for this work.

5.1. Corrosion Control Diagrams

Development of Corrosion Control Diagrams (CCD) is a required RBI project deliverable per company guidelines. A CCD is essentially a marked-up version of the respective process flow diagram (PFD). The marked-up information includes process stream operating conditions (temperature, pressure) and fluid composition. In addition information regarding corrosion damage mechanisms, material of construction, use of cladding, post-weld heat treatment, operating limits (e.g. pH range) and significant prior inspection findings are also noted on CCDs.

The CCD serves many purposes. First, comments relevant to equipment performance, and/or assumptions used for assigning susceptibility to corrosion damage mechanisms are noted on the CCDs. For example it is appropriate to note that a process stream may contain hydrogen sulfide during a process upset. The above-mentioned process and corrosion related information noted on CCDs are basically inputs for the RBI model.

Second, the CCD can serve as a RBI data quality control tool. Critical inputs that are used in the RBI model are easily reviewed using the CCD and can be updated as needed. It is not necessary for operations personnel to understand the RBI model or use the software to offer feedback.

Third, besides serving as a repository of process and corrosion related information the CCD should contain RBI analysis outputs for select equipment. The risk profile for equipment that requires follow-up risk mitigation plan, along with information on probable corrosion damage mechanisms is noted on the CCD.

Fourth, plant personnel can then use these drawings as indoctrination to the unit and potential concern areas.

Finally, the CCD serves as the starting point for the annual RBI reassessment process.

Appendix I is an example of the MDEA amine unit corrosion control diagram.

5.2. Risk driver analysis

The complete scope of the GPP RBI analysis covered approximately eight-hundred sixty three vessels and piping circuits. In order to effectively manage fixed equipment risk an understanding of risk origin, its distribution among process units and ability to project future risk value is needed. The RBI analysis offers the ability to understand fixed-equipment risk.

The API-RBI database is a repository for information relevant to fixed equipment health. It contains data inputs regarding equipment design information, process conditions, equipment inspection history and equipment condition. In addition, it is an intelligent database that utilizes industry recognized best engineering practices for calculations necessary to risk management. For example, it provides calculated output values for equipment risk, equipment LOF drivers, COF drivers, and risk projections for the evaluation period. Using these software calculated outputs, it can be determined if equipment risk is unacceptable and what are the risk drivers. Is the risk for a particular equipment item consequence driven, likelihood driven or both risk dimensions are responsible?

Table 1 shows the calculated damage subfactors IDF and CDF the Total Damage factor, the risk components (LOF, COF) and the risk value extracted from the RBI software for several higher risk equipment items. Information similar to Table 1 is available for all plant fixed equipment items and it is sorted in descending risk value order. A risk driver analysis is necessary for the equipment items that exceed acceptable risk criteria.

In equation (1), the LOF can be expressed as probability of events per year while the consequence of failure can be expressed as area impacted from event (ft^2/event). The resulting risk units, from equation (1), are then expressed in ft^2/year. The Total Damage Factor (TDF) is an expression of active corrosion damage mechanisms for the equipment (equation 3). The Internal Damage Factor and the Cracking Damage Factors reflect the relative level of concern about the equipment susceptibility to internal thinning corrosion and environmental cracking degradation. The EDF and the MDF had zero values for all equipment items listed below in Table 1. For this work, a value of one was used for the Management System Factor, while the default API-RBI software Generic Failure Frequencies (GFF) were used.

As mentioned earlier, Total Damage Factor (TDF) is the sum of internal, external, cracking and mechanical damage factors (Equation 3). Table 1, lists the TDF, internal and cracking damage factor along with the LOF, COF and risk values for several high-risk equipment items. From Table 1, Vessel-2 has a TDF value that equals the cracking damage factor. Therefore Vessel-2 is a high-risk vessel with its LOF mainly driven by susceptibility to environmental cracking mechanisms. Similarly, RBI software calculations for Vessel-1 show that the TDF is driven mainly from internal corrosion issues with minor contribution from a susceptibility to environmental cracking mechanisms.

Development of the appropriate risk mitigation measures is based on the understanding of the above risk driver analysis. Based on the risk-driver analysis, a set of detailed inspection recommendations specific to anticipated damage mechanisms is provided for all fixed equipment.

The risk driver analysis may identify that inspection alone can not effectively manage risk for a particular equipment item and alternative or additional measures are necessary. For example, in the case of a high risk pressure vessel that has consequence as the main driver, one will need to consider the following:

- isolation devices (e.g. remote operated valves, automatic isolation devices),
- detection systems (e.g. hydrocarbon sensors, cameras),
- process design changes (e.g. reactants, temperature, pressure, catalyst),
- equipment location within facility or mitigation system enhancements (e.g. deluge system, secondary containment).

Replacement with a brand new pressure vessel will not reduce a consequence driven risk value as it would with a likelihood driven risk.

Table 1. Equipment risk driver values

Equipment ID	Component Description	Equipment type	Internal Damage Factor (IDF)	Cracking Damage Factor (CDF)	Total Damage Factor (TDF)	Likelihood-of- Failure (LOF) (events/year)	Consequence of Failure (COF) (ft^2/event)	Risk (ft^2/yr)
Exchanger 1	Amine unit exchanger 1- shell side	Exchanger	0	4705.5	4705.5	14 10^{-2}	11,825.9	1702.8
Exchanger 2	Amine unit exchanger 2 – tube side	Exchanger -channel	0	1057.8	1057.8	3.2 10^{-2}	28,528.8	923.5
Pipe 1	Amine unit pipe circuit 6”	Pipe-6	672.8	0	672.8	2 10^{-2}	39,089.5	804.8
Pipe 2	Amine unit pipe circuit 6”	Pipe-6	465.9	0	465.9	1.4 10^{-2}	33,398.6	476.2
Vessel 1	Amine unit vessel - bottom side	Column-Btm	5000	27.9	5027.9	15.3 10^{-2}	2,884.2	443.8
Pipe 3	Amine unit pipe circuit 2”	Pipe-2	4269.4	0	4269.4	13 10^{-2}	3,133.2	409.3
Pipe 4	Amine unit pipe circuit 6”	Pipe-6	5000.0	0	5000.0	15.3 10^{-2}	2,251.3	344.4
Pipe 5	Amine unit pipe circuit 6”	Pipe-6	236.5	0	236.4	0.7 10^{-2}	33,414.6	241.8
Vessel 2	Amine unit vessel	Vessel	0	128.8	128.8	0.3 10^{-2}	44,679.2	176.1
Vessel 3	Process vessel	Vessel	0	470.5	470.5	1.4 10^{-2}	4,305.5	62.0
Vessel 4	Amine unit vessel	Vessel	188	28	215.8	0.6 10^{-2}	2,636.5	17.4

5.3. Inspection plan

The inspection plan is one of the main RBI project deliverables. Based on the risk-driver analysis a set of detailed inspection recommendations specific to the anticipated damage mechanism is provided for all GPP fixed-equipment. The inspection plan covers the types of probable failure modes, equipment areas to look for probable failure modes, the recommended type of inspections, and the extent of inspection coverage. Non-intrusive, on-stream inspection recommendations can be offered as an alternative to internal inspection.

For example, inspection for Vessel-3 should focus in assessing environmental assisted cracking damage. Welds and their heat–affected zones are areas that cracks may appear in a deaerator. In particular the head-to-shell circumferential weld and welds in the liquid zone are considered higher potential for crack initiation [(9)]. The extent of weld inspection and therefore the associated inspection cost is based on the equipment risk value. Since the deaerator is among the highest plant equipment risk contributors, extensive inspection coverage is justified.

6. Inspection findings

Several equipment inspections were completed during the year 2009, following the RBI plan recommendations. Implementation of the recommended inspection plan has strengthened the GPP mechanical integrity program and confirmed the need to ensure that the RBI process is closely followed. The following discussion will focus on the lessons learned from two examples, a success and a failure of the RBI inspection plan. In both cases, the outcome is directly related with the degree of adherence to the RBI methodology.

6.1. NGL storage tanks

Five identical bullet type NGL product storage drums are in service at the GPP. These drums containing light hydrocarbon product, are carbon steel construction, approximately 115 ft long and 11 ft inside diameter, operating at 170psig, and 100°F. The GPP process is designed to remove water and hydrogen sulfide from the NGL product prior to storage in these drums. The above information reflecting the GPP design was entered in the RBI software and documented in the corrosion control diagrams. In absence of water and hydrogen sulfide, hydrocarbon product in ambient temperature is not considered corrosive to carbon steel.

The RBI process is a multi-discipline, cross functional exercise that requires inputs and an organized interaction among plant specialist resources. The RBI specialist develops the initial corrosion control diagrams by extracting and documenting the existing in various plant files knowledge. The next RBI process step requires a review of the corrosion control diagram with appropriate plant personnel to validate and update as necessary the documented information. The updated information is then stored in the RBI software database and used for the risk calculations.

During this plant review, it was pointed out that both water and hydrogen sulfide can be present in the storage drums as the result of process upset conditions. Process upsets result in NGL product that does not conform to specifications. The storage drums are utilized to blend the out of specification product back in the GPP process stream for reprocessing.

Carbon steel equipment processing hydrocarbons in presence of hydrogen sulfide and moisture, is susceptible to hydrogen induced environmental degradation mechanisms. Therefore, these storage drums were assigned a susceptibility to hydrogen induced environmental damage and combined with their calculated consequence of failure, their calculated risk resulted in RBI recommendations for an inspection.

Inspections revealed hydrogen blistering and environmental degradation mechanisms. As a follow up of a fit-for-service engineering analysis, per API-579, it was decided to retire two of the five storage drums. The storage drum internal surface was coated as a mitigation measure to prevent further hydrogen induced damage. Coating acts as a barrier that stops the corrosion reaction of hydrogen sulfide with the steel surface. Without hydrogen sulfide reacting with the steel substrate, there is no free hydrogen produced and therefore the possibility for hydrogen induced damage of the steel surface is eliminated. The above mitigation measures are complimented with an inspection monitoring program that is designed to ensure continued integrity of the coating system and the carbon steel drums.

6.2. Inlet surge drums

The GPP has three inlet surge drums, two were part of the original plant construction and are carbon steel construction with 316L grade stainless cladding .The third was placed in service seven years ago is 304L stainless steel construction.. The surge drums operate under similar conditions, processing the same feed streams from production.

The RBI review for these vessels focused on the potential for internal thinning and chloride stress cracking. As described in section 4.2 of this paper, the inlet surge drums operating temperature of 75°F is below the minimum necessary temperature threshold for chloride stress cracking initiation. According to API standards, below operating temperatures of 100°F stainless steel is not susceptible to chloride stress cracking.

There were no findings related to corrosion thinning or chloride crack initiation sites in the extensive inspection files that document the performance records of these surge drums. As a result, during the plant review of the RBI model, it was concluded that the equipment was suitable for service within the GPP operating envelope and had a low probability for an active corrosion mechanism. Inspection deferral is justified in the above-described scenario, but since the 304L construction drum was recently placed in service an internal inspection was performed. It is good practice and necessary to validate the theoretical RBI model with actual field experience data obtained from an initial vessel inspection. There was extensive inspection history, over 25 service years worth, for the drums with 316L cladding but no inspection data on the recently installed 304L drum.

The 304L drum internal inspection identified pitting corrosion in numerous locations throughout the bottom vessel section, the normal liquid level area. Microbiologically induced corrosion is considered as responsible for the observed pitting damage. This was not predicted by the RBI because no Production personnel were involved. The 316 cladded drums had also pitting corrosion but the pit depth was lower than what was found in the 304L drum.

There was no need to weld repair, and the 304L inlet drum is fit for continued service with proper mitigation measures to ensure its long-term integrity. The source of microbes in the system is under investigation and recent process changes in the upstream production field associated with the GPP are currently being reviewed. Bacteria count samples are collected in additional production locations aiming to track their source in the feed stream. Chemical treatment program changes are planned to eliminate the bacteria in the system and effectively control the corrosion.

The pitting corrosion resistance of the stainless steels is dependent on their alloying composition, with chromium, molybdenum and nitrogen being the main elements used to strengthen austenitic grade steel resistance to pitting. A Pitting-Resistance Equivalent Number (PREN) [(10)] can be used to predict the pitting resistance of an alloy. The PREN is given by an equation:

$$\mathbf{PREN = \%Cr + 3.3\%\ Mo + 16\%\ N} \qquad \textbf{[Eq. 4]}$$

The larger the numerical value of PREN the higher the expected resistance of the alloy. For example, the nominal composition of the 304L grade results in a PREN value of nineteen while the 316L PREN value is higher than twenty-five. These PREN values can explain the observed pit depth differences between the 316L and the 304L construction inlet drums.

6.3. Lessons learned

The above two examples provide background information on instances where RBI successfully predicted corrosion, and where it didn't. . A failure of RBI to successfully predict equipment condition can be usually traced back to one or more of three category pitfalls. This section provides a brief review of these three main failure causes. In addition, we will tie both abovementioned GPP examples into the appropriate category of RBI implementation pitfalls.

The three main RBI implementation project pitfalls, listed in order of increasing difficulty to address, are a) data entry quality, b) inspection recommendation and c) operating envelope review.

A. Data Entry Quality Each equipment item in a RBI database requires a minimum of twenty separate entries and depending on the type of active corrosion damage mechanisms the number of entries can exceed thirty parameters. These entries range from fabrication information such as thickness to operating conditions such as temperature or pressure. Information is collected from various sources including U1 forms, equipment drawings, process flow diagrams and electronic databases. Simple typing errors such as transposing numbers are possible. Interpretation errors are common and directly related to the experience level of the RBI analyst entering the values in the database. Potential pitfalls in this category can be easily handled with appropriate quality checks performed by personnel experienced in utilizing the RBI software build-in query capabilities.

B. Inspection Recommendation A second category of RBI pitfalls is related to the inspection plan recommendations. It is possible that a pressure vessel or piping system has been properly evaluated and risk assessed but the follow-up inspection recommendation is incorrect. The recommendation error can be regarding the type of inspection method proposed, the location within the equipment and/or the extent of the inspection coverage. Examples of incorrect inspection recommendation follow-up to the RBI analysis are provided for clarity:

- Using wet fluorescent magnetic particle testing (WFMT) on stainless equipment for crack detection is an example of mismatch between inspection technology and application. The WFMT technology is used for crack detection in ferromagnetic materials, while for stainless equipment liquid penetrant is the appropriate inspection choice.
- If an inspection is recommended for corrosion under insulation, then checking at nozzles protrusions through the insulation is a suggested location within the equipment. Therefore, inspecting the least susceptible area of a vessel (i.e. shell under good condition insulation) can result in misleading evaluation.
- The extent of inspection coverage needed to reveal potential localized thinning requires more than couple of ultrasonic thickness readings on a pipe elbow. Obtaining an x-ray of the pipe elbow is an appropriate inspection method to ensure that we have a clear understanding of its condition.

The effectiveness of an RBI analysis to reveal mechanical integrity improvement opportunities in a facility is traced to the quality of the operating envelope review. It is critical for a successful review to engage all plant disciplines and appropriate engineering resources (e.g. process engineer, inspector, operations and maintenance representatives, plant management). The starting point for the review is the plant design information that includes the operating description and materials of construction. Over years, continuous improvement efforts and/or feed stream changes can result in changes to process stream composition, operating parameters (i.e. temperature, pressure) even material upgrades and equipment changes. Updated information should be documented in plant files. In addition, process upsets can result in a substantially different equipment operating envelope that must be incorporated in the RBI analysis and evaluated properly.

C. Operating Envelop Review The plant review serves as a quality assurance measure, ensuring that the RBI analysis considers current facility information including abnormal operations. Involving all appropriate personnel and having a review under the guidance of an experienced corrosion specialist, able to understand the various possible damage mechanisms that can be active as the process parameters change, is critical for the RBI analysis success.

The hydrogen induced damage potential for the GPP's storage drums was identified as a result of the plant review. Based on given information the storage drums were supposed to process NGL product. Carbon steel is not susceptible to corrosion attack from light hydrocarbons in the given temperature range. Because the upstream equipment is processing hydrocarbons containing hydrogen sulfide, the question was raised regarding potential presence in the storage drums. During the plant operating envelope review meeting it was pointed out that process upsets result in out of specification product containing hydrogen sulfide to be stored in these drums and blended back in the process stream.

The failure to predict the potential for the inlet surge drum pitting corrosion damage is also traced to the same plant operating envelope review meeting. In retrospect, while in the storage drum case questions regarding process upsets were raised, for the inlet drums the imaginary plant boundary line to the upstream production field was not crossed. The fact that inspection records from the previous twenty-five years of operation were not indicating any issues did not help raise "what-if" scenarios regarding the plant inlet stream.

As stated earlier, success for the plant review meeting rests with involving all appropriate personnel and reviewing all pertinent information. The inlet surge drums are the boundary entrance point from upstream operations. We believe that process changes resulting in bacteria activity or oxygen entering the process stream are the cause for the observed pitting corrosion attack. In hindsight, the inlet process changes and their potential impact in corrosion rates could have been identified during the plant review meeting.

7. Risk criteria

Absolute risk value calculations have high resource requirements and often not even possible to complete due to involved uncertainties with input variables. The RBI process is focused on a systematic determination of relative risks. The relative equipment risk ranking is used to focus the risk management activities on the higher risk items. If a quantitative RBI study is conducted rigorously and properly, the resultant relative risk number should be a fair approximation of the actual risk of containment loss due to equipment failure [(11)].

In this context of relative risk ranking it is then necessary to establish the risk criteria that are used for the RBI risk assessment and management process. Suggestions for developing risk criteria are offered in this section. It is important to note that the risk criteria are supporting instead of managing the RBI process and this distinction will be further discussed.

The starting point for setting risk criteria for a RBI analysis is the existing corporate targets. Alignment of RBI process targets with corporate targets is necessary. It is common practice for corporations to set targets defining acceptable and prudent levels of safety, environmental and financial risks. These targets are company specific reflecting the particular operating philosophy.

If the corporate risk targets are expressed in a quantitative format then they can be directly applied to the RBI analysis. If the corporate risk targets are expressed in qualitative terms then a translation to quantitative RBI targets is needed. Risk in qualitative terms is usually described in terms of high, medium and low. Similarly the consequence and probability of failure risk dimensions are given in qualitative terms. For example potential consequences of failure can be described as "catastrophic", "major", "significant" or "minor". The probability of failure in qualitative terms can be expressed as "rare", "unlikely", "possible" or "almost certain". The API-580 RBI recommended practice provides guidance that can be used as starting point for translating these qualitative terms into quantitative RBI targets (11).

Similarly with all technical disciplines it is expected that the RBI specialist maintain effective industry networking relationships. Participation in API committees responsible for maintaining and developing industry standards is necessary. RBI software user-group meetings are scheduled regularly each year and provide excellent opportunities for sharing information, including risk target levels used by operating companies.

The established risk targets serve as a trigger in the RBI analysis, initiating the development of mitigation plans for the equipment with risk values above the acceptable targets. Feedback from the performed risk mitigation plans and inspection activities is important for the facility's continuous improvement efforts. The risk targets should be re-evaluated based on the findings and feedback of the equipment inspections. These are not meant to be absolute, unchangeable values.

Specific inspection plans are developed for each equipment item that exceeds the acceptable risk targets. The inspection results can either reveal equipment integrity issues or confirm that there is no corrosion related damage. Feedback from the field inspection work must be entered in the RBI model and may result in reassessment of the prior analysis regarding equipment condition. In addition, the risk targets should be evaluated based on the inspection feedback. If integrity issues are identified for equipment with calculated risk value slightly above the acceptable target then the inspection scope must be extended to cover lower risk value equipment. In this scenario, the risk acceptable target is effectively reduced to reflect inspection findings. The risk targets are a necessity for initiating the risk assessment and management process but the process itself is the critical element. The focus needs to be with the process and not with a strict acceptance of the risk targets or criteria.

CONCLUSIONS

The presented approach can be applied to any process facility across the oil and gas value chains. RBI analysis offers the ability to understand fixed-equipment risk and establish proper risk priorities for likelihood-of-failure mitigation plans. RBI implementation is an intelligent database system, compliant with industry best practices for fixed-equipment risk management that enables plant decisions for asset management based on quantitative data analysis (e.g. project priorities for allocating resources within corporate portfolio and/or within plant unit)

Successful implementation of RBI analysis is defined by the quality of process implementation. Risk targets used in RBI analysis must align with corporate targets and be viewed as only one part of the process that can be updated as based on the inspection results.

Acknowledgements
The author would like to thank Hess Corporation for granting permission to publish this paper and Bill Capdevielle, Senior Facilities Engineering Advisor, for his thoughtful insights and suggestions to the final paper.

REFERENCES

1. NACE conference paper, March 23rd 2009 Atlanta, Tony Poulassichidis, "A Gas plant perspective on utilizing risk-based inspection analysis to improve mechanical integrity",
2. Gas Processors Suppliers Association (GPSA) Engineering Data Book, 12th edition (electronic version) page 16-32
3. Corrosion in natural gas processing plants, Gas Research Institute report GRI-90/0050, page 104
4. API recommended practice 571, 1st edition 2003, "Damage mechanism affecting fixed equipment in the refining industry" section: 5.1.1.1
5. API Publication 581, "Risk-Based Inspection, Base Resource Document". Second edition, October 2000, Appendix H-11, page 290
6. API recommended practice 571, 1st edition 2003, "Damage mechanism affecting fixed equipment in the refining industry" section: 4.5.1.3.g
7. R.B Nielsen, K.R. Lewis, J.G. McCullough, D.A. Hansen "Controlling corrosion in amine treating plants", NACE 1995, paper 571.
8. API recommended practice 571, 1st edition 2003, "Damage mechanism affecting fixed equipment in the refining industry" section: 5.1.2.2.3.c
9. NACE RP0590-96, Recommended practice for prevention, detection and correction of Deaerator cracking, section 2.2, page 9, 1996 edition
10. ASM Handbook, Volume 13A, Corrosion: Fundamentals, Testing, and Protection, page 697, 2003 edition
11. API Publication 580, "Risk-Based Inspection". Second edition, November 2009, page 25, 51, 60-61.

Appendix I: Example of corrosion control diagram

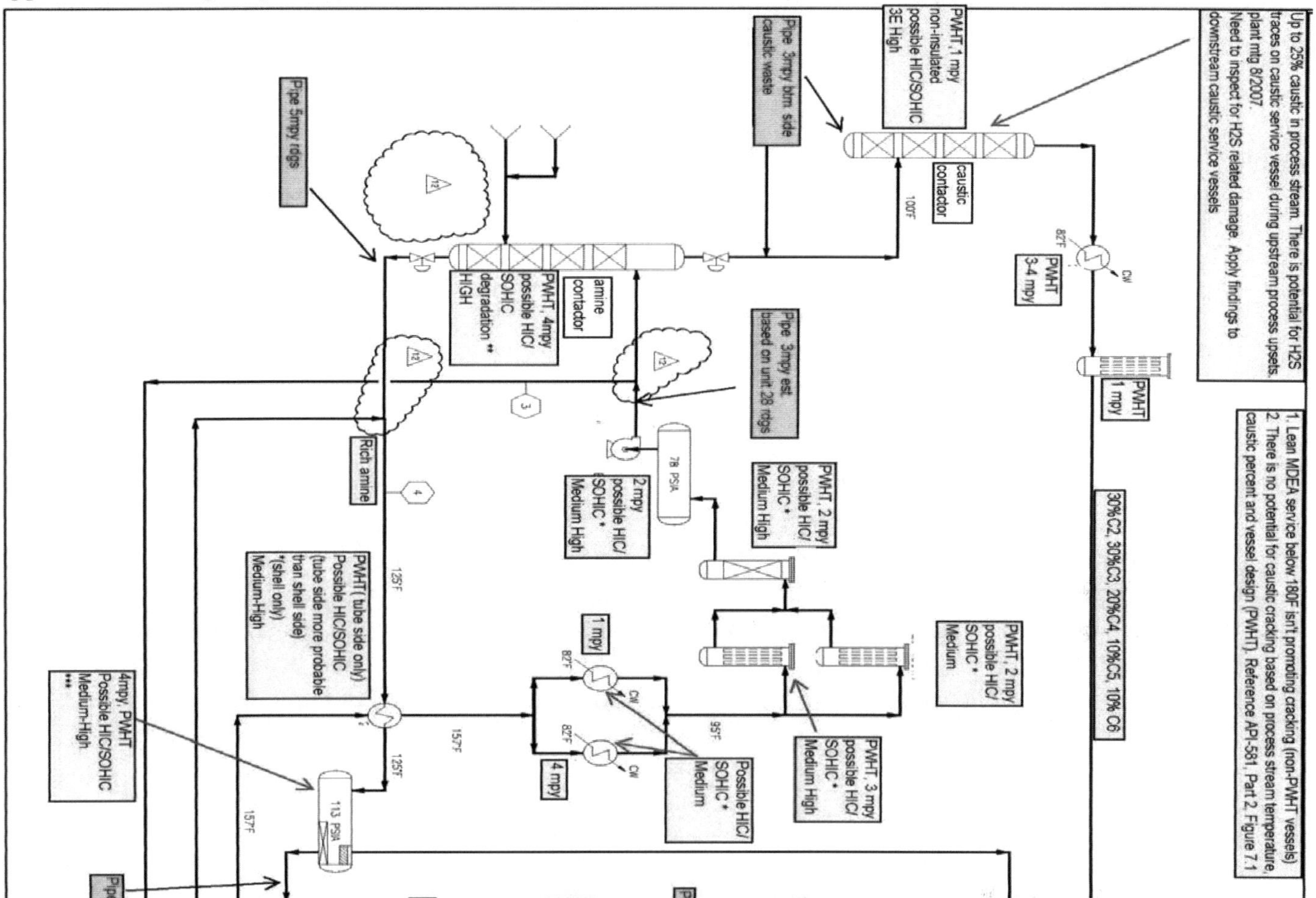

The Importance of MI Inspection for Successful Capital Projects

Paul F. Schubert
SGS North America Inc.
drpaulschubert@gmail.com

Kyle Stell
SGS North America Inc.

Keywords: mechanical integrity, asset integrity management, capital project inspection, Process Safety management, PSM, API-510, API-570, PMI

Abstract

It has been a common practice in the industry to wait until a new process unit has been placed into service before performing the initial inspection to establish corrosion rates of new piping systems and vessels. One problem with delaying inspection is that use of nominal thickness rather than measured thicknesses can result in significant errors in calculated corrosions rates as a result of the mill tolerances. This can mask potentially hazardous conditions, or cause concern where none is really warranted. Involvement of the inspection department in a capital project can significantly improve quality, reduce cost, and ensure compliance. The objectives of this paper are to provide owner/user inspection agencies with: 1) justification for involvement in capital projects; 2) effective roles for inspection departments in capital projects; 3) justification for performing vendor surveillance in capital projects; and 4) the technical advantages from performing pre-service baseline inspections.

1. The Need for Inspection in Capital Projects

In a recent project involving initial on-stream inspection of four new process units which had been on stream for less than two years we inspected approximately 150 pressure vessels and heat exchangers. Out of these equipment items, we found 22 instances where the wrong thickness of material was utilized during fabrication. In most cases these equipment items were re-rated. In a few cases the equipment items had to be removed from service and replaced. A thickness survey at the fabrication facility would have probably prevented these fabrication errors, and would have probably cost only a fraction of the cost of shutting down the operating units to remove and replace the off-spec equipment . This discrepancy between the specified equipment and the actual installed equipment highlights the need for inspection as part of the original capital project for process unit construction. In fact, there are several drivers for including inspection in capital projects including regulatory requirements, improved QA/QC, more effective vendor surveillance programs, and more effective budgeting.

2. Regulatory Requirements for Inspection in Capital Projects

In 1992 OSHA issued the Process Safety Management standard (29 CFR 199.1910) in response to a series of catastrophic failures in the chemical process industry. Within the standard, there are places where inspection is required to meet the standard. In addition, the standard established the concept of Recognized and Generally Accepted Good Engineering Practices (RAGAGEP). RAGAGEP was intentionally left undefined to allow industry to define what constituted good engineering practice.

2.1 PSM Standard 1910.119 Appendix C Compliance Guidelines Paragraph 9

PSM Standard 1910.119 Appendix C Compliance Guidelines Paragraph 9 states, "The use of appropriate gaskets, bolts, valves, lubricants and welding rods need to be verified in the field." Clearly, the requirement for "verification in the field' cannot be met without inspection at the project's site during construction. The paragraph further states that "It may be appropriate to conduct audits of the equipment supplier's facilities to better assure proper purchases of required equipment which is suitable for its intended purpose." The key phrase, "suitable for its intended purpose," constitutes justification for vendor surveillance.

2.2 OSHA Refinery NEP Audit Protocol

OSHA is in the final stages of completing an audit of the refining industry's compliance with the PSM Standard. As part of this National Emphasis Program (NEP), OSHA issued a protocol with a list of questions and possible violations related to the questions. Question 9 in the Piping section of the protocol states, "Considering the five piping circuits randomly selected, can the employer demonstrate that the piping was installed according to design specifications?" The key words in this being "was installed." The document goes on to list possible violations if the employer cannot demonstrate compliance:

a. "The employer did not conduct appropriate check and *inspection* to ensure piping was *installed* according to design specifications[1]."
b. "The employer did not ensure that piping construction and equipment was in accordance with its design specifications *before startup* ."
c. The employer did not follow RAGAGEP when it failed to *conduct PMI testing*"

One of the most important phrases in these possible violations is "before start-up." Therefore the question becomes, "Is your inspection and verification documentation of material composition and thickness dated prior to unit start-up?" API-578 Material Verification Standard (PMI) is listed as RAGAGEP, and would be the needed inspection protocol to verify material composition. In the absence of specific instructions to an E&C organization, their PMI documentation typically is a spread sheet and a marked up construction isometric drawing. We recommend that a separate layer be added to either the AutoCAD piping inspection isometric or equipment drawings. The PMI results can then be entered into a data base, and then linked to the AutoCAD drawing.

2.3 Verification that Piping Meets Specifications

Verification that the piping meets the design specification prior to start-up would come under the API-574 inspection practices for piping systems. Section 12.1.c of API-574 states" "*All inspection records* should contain as a minimum (emphasis added):

a. "The original date of installation."
b. "The specifications and strength levels of the materials used."

c. "The *original thickness.*"

Experience has shown that in many cases even though "all inspection records" is specified, this data is not entered into the corrosion monitoring database.

Furthermore, there is a significant issue with what is used as the "original thickness." Figure 1 illustrates the problem with using nominal thickness as the original thickness. For 6 inch schedule 40 pipe the nominal thickness is 0.280 inches. The mill tolerance is 12.5%, which can give an actual and acceptable thickness out of fabrication of 0.245 inches, fully 0.035 inches less than the nominal thickness. If inspection is then conducted a year later, and an ultrasonic thickness reading of 0.245 inches what is the corrosion rate? If the nominal thickness is used for the original thickness, then the corrosion rate is calculated to be 35 mils per year, indicating severe corrosion. However, if the pipe's original thickness was really at the minimum allowed thickness to meet the mill tolerance (0.245 inches), then the corrosion rate would be 0 mils per year. In the absence of a actual thickness measurement prior to unit start-up there is no way to determine what the real corrosion rate is.

Figure 1.
The problem with using nominal thickness for original thickness.

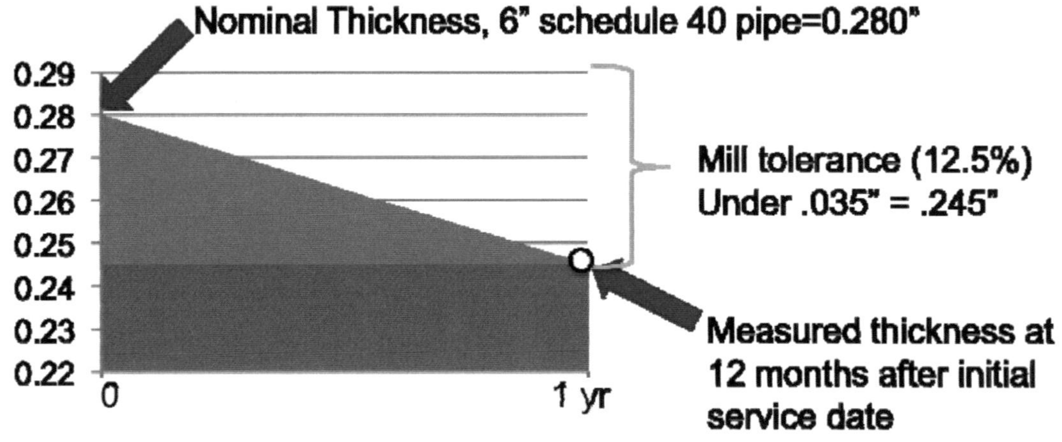

2.4 RAGAGEP for Pressure Vessels

API-510 is considered RAGAGEP for pressure vessel inspection practice and requires that a corrosion rate be established for each pressure vessel. It provides two options. A probable corrosion rate can be established, or a corrosion rate can be established based on actual measurements. The three ways a probable corrosion rate can be established are: "similar service," "user experience," or "published data." One of the key problems in using a probable corrosion rate is being able to support it in an audit. Typically there is little or no documentation of how the probable rate was established. In addition, there is usually not a written procedure in the owner-operator's mechanical integrity manual stating how the rate will be established.

If a probable corrosion rate cannot be or was not established, Section 5.3.c of API-510 states that, "on-stream determinations shall be made after 1000 hours of service by using suitable corrosion monitoring devices or *actual nondestructive thickness measurements* of the vessel or the system" The 1000 hours of service requirement means that the measurements must be made within 41 days.

As with the piping, it is important to determine the actual original pressure vessel wall thickness to established reliable corrosion rates for new units. We will consider an example of a vessel with walls nominally 1.5 inches thick. Under ASME-SA 20 Section IIA acceptable over tolerance for a 1.5 inch thick plate would be 0.060 inches, and an under tolerance for the plate would be 0.010 inches. If we further assume that we have a corrosion allowance of 0.125 inches, then the minimum thickness (t_{min}) for the vessel in service would be 1.375 inches.

Assume that after 1 year of service the actual thickness is measured as 1.490 inches. If the plate was originally at the over tolerance, then the actual corrosion rate would be calculated as 1.560 minus 1.490, resulting in an actual corrosion rate of 70 mils per year. This would be severe corrosion that would be missed in the absence of an actual measurement. The interval for the next inspection would be calculated by determining the time required to reach half of the remaining life. That would be the time required to go from 1.490 inches (the measured thickness) to 1.350 inches (T_{min}). At 70 mils per year, it would take 1 year to reach half of the remaining life.

Using the nominal thickness, the severe corrosion would be entirely missed, and the inspection interval would be well past the date when T_{min} was reached. Again, assuming after a year the thickness is measured at 1.490 inches. Using the nominal thickness (1.500 inches) to calculate the corrosion rate gives a corrosion rate of 10 mils per years. Using this corrosion rate, the calculated next inspection due date would be 11 years. In this case the owner-operator would believe the vessel to be safe when the true situation was a highly hazardous situation.

3. Improving QA/QC

The inspection organization can significantly improve the QA/QC of a capital project. We recommend that the inspection group develop and maintain the QA/QC manual, field new construction QA/QC questions in the case of an OSHA audit, and be responsible for the following essential steps during the early stages of the project:

- Review of the proposed engineering specifications to be used during procurement and construction.
- Review the QC manuals of construction contractors.
- Establish a Project QA/QC organizational chart.
- Establish a system turn-over protocol.
- Establish that a baseline mechanical integrity inspection be included as part of the E&C firms responsibility.

3.1 Site Specific Engineering Specification Manual

A major cause of equipment failure is a result of insufficient materials specifications and system understanding. Much of this is a result of what was done in regards to process hazards analysis (PHA), mechanical integrity, and process safety information during the capital project. In order to improve the specifications, we recommend that a site specific specification manual be developed and be the authoritative source for specifications.

Typically, an E&C firm's specification manual will not include all of the needed specifications. The inspection organization should review what is provided in the E&C firm's specification manual, and determine what is needed. Among the items most frequently needed are:

- Site specific engineering specifications.
- Supplemental site specific mechanical integrity procedures such as PMI practices and pre-service testing of Pressure Safety Valve (PSV) requirements.
- Site specific approved manufacturer's list (AML). Most E&C firms have an AML, and many owner-operators have their own AML, but the two typically don't match.
- Site specific Approved Fabricators List (pressure vessels, exchangers, pipe fabricators). Most E&C firms have their own AFL. However, their AFL is typically cost driven rather than quality driven. A site specific AFL provides leverage for quality and schedule demands.

3.2 Site Specific QA/QC Manual

A similar strategy should be followed for a site specific QA/QC manual as was followed for the specification manual. All major E&C firms have QA/QC manuals, but few get the owner-operator inspectors involved. A site specific QA/QC manual allows the owner-operator to establish their hold points.

Essential items to include in the QA/QC manual are:

- The owner-operator inspector as the reviewer and approver of non-conformance reports (NCR's). Typical E&C QC manuals do not require client approval of non-conformance reports written by the E&C QA inspectors.
- A pre-hydro check list should be established. It should ensure that:
    - It is properly supported
    - All welding is complete (shoes, guides, high and low point drains)
    - All non-destructive testing (NDT) is complete and documented
    - All PMI is completed (welds, fittings, pipe, valves, plate, instrumentation)
- The turn-over requirements are specified (see discussion below).

3.3 Project QA/QC Organizational Chart

It is essential to have an up front understanding with the E&C firm's project manager, the operating company's project manager, and the QC personnel that the owner-operator's QA position makes all decisions concerning quality and mechanical integrity. In this case, the project manager only makes administrative decisions concerning QC personnel. This ensures that decisions are quality and integrity driven rather than time and cost driven. This does not mean that the owner-operator should be insensitive to those very real issues. However, in the absence of a strong owner-operator QA position, quality typically suffers significantly.

3.4 System Turn-Over Protocol

Unless the owner-operator is specific about the documentation that is needed in order to accept turn over of the plant and equipment, the owner-operator is unlikely to have all of the documents needed to ensure PSM compliance. According to the PSM standard section 1910.119(j)(6)(i):

> "In the construction of new plants and equipment, the employer shall assure that equipment as it is fabricated is suitable for the process application for which they will be used. "

A check off list should be developed which requires owner-operator QA sign-off that includes the following documentation in the turn-over package are:

- Initial base line UT data
- AutoCAD inspection isometrics and drawings
- Corrosion monitoring data base established, complete, and linked to the drawings
- Listing of specifications used
- Final shop QC plan that lists hold points and sign offs.
- NCR log and individual NCRs. Acceptance signature of QA inspector.
- Release forms signed by QA inspector.

4. Improved Budgeting

The case has been made for including inspection, including an initial pre-start up base line inspection as part of the capital project. This is advantageous for all parts of the organization. It ensures that the inspection needed for compliance is budgeted. Since this compliance driven testing is not truly an operating expense, it should be included in the capital budget. Furthermore, it ensures that no hidden post-construction administrative costs are placed upon the operating budget. If a hard dollar inspection bid is required, then at a minimum condition monitoring locations (CML) placement, P&ID's, piping specifications, line test, equipment fabrication drawings, and construction isometrics will be needed. In addition obtaining the baseline inspection will require the following:

- Coordinating the bid walk down
- Developing the bid evaluation and contractor selection process
- Developing the AFE
- Writing the PO
- Coordinating the inspection project kick-off
- Daily supervision of the inspection project.

These activities require a substantial time investment, and should not be placed on the operating budget.

5. Conclusion

Full compliance with OSHA's PSM standard requires that inspections are made during the construction to verify that the equipment as constructed is suitable for the intended use, and that there is sufficient documentation to support that under OSHA audit. Inspection is needed not just to verify that incoming parts have the right alloy composition and wall thickness, but that they are present in the as fabricated equipment. Inspection is also needed prior to start-up to establish corrosion rates, as mill tolerances make it impossible to use nominal thicknesses and calculate valid rates.

However, inspection as part of the capital budget provides benefits beyond just regulatory compliance. If effectively used, owner-operator inspection involvement can ensure that the site specific specifications and needs are not neglected as a result the tendency to use more generalized specifications and QA/QC manuals. This improves the level of materials specification and system understanding, and thereby can substantially reduce one of the major reasons cited for equipment failure. In addition, it helps to ensure that documents essential to demonstrate compliance are successfully transferred from the E&C firm to the owner-operator.

Finally, inclusion of inspection in the capital project budget places the cost where it should be born. Often inspection resources from the operations group are required to conduct the baseline inspection for new equipment. Typically they do not have the human resources or the budget to perform the inspection in a timely manner, nor to oversee the inspection process if it is outsourced. Inclusion of these costs in the capital budget places them where the required work is generated.

Notes

1. Italics were used in various locations to bring emphasis to the printed data.

GOOD ENGINEERING PRACTICES:

INSPECTION AND TEST FREQUENCIES

Richard W. Prugh
Chilworth Global
250 Plainsboro Road
Plainsboro, New Jersey 08536
rprugh@chilworth.com

Keywords: Inspection, Test, Frequency, Mechanical Integrity, Leading Indicator

Abstract

The "Process Safety Management" standard of the Occupational Safety and Health Act [OSHA] requires that "The frequency of inspections and tests of process equipment shall be consistent with applicable manufacturers' recommendations and good engineering practices, and more frequently if determined to be necessary by prior operating experience." Guidance concerning inspection and test frequencies is provided in publications of the Center for Chemical Process Safety, and an objective of this presentation is to compile sources for the available data.

Objectives of preventive-maintenance and mechanical-integrity programs are to ensure that devices, systems, and equipment will perform satisfactorily during the interval between inspections or tests [I&T]. Thus, "leading indicators" (such as corrosion rates, and detection of incipient failures) are important in establishing appropriate I&T frequencies, rather than "present conditions" (such as hydrostatic tests, and detection of existing failures) or "lagging indicators" (such as historical failure rates). Of critical importance is the recording of "as-found" conditions, prior to repairs, calibrations, or replacements. Suggestions are provided for adjusting inspection-and-test frequencies and intervals, both upward and downward, as based on a "predictive-maintenance" program.

Of current interest is the testing of the shut-off devices on blowout preventers at wellheads beneath oil platforms. Typically, the shut-off devices are of a fail-safe design, where loss of air, gas, or hydraulic pressure would cause the valves to close. Frequent testing is required to ensure that emergency shutdown devices [ESD] will function properly when needed to protect personnel and property. Although the ram preventer (pinch device) cannot be tested without destroying the piping, the actuating systems need similar frequent testing.

Sinking of the "Deepwater Horizon"

1.0 Introduction

Most chemical companies routinely inspect and test process-control and emergency devices, to provide some assurance that these devices will operate properly if needed. Other "safety-critical" equipment – such as pressure vessels, storage tanks, gas detectors, and sprinkler systems – are also subjected to inspections and tests, to aid in determining the "lifetime" and/or operability of such equipment. Fewer companies inspect and test other equipment that could pose threats to personnel and property, such as piping, pumps, hoists, and plug-in electrical equipment, thus following a "fix-it-when-it-breaks" philosophy.

OSHA's "Process Safety Management" standard requires that inspections and tests be performed as an important part of a mechanical integrity program. The frequency of such inspections and tests is to be guided by manufacturers' recommendations and good engineering practices, and more frequently if determined to be necessary by prior operating experience." [1] Recent publications of the Center for Chemical Process Safety [2,3] and the American Petroleum Institute [4,5] provide good and alternative guidance for the frequency of inspections and tests. The objective of this paper is to provide methods for establishing and adjusting the inspection and test frequencies.

2.0 A Guiding Principle

If a component or system is critical to safety, it is important to know if that component or system will operate properly when needed. There are only two opportunities to determine proper operability: (1) when the component or system must operate [a "demand"], and it fails [incident investigation] or it functions properly [often, no investigation or report]; and (2) when the component or system is tested, during a brief interval when operation is not required. [6] Obviously, failure may be intolerable,[a] so inspections and tests of safety-critical devices are important preventive-maintenance and mechanical-integrity activities.

The primary objective of inspections and tests is to ensure that the equipment or device will not fail during the interval between such inspections and tests. For the statisticians among us, this should be modified to "ensure that there is a very low probability of failure between such inspections and tests".

Thus, an inspection or test must provide sufficient information about the condition of equipment to allow prediction of performance during the upcoming interval before the next inspection or test. A corollary of the primary objective is that the result of an inspection or test should be a "leading indicator" of future performance. This applies particularly to devices or systems that are the "last line of defense", such as rupture disks, relief valves, explosion vents, and shutdown-interlock systems, since they *must not fail* if needed.

[a] Some "demand" failures are "repairable", if the failure can be detected and corrected before life-threatening and/or property-threatening conditions develop. [7]

Further, when the results of inspections and tests can be used to predict future performance – as a leading indicator – and deterioration in performance is detected, plans can be made for repairs or replacements can be ordered. Also, the observed deterioration may provide some "lead time" for a change in design or operation of the equipment or process.

The Process Safety Management standard of the Occupational Safety and Health Act requires an Inspection and Test program for sites that handle highly-hazardous chemicals. [1] The program is to establish inspection and test frequencies, as guided by "recognized and generally-accepted good engineering practices". Additional guidance is provided in Section 9. of Appendix C in the OSHA standard, which states "Mean time to failure . . . would influence the inspection and testing frequency." This requirement is a Federal recognition of the importance of inspections and tests in minimizing the frequency of equipment failures.

3. A Quandary

Some performance tests provide little or no predictive information. For example, if an interlock system functions properly, that information shows that the system worked properly when tested and that the system very probably would have functioned properly prior to the test. However, the system might not function properly in any subsequent test or in the future. Similar questions arise following hydrostatic tests on pressure vessels and following human-performance evaluations. How then can good performance in the future be ensured, and how can the results of inspections and tests provide information concerning future performance?

4. Past Performance as a Guide to Future Performance

Good performance of a component or system in the past can give an indication of good future performance, (1) provided that an estimate can be made of the failure frequency or the average interval between failures *and* (2) provided that the prior good performance is well-documented. If there are many identical units at a site, and they are in identical service conditions, their observed frequency of failure can be readily determined. [8] If there are many similar units at a site, it may be possible to establish a "generic" failure frequency, with corrections for dissimilar service conditions. [9] When data are obtained by these methods, it may be a result of "breakdown" maintenance. It is a manifestation of "no good deed goes unpunished" that good preventive-maintenance practices give meager failure-rate data. Thus, during scheduled inspections and tests, it is particularly important to note and record – and forward to the preventive-maintenance staff, for tabulation and analysis – any "as-found" information (including good test results) that can lead to estimation of future service lifetime.

If there have been no failures in a particular interval, there are methods for obtaining a rough estimate of the failure frequency, for later use in establishing inspection and test frequencies. The following are two examples:

4.1 *The "One-Third" Rule*

In the absence of any failure-rate data for a component or system, but with a history of no failures in a significant interval, the "one-third rule" provides a crude estimate of the likely failure frequency. [10] This "rule" presumes that over a period equal to three times the failure-free interval, a failure is likely to occur. The applicable relationship is:

$$\lambda = 1 / (3\,t) \qquad \text{failures per year}$$

where <u>t</u> is the interval within which no failures occurred.

<u>Example.</u> If there is evidence that a component has not failed in ten years of operational demands, this method would give a failure-frequency estimate of 0.03 failure per year. Thus, it is likely that a failure would occur in the next twenty years of similar demands.

This concept can be used to establish inspection-and-test intervals. If – for example – a test shows that a component or system is in good condition, then there is a very high probability that it has not failed since the last inspection or test [or repair]. Thus, by the "one-third" rule, it is unlikely that the component or system would fail in the next two such intervals. In this case, it might not be unreasonable to extend the inspection-and-test interval to three times the previous interval. However, this "adjustment" might be overly optimistic, and a suggested practice is to wait until three consecutive good inspection-and-test results have been obtained [and documented], and then double the inspection interval. Of course, if the inspection or test reveals unsatisfactory performance, the inspection interval should be reduced [by one-half, for example]. If this "halving" occurs as many as three times, then "re-engineering" the component, system, or process would be needed.

4.2 *Upper Confidence Bound Method*

Again, in the absence of any failure-rate data for a component or system, but with a history of no failures in a significant interval, an upper limit for the failure frequency can estimated with the "chi-squared" method. [11] The applicable relationship is:

$$\lambda \geq X^2_{(2n+2,a)} / (2\,t) \qquad \text{failures per year}$$

where <u>n</u> is the number of failures in <u>t</u> years;
<u>a</u> is equal to the decimal value of (100% – Confidence); and
$\underline{X^2}$ is the "chi-squared" value that corresponds to <u>n</u> and <u>a</u>.

Values of $\underline{X}^2$ for confidence level of 10%, 50%, and 90% – and some examples of estimated failure frequencies – are given in the following Table: [12]

Number of Failures in 10 Years	0	1	2	3	5	10
Failure Freq. [90% Confidence]	>0.01/yr. [<95 yrs.] X^2=0.211	>0.05 /yr. [<19 yrs.] X^2=1.06	>0.11 /yr. [<9 yrs.] X^2=2.20	>0.17 /yr. [<5.7 yrs.] X^2=3.49	>0.32 /yr. [<3.2 yrs.] X^2=6.30	>0.70 .yr. [<1.4 yrs.] X^2=14.0
Failure Freq. [50% Confidence]	>0.07 /yr. [<14 yrs.] X^2=1.39	>0.17 /yr. [<6 yrs.] X^2=3.36	>0.27 /yr. [<3.7 yrs.] X^2=5.35	>0.37 /yr. [<2.7 yrs.] X^2=7.34	>0.56 /yr. [<1.8 yrs.] X^2=11.3	>1.07 /yr. [<0.9 yr.] X^2=21.3
Failure Freq. [10% Confidence]	>0.23 /yr. [<4 yrs.] X^2=4.61	>0.39 /yr. [<2.6 yrs.] X^2=7.78	>0.53 /yr. [<1.9 yrs.] X^2=10.6	>0.67 /yr. [<1.5 yrs.] X^2=13.4	>0.92 /yr. [<1.1 yrs.] X^2=18.5	>1.5 /yr. [<0.6 yrs.] X^2=30.8

Example: If there have been 2 failures of a component in the past ten years, there would be 90% confidence that the failure frequency would be at least 0.11 per year and that at least one failure would occur in the next ten years. In other words, it is very unlikely that the component could "survive" the next nine years without a failure. The "50/50" failure frequency would be at least 0.27 failure per year, or at least one failure in the next 3.7 years. There would only be a 10% likelihood that there would be more than 5.3 failures in the next ten years, if likelihood and probability can be assumed to be equivalent to confidence. These estimates presume that the component is as well maintained in the future as it has been in the past, and that it is no more stressed in the future that it has been in the past.

Working backwards with this method, it can be shown that the confidence that the "one-third rule" gives a "good" upper-limit failure frequency would be about 70%. That is, there is a 70% probability that at least one failure would occur within an interval that is three times the no-failure interval.

As indicated above, the inspection-and-test interval could be doubled following three consecutive good results, and this "chi-squared" method can be used to estimate the confidence that accompanies a change in the inspection-and-test interval.

4.3 Bayesian Estimation of Failure Frequency

Past performance can be used to adjust failure frequencies, using the Bayesian method for combining available "generic" data with recent plant-specific data. [13,14] Although this method gives a good frequency of failure – *provided that there is good documentation of test results* – the method is complicated and is beyond the scope of this paper.

5. The Effect of Inspection & Test Frequency on "Fractional Dead Time"

Most – if not all – components and systems are likely to fail to perform as intended during their service lifetime. The time-fraction during which the component or system is in a failed condition is the "fractional dead time". Ideally, this fractional dead time is a small part of the service lifetime. However, if a component or system is not inspected or tested, and a failure occurs early in the service lifetime – as a result of "infant mortality" [15] – the component or system might be inoperable for most of its service lifetime. In Fault Tree Analysis, a component is considered to be in a failed state if it has never been tested since installation. [16]

An approximation of fractional dead time [FDT] is one-half of the inspection interval. [17] Thus, if the inspection interval is short, then the FDT is also short. Since FDT also includes the repair time – if failure is detected – there is a practical limit to inspection frequency.

Many modern instrumentation systems incorporate self-testing and self-diagnosis, [18] such that incipient failures are revealed and alarmed. This includes pressure and temperature sensors, programmable-logic controllers with "watch-dog" timers, Automatic-start-and-test fire pumps might also be included in this category. However, many items of equipment are not automatically tested and are not tested routinely by upsets in a process. Thus, inspections and tests are needed to detect these "unrevealed" failures.

6. A Recent Incident Involving Failure to Test

The "blowout" of the "Deepwater Horizon" oil well platform, at 9:45 PM on April 20, 2010, involved failures of components in the Blow-Out Preventer [BOP] systems that were intended to prevent such an incident. [19] This included two manually-actuated "variable-bore rams" [VBR] that would seal the annulus around the drill pipe and a fail-safe "blind shear ram" that would pinch-shut the casing and shear the drill pipe. The "automatic mode function" [AMF] system was designed so that the "blind shear ram" would operate to close the well with pressure from a hydraulic accumulator, upon loss of electrical signals or hydraulic pressure, and thus stop the upward flow of fluids from the well. [20]

Investigations indicate that the exceptionally-dense drilling mud was being removed from the borehole – and was being replaced with seawater – when the incident occurred. [20] The pressure in the oil-and-gas reservoir blew the remaining mud and seawater out of the borehole, and attempts to close the blowout-preventing systems were unsuccessful. It was later discovered that a battery was "dead" in the electrical control circuit of one [blue] of the two independent "control pods" that would actuate the blind shear ram, and that a solenoid valve was defective in the other [yellow] "control pod". The incident-investigation team found that batteries were not being replaced annually as was recommended by the equipment manufacturer and that battery failure had been reported previously. Although functional testing of the blind shear ram is not required by pertinent standards, testing of the VBRs and the BOP actuation systems prior to subsea deployment was not performed as required, and the scheduled weekly testing of the AMF system also was not being performed. [21] A conclusion of the investigation team was that these failures prevented operation of the blind shear ram, which was the "last defense" against well blowout.

7. Inspection and Testing Schedules

The interval between inspections and tests should be "short" – by comparison with the expected time between failures – for "safety-critical" equipment and components. This would allow detection of deterioration as the end of the service lifetime approached, so that replacements could be ordered or major repairs could be scheduled. A risk assessment – based on the consequences of failure and the expected time between failures – might be needed [22] to establish which of many items of equipment are "safety-critical" and which are not.

Most inspections and tests require that the component or system be removed from service or bypassed. Thus, there is a risk that safety would be compromised during the inspection and test interval, or it would be necessary to suspend operations. Ideally, such testing would be performed during shutdown periods, or there would be some "redundancy" or "diversity" in the system that would function to provide protection while the testing was underway.

Ideally, the inspection or test interval should be set so that the equipment would be found in good condition, but perhaps with some evidence of deterioration. This could include corrosion, erosion, or cavitation; partial pluggage or fouling; leakage through seals, gaskets, or other fillings; slow or erratic response to test conditions; evidence of overheating, overloading, or friction; mechanical defects such as cracking or warping; and loose electrical or power-media connections. If the deterioration is so severe that repairs must be made, the interval between inspections and tests should be reduced.

The probability of finding a component or system in a failed condition is related to the failure frequency and the time interval of interest, by: [23]

$$P = (1 - e^{-\lambda t}) = (1 - e^{-t/\tau})$$

where λ is the failure frequency, in failures per year, for example;
τ is the average interval between failures, in years, for example; and
t is the interval of interest, in years, for example.

Thus, for a "statistical" failure frequency of one per year, the probability of finding the component in a failed state after two years would be about 86%. However, if this component was checked after one month, there would be about a 92% chance of finding the component in operable condition, but perhaps with some indications of deterioration. Thus, in order to find a component in good condition during a test or inspection, the λt product should be about 0.1; that is, the inspection or test interval should be about one-tenth of the average interval between failures, to obtain a failure probability of about 9.5%. If this is not "good enough", an interval of one-hundredth of the failure interval would give a 99% probability of finding the device in good condition.

If an item of equipment is found to be in very good condition, it may be possible to increase the interval between inspections and tests. For example, if the inspection interval is one-tenth of the expected interval between failures, the probability of failure would be about 9.5%. If three consecutive good results are obtained, it is likely that the actual failure rate might be only one-third of the expected failure rate. Thus, doubling the inspection interval would result in a probability of failure of about 6.4%, or about 67% of the initial expected probability of failure.

8. Risks of Testing

In addition to possible compromised safety during tests, there are some undesirable risks that may accompany "invasive" testing. They include: [24] (1) not correcting the failure; (2) leaving the safety component bypassed; and (3) testing introduces more demands and aggravates "wear-out" or "mortality" failure modes. Thus, there may be a practical limit for the frequency of tests, particularly where the testing severely stresses the component(s) being tested. This could include the electrical circuits and drivers for heavy-duty pumps and blowers, for example. The equipment manufacturer may be able to provide a maximum number of "demands in a lifetime" that was used as the basis for design, to aid in determining the frequency or number of tests that may be excessive.

9. Predictive Maintenance, or Leading Indicators

Inspections and tests typically give "go/no-go" results and, as such, only give an indication of the present condition or state of operation. However, inspections and tests also can reveal incipient failures, which can be a "leading indicator" of future performance. [25] Thus, it is important to look for *and document* any evidence of age-related or stress-related deterioration; unexpected corrosion or erosion; slower-than-normal or erratic functioning speed; metal-to-metal contact; exposure to high temperatures; and abnormal bending, warping, deformation, or mis-fitting of components. Several forms of non-destructive testing also can give "leading indicators" of age or chemical-related deterioration, and this includes acoustic-emissions [during pressure tests] and ultrasonic wall-thickness tests, This evidence should be documented before any repairs, adjustments, calibrations, etc. are performed, so that this "as-found" data can be used to adjust the frequency and/or type of inspections and tests.

Documenting trends in performance also can be a leading indicator of impending failure. This would include trends in vibration for rotating or cyclic-operation machinery; trends in pressure for pumps and blowers; and trends in power consumption or amperage for motors. Also, a decreasing trend in plant throughput or productivity should prompt a study into possible equipment failures, if there is no other obvious cause. Of course, someone needs to review the inspection-and-test and trend reports, and then determine if the inspection-and-test, repair, or replacement interval needs to be adjusted. [26]

Another application of predictive maintenance is to determine the average interval between failures and then replace – or remove and repair – the component before it is likely to fail. If there is a large "population" of similar components at a site, the inspection-and-test program may be able to provide a good value for failure rate. Alternatively, records of failures, the durations of good performance prior to the failures, and the "as-found" conditions that caused the failures, can be used to determine failure modes and failure rates. Also, disassembly of components – or destructive testing to failure – can provide information for adjustment of failure rate and replacement frequency, if such actions can be justified by a "cost-benefit" analysis.

A major source of leading indicators is "inspection" action by operators, mechanics, and members of the site staff in performing frequent patrols and safety audits of operating areas. Noting unusual noises and odors, observing leaks and spills, and hearing comments concerning operating problems can be symptoms of impending failure and can lead to investigation and correction. The frequency of such patrols and inspections should be adjusted according to the number and seriousness of the findings. Also, "near misses" often are tests of emergency systems and usually involve human or equipment failures. If they are reported promptly, investigation of such incidents can lead to appropriate corrections before "near misses" become life-threatening or property-threatening incidents.

10. Conclusion

If there are any safety-critical components or systems at a site, that site should have a good mechanical-integrity program, and the program should contain a schedule and procedure for inspections and tests of those components. It is very important to record the results of such inspections and tests – and particularly to record the "as-found" conditions – and then analyze the results, in order to predict future performance and to adjust the inspection-and-test frequency, if warranted. Without such a program, site productivity and safety performance could be severely impacted by essentially-random equipment failures.

11. References

1. U. S. Department of Labor, Occupational Safety and Health Administration, "Process Safety Management of Highly Hazardous Chemicals", 29 CFR 1910.119(j)(4) (1992).

2. Center for Chemical Process Safety [AIChE], "Guidelines for Mechanical Integrity Systems", (2000).

3. Center for Chemical Process Safety [AIChE], "Guidelines for Chemical Process Quantitative Risk Analysis", 2nd Ed. (2000).

4. American Petroleum Institute, "Risk-Based Inspection", API Recommended Practice 580 (2002).

5. American Petroleum Institute, "Risk-Based Inspection Technology", API Recommended Practice 581, 2nd Ed. (2008).

6. Center for Chemical Process Safety [AIChE], "Guidelines for Chemical Process Quantitative Risk Analysis", 2nd Ed., page 530 (2000).

7. Center for Chemical Process Safety [AIChE], "Guidelines for Chemical Process Quantitative Risk Analysis", 2nd Ed., page 679 (2000).

8. Center for Chemical Process Safety [AIChE], "Guidelines for Chemical Process Quantitative Risk Analysis", 2nd Ed., page 507 (2000).

9. Greenberg, H. R., and J. J. Cramer [Stone & Webster Engineering Corporation], "Risk Assessment and Risk Management for the Chemical Process Industry", page 222 (1991).

10. Center for Chemical Process Safety [AIChE], "Guidelines for Chemical Process Quantitative Risk Analysis", 2nd Ed., page 505 (2000).

11. Center for Chemical Process Safety [AIChE], "Guidelines for Chemical Process Quantitative Risk Analysis", 2nd Ed., page 505 (2000).

12. "Handbook of Chemistry and Physics", 77th Ed., page A-105 (1996).

13. Center for Chemical Process Safety [AIChE], "Guidelines for Chemical Process Quantitative Risk Analysis", 2nd Ed., page 717 (2000).

14. Mannan, S. [editor], "Lees' Loss Prevention in the Process Industries", 3rd Ed., pages 7/10 and 7/38 (2005).

15. Center for Chemical Process Safety [AIChE], "Guidelines for Chemical Process Quantitative Risk Analysis", 2nd Ed., pages 482 and 495 (2000).

16. Center for Chemical Process Safety [AIChE], "Guidelines for Chemical Process Quantitative Risk Analysis", 2nd Ed., page 677 (2000).

17. Center for Chemical Process Safety [AIChE], "Guidelines for Chemical Process Quantitative Risk Analysis", 2nd Ed., pages 530 and 532 (2000).

18. Center for Chemical Process Safety [AIChE], "Guidelines for Chemical Process Quantitative Risk Analysis", 2nd Ed., page 541 (2000).

19. Wikipedia, "Deepwater Horizon" and "Blowout Preventers" (2011).

20. BP Exploration & Production, Inc., "Deepwater Horizon Accident Investigation Report", pages 11, 12, 47, 49, 152-154, 169, 171, 172, 178, and 186 (September 8, 2010). [a]

21. BP Exploration & Production, Inc., "Deepwater Horizon Accident Investigation Report", pages 10, 32, 41, 79, 83, and 175 (September 8, 2010).

22. American Petroleum Institute, "Risk-Based Inspection Technology", API Recommended Practice 581, page 1-15 (2002).

23. Center for Chemical Process Safety [AIChE], "Guidelines for Chemical Process Quantitative Risk Analysis", 2nd Ed., page 481 (2000).

24. American Petroleum Institute, "Risk-Based Inspection", API Recommended Practice 580, pages 8 and 32 (2002)

25. Center for Chemical Process Safety [AIChE], "Guidelines for Mechanical Integrity Systems", page 239 (2006).

26. Center for Chemical Process Safety [AIChE], "Guidelines for Mechanical Integrity Systems", pages 49, 50, 123, and 127 (2006).

*

http://www.bp.com/liveassets/bp_internet/globalbp/globalbp_uk_english/incident_response/STAGING/local_assets/downloads_pdfs/Deepwater_Horizon_Accident_Investigation_Report.pdf

APPENDIX A

Additional References

for

"Recognized and Generally-Accepted Good Engineering Practices"

Loss Prevention Symposia

9: Kletz	13: Puyear	13: Schwarz	15: Anonymous
17: Woolfolk	17: Livingston	18: Munson	20: Arnold
22: Burk	27: Sadler	29: Remson	31: Freeman
32: Stellmach	33: Summers	33: Scheier	33: Pratt
34: Roe	34: Hanauska	35: Weber	35: Summers
36: Renshaw	38: Frank	38: Rodante	38: Woycheese
40: Remson	43: Harris	43: Lay	43: Broadribb
44: Harris	44: Patschke		

Center for Chemical Process Safety Conferences

1996: Mohindra
1998: Leonard
1998: Farquharson
2001: Behie
2003: Urbanik

Factory Mutual Engineering Data Sheets

7-17: Explosion Protection
7-43: Loss Prevention
7-45: Instrumentation
7-47: Physical Operations
7-49: Emergency Venting
7-76: Dust Explosions
9-0 : Maintenance/Inspection
17-4: Rotating Machinery

An Update on OSHA's PSM National Emphasis Programs

Lisa A. Long (Presenter)
U.S. DOL-OSHA
200 Constitution Ave., NW
Room N3119
Washington, DC 20210
long.lisa@dol.gov

Michael L. Marshall, P.E. (Presenter)
U.S. DOL-OSHA
200 Constitution Ave., NW
Room N3119
Washington, DC 20210
marshall.mike@dol.gov

James Lay, P.E.
U.S. DOL-OSHA
200 Constitution Ave., NW
Room N3119
Washington, DC 20210
lay.jim@dol.gov

Keywords: Refinery NEP, chemical NEP, chem NEP, OSHA, national emphasis program

Abstract

Since 2007, OSHA has initiated two major National Emphasis Programs (NEPs) in process safety. The Petroleum Refinery Process Safety Management NEP (Refinery NEP) was effective on June 7, 2007; the PSM Covered Chemical Facilities NEP (Chemical NEP) was effective on July 27, 2009. This presentation will provide an update on the findings of these NEPs, including specific items cited and challenges encountered.

1. Introduction / Background

An NEP is a type of OSHA Special Emphasis Program. OSHA develops NEPs to focus outreach efforts and inspections on specific hazards in a workplace. The Refinery and Chemical NEPs provide OSHA inspection tools for evaluating catastrophic hazards (i.e., fire, explosion, and toxic releases) as they relate to employers' compliance with PSM standards.

This paper builds on a paper presented at the 2010 Global Congress on Process Safety titled, "OSHA Refinery and Chemical NEP Findings."

Refinery NEP Background:

OSHA evaluated data from 1992 and 2005 and determined that serious PSM incidents occurred most often in the petroleum refining industry (SIC 2911/NAICS 32411). This evaluation defined serious PSM incidents as those involving an employee fatality, or the hospitalization of three or more workers, where the incident involved a PSM covered process. As a result of the incident history in petroleum refining, including the major catastrophe at the BP Texas City Refinery in 2005, OSHA developed the Refinery NEP, which became effective on June 7, 2007.

The Refinery NEP outlined policies and procedures for OSHA Compliance Officers to use in conducting inspections of petroleum refineries, and mandated that OSHA inspect all petroleum refineries, excluding those in the Voluntary Protection Program (VPP), within federal jurisdiction. As of January 1, 2011, all but two Refineries under federal jurisdiction have been inspected. The remaining inspections should be complete by October 1, 2011.

The Refinery NEP emphasizes the *implementation* of PSM programs. Compliance Officers use a list of static questions, which is published in the Refinery NEP directive, and a smaller list of dynamic questions, which is unpublished and periodically changed. The questions focus on issues identified as contributing to previous incidents and contain compliance guidance to assist Compliance Officers in evaluating PSM implementation in a unit(s) selected by the inspection team. Contractors working in or adjacent to the selected unit are also inspected.

Due to the nature and large number of static and dynamic questions addressed during each Refinery NEP inspection, these inspections are both intensive and wide-ranging. As a result, the time to complete them is relatively long and the OSHA resources committed relatively large compared to the typical General Industry workplace inspections.

Chemical NEP Background:

The Chemical NEP is an OSHA enforcement initiative for inspecting facilities with PSM covered processes (excluding petroleum refineries), i.e., those which contain highly hazardous chemicals (HHCs) in amounts at or greater than the threshold quantities listed in 29 CFR 1910.119. It was started as a pilot program for programmed inspections in OSHA Regions 1, 7, and 10, and unprogrammed inspections in all regions. (Unprogrammed inspections are not scheduled; they are inspections initiated as a result of an incident, employee complaint, media referral, etc..) At the end of the pilot period, the Chemical NEP was continued while a permanent Chemical NEP is being developed. A permanent Chemical NEP is expected to be issued in early 2011, and will likely extend the Chemical NEP nationwide.

The Chemical NEP outlines a targeted strategy for programmed inspections. The pilot target inspection pool includes facilities with PSM covered processes including:

- Facilities listed in EPA's RMP database as Program 3;
- Explosives manufacturing facilities;
- Facilities listed in OSHA's IMIS database as having been previously cited for PSM; and
- Facilities included through OSHA Area Offices' local knowledge.

The Chemical NEP's approach for inspecting PSM covered facilities seeks to better allocate OSHA's resources to perform a large number of less resource-intensive inspections. These are intended to encourage facility compliance by increasing the probability that a facility will be inspected. To accomplish this, the Chemical NEP focuses on a few very specific items to address during an inspection, using a dynamic list of questions developed by OSHA's National Office. The dynamic list is not in the public domain; consequently, employers will not know which areas of PSM will be the focus of their inspection. The dynamic list is periodically updated to assure the integrity of the inspection process. In order to ensure the most appropriate questions, lists have been developed for the following types of facilities:

- PSM General;
- Ammonia Refrigeration;
- Chlorine for Water and/or Wastewater Treatment; and
- Chemical Processing

Chemical NEP refinery inspections have averaged 125 OSHA man hours per inspection compared with Refinery NEP inspections that have averaged over 1000 man hours per inspection.

2. Petroleum Refinery NEP Results

2.1 Inspection Statistics

As of December 31, 2010, OSHA has opened 248 inspections under the Refinery NEP. These include inspections in 60 of the 62 non-VPP (Voluntary Protection Program) refineries that are under Federal OSHA's jurisdiction. OSHA expects the remaining inspections to be opened in fiscal year 2011. In addition to the refinery inspections, OSHA has inspected an average of three contract employers at every refinery inspected.

Citation Statistics

OSHA has found a significant number of PSM violations in the refining industry. As a result of this NEP, OSHA has issued citations in 58 of the 60 Refinery NEP opened inspections, resulting in an average of 17 citations per inspection with an average penalty of $162,000 per inspection. Two other inspections are ongoing, and citations have not yet been issued.

2.2 Frequency of Citations by Standard Element

Violations broken down by Element and Sub-element of the PSM Standard

OSHA inspection data was analyzed to determine which element and sub-element of the PSM standard were most frequently violated. The following tables show the top elements and sub-elements.

Table 1 – Refinery NEP Most Frequently Cited PSM Elements

Element	Description	Percentage of Total Violations
J	Mechanical Integrity (MI)	19%
D	Process Safety Information (PSI)	18%
F	Operating Procedures (OP)	17%
E	Process Hazard Analysis (PHA)	16%
L	Management of Change (MOC)	10%
M	Incident Investigation (II)	7%
O	Compliance Audits (CA)	4%
H	Contractors	3%
G	Operator Training	2%
C	Employee Participation	2%

Table 2 - Refinery NEP Most Frequently Specific PSM Standard

Sub-element	Description	Percentage of Total Violations
(d)(3)(ii)	PSI – equipment complies with RAGAGEP	4.7%
j(5)	MI – equipment deficiencies	4.1%
e(5)	PHA – system to promptly address recommendations	3.5%
d(3)(i)(B)	PSI – information pertaining to the equipment in the process – P&IDs	2.6%
(l)(1)	MOC – establish & implement MOC	2.6%
j(2)	MI – written procedures	2.4%
(f)(4)	OP – safe work practices	2.0%
e(3)(v)	PHA – facility siting	1.9%
(d)(3)(i)(D)	PSI – relief system design and design basis	1.7%
f(1)(i)(D)	OP – emergency shutdown procedures	1.7%
j(4)(i)	MI – inspection and testing	1.7%

2.3 *Refinery NEP Challenges and Learnings*

The major challenge presented by the Refinery NEP is the significant resource commitment needed to complete NEP inspections. This resource issue impacts both OSHA and refinery employers.

OSHA is analyzing the information from the Refinery NEP inspections to find trends or lessons learned that will be helpful to the refining industry. As a starting point, the violations from the top cited sub-element 1910.119 (d)(3)(ii), failure to document compliance with RAGAGEP (Recognized and Generally Accepted Good Engineering Practice) were analyzed.

The predominant issues highlighted in the (d)(3)(ii) violations are valves and facility siting. Violations cited in association with valves mainly concern relief valves. Twenty different refinery inspections resulted in citations for failure to document that relief valves comply with RAGAGEP. Below is a summary of the most common issues leading to regarding relief valve citations:

- The employer failed to provide adequate controls to ensure that intervening valves to/from relief devices remained in the open position during pressure vessel operation. (citations in 16 refinery inspections).
- The employer failed to document that the relief valves did not have excessive back pressure. (citations in 4 refinery inspections).
- The employer failed to document that relief valves did not have inlet pressure drop that could lead to valve chattering. (citations in 3 refinery inspections).
- The employer failed to install required relief valves on pressure vessels. (citations in 3 refinery inspections).
- The employer failed to document that relief valves were properly sized for the given service. (citations in 2 refinery inspections.)

In approximately 25% of the refinery inspections, citations for employers not following RAGAGEP referenced facility siting. The majority of these violations were for facilities out of compliance with RAGAGEP which, therefore, left workers unprotected by either adequate separation distance or building construction. In many cases, there was more than one instance of this in a given facility. Other facility siting violations resulted from a lack of documentation of structure vulnerability altogether, making it impossible to determine whether structures provided employees with adequate protection.

After the first year of the Refinery NEP, OSHA published a letter to Refinery Managers that included appendices with paraphrased language of all citations that had been issued at that time. A copy of this letter is available on the PSM topic pages of OSHA's website (1). Employers may find the specific citation language helpful in reviewing their own programs.

3. Chemical NEP Results

3.1 Inspection Statistics

The pilot Chemical NEP has been effective since July 27, 2009. As of December 31, 2010, OSHA had opened approximately 151 Chemical NEP inspections. Ninety-six of these inspections have resulted in either citations issued or have been closed without citations. Another 19 inspections were opened, but resulted in no citations because there was no PSM covered process. On average, Chemical NEP inspections have resulted in 9 citations per inspection with total proposed penalties of $25,000.

For the purposes of the Chemical NEP, OSHA divided PSM covered facilities into three categories: ammonia refrigeration, chlorine for water/wastewater treatment, and all other. The Chemical NEP instructed OSHA Regions, where possible, to conduct approximately 25% of programmed inspections in workplaces that contain PSM covered ammonia refrigeration process, 25% in workplaces that use chlorine for water/wastewater treatment, and the remaining 50% in all other facilities. This was done to ensure that OSHA inspectors visit a variety of covered processes. OSHA recognized that without this guidance, pilot Regions with large numbers of food processing facilities might conduct many more inspections in ammonia refrigeration and water treatment facilities, rather than facilities with other types of PSM covered processes.

The following graphs show the breakdown of facility categories for Chemical NEP inspections in programmed, unprogrammed, and all inspections. Inspections of contract employers were broken out separately because the data makes it difficult to tell what kind of facility the inspected contract employer was working in. The data illustrates that more Chemical NEP inspections took place in ammonia refrigeration process than OSHA had anticipated; this is largely due to the fact that the pilot Regions have more ammonia refrigeration facilities than other PSM covered chemical processes. OSHA has also found that many PSM covered facilities that used chlorine for water or wastewater treatment have either switched to non-PSM covered chemicals or are run by municipalities, and are, therefore, outside of Federal OSHA jurisdiction.

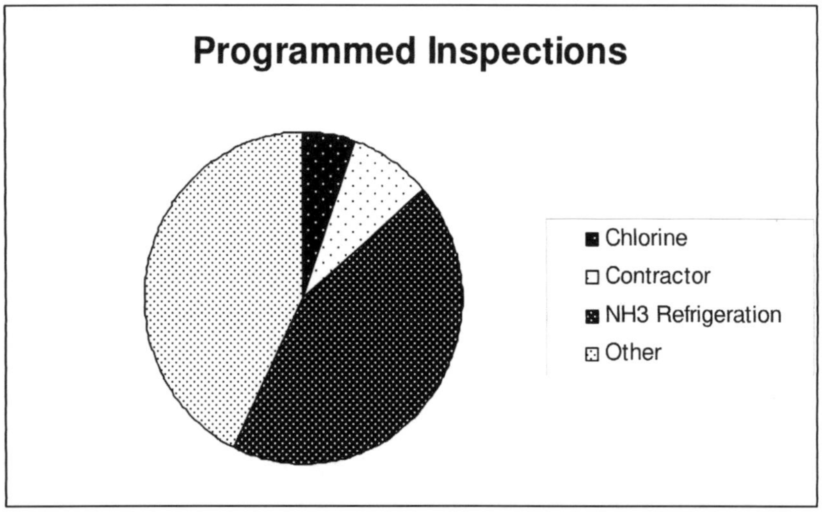

Figure 1: Facility Categories for Programmed Chemical NEP Inspections

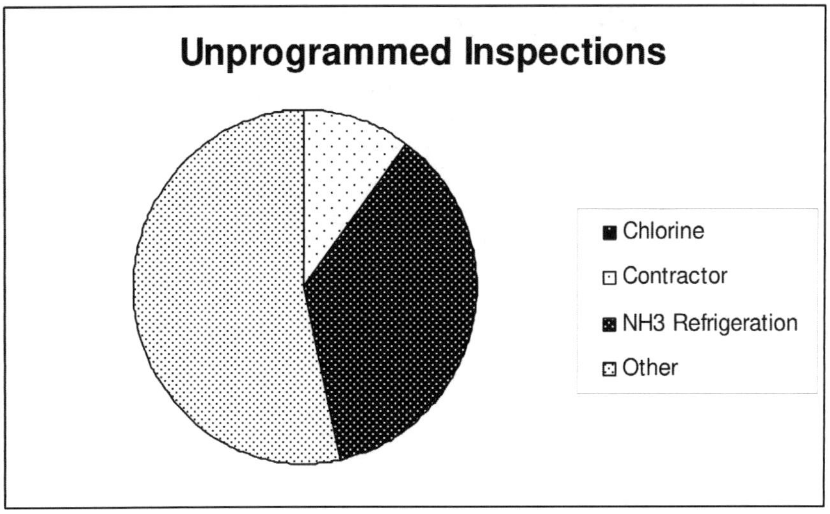

Figure 2: Categories for Unprogrammed Chemical NEP Inspections

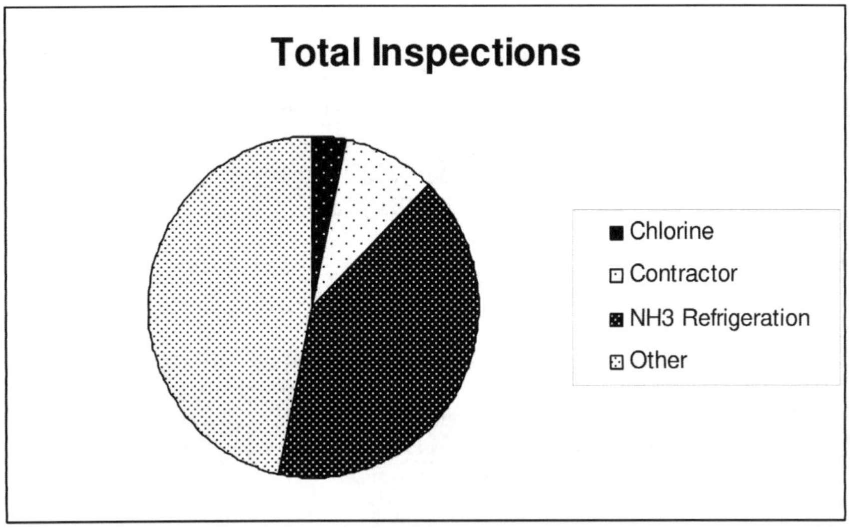

Figure 3: Categories for All Chemical NEP Inspections

3.2 Frequency of Citations by Standard Element

Violations broken down by Element and Sub-element of the PSM Standard

OSHA inspection data was analyzed to determine which element and sub-element of the PSM standard were most frequently violated. The following tables show the top 10 elements and sub-elements.

Table 3 – Chemical NEP Most Frequently Cited PSM Elements

Element	Description	Percentage of Total Violations
J	Mechanical Integrity (MI)	22.5%
D	Process Safety Information (PSI)	21.6%
E	Process Hazard Analysis (PHA)	17.9%
F	Operating Procedures (OP)	12.1%
G	Operator Training	4.4%
L	Management of Change (MOC)	4.3%
H	Contractors	3.5%
O	Compliance Audits (CA)	3.5%
C	Employee Participation	3.0%
M	Incident Investigation (II)	2.5%
N	Emergency Planning & Response	2.3%
I	Pre-startup Review	1.6%
K	Hot Work	0.7%

Table 4 – Chemical NEP Most Frequently Cited OSHA Standard Sub-element

Sub-element	Description	Percentage of Total Violations
119(d)(3)(ii)	PSI – equipment complies with RAGAGEP	3.8%
119(j)(2)	MI – written procedures	3.8%
119(j)(4)(i)	MI – inspection & testing performed	2.2%
119(e)(5)	PHA – findings & recommendations	2.0%
119(j)(4)(iii)	MI – inspection & test frequency	2.0%
29(b)(1)	Mobile ladders and scaffolds	1.9%
119(j)(5)	MI – equipment deficiencies	1.8%
119(d)(3)(i)(B)	PSI P&IDs	1.6%
119(j)(4)(iv)	MI – inspection & test documentation	1.6%
119(n)	Emergency planning & response	1.6%
119(o)(1)	Compliance audits performed	1.6%
119(e)(1)	PHAs performed	1.5%
119(e)(iii)	PHA engineering controls	1.5%
119(f)(3)	OP - certification	1.5%
119(f)(4)	OP – safe work practices	1.5%
151(c)	Eyewash stations	1.5%

3.3 Chemical NEP - Challenges and Learnings

Chemical NEP inspections have taken place in facilities with a wide range of NAICS codes. In fact, there have been inspections in 65 distinct NAICS codes. This wide range of facility types makes it much more difficult to narrow in on specific industry issues than for refineries, which are all very similar.

The variety of facility types also makes it difficult to analyze inspection data for trends or to draw conclusions. However, the data presented above does reveal two important findings. The top four PSM elements cited are the same for both Refinery and Chemical NEP inspections, indicating that employers with PSM covered processes should focus on these elements. Also, Chemical NEP inspections have resulted in a number of citations for OSHA standards other than the PSM standard. This may not be completely unexpected as the types of hazards are diverse in these facilties.

4. What is the Future of PSM NEPs?

Both the Refinery and Chemical NEPs have proven to be successful tools for determining and improving compliance with OSHA's PSM standard. Based on inspection results, OSHA feels that this effort has been very beneficial to worker safety. OSHA is currently revising the Chemical NEP and expects to expand it nationwide. Refinery NEP inspections are nearing completion, and OSHA will revisit its compliance program for refineries. OSHA plans to continue to analyze and present information from the findings of these PSM inspections to help employers identify weaknesses in their process safety management systems.

5. Disclaimer

This paper represents the views of the authors. Although it is based on OSHA programs, it is not an OSHA policy document.

6. References

(1) Refinery NEP First Year Summary, Safety and Health Topics: Process Safety Management (PSM), Accessed January 2011.

(2) Occupational Safety and Health Administration, *Directive Number: CPL 03-00-010, Petroleum Refinery National Emphasis Program,* August 18, 2009, http://www.osha.gov/pls/oshaweb/owadisp.show_document?p_table=DIRECTIVES&p_id=4014, Accessed January 2010.

(3) Occupational Safety and Health Administration, *Directive Number 09-06 (CPL 02), PSM Covered Chemical Facilities National Emphasis Program,* July 27, 2009, http://www.osha.gov/pls/oshaweb/owadisp.show_document?p_table=DIRECTIVES&p_id=4008, Accessed January 2010.

(4) Long, Lisa et. al., "OSHA Refinery and Chemical NEP Findings," 2010 AIChE Spring Meeting Conference Proceedings.

(104b) Chemical NEP Inspection Preparation

Mark S. Dreux, Arent Fox PLLC, Washington, DC

On July 27, 2009, OSHA issued a National Emphasis Program for the chemical industry ("the Chemical NEP"). Pursuant to the Chemical NEP, OSHA has been and will be conducting PSM inspections of numerous chemical facilities. This presentation will review lessons learned from OSHA's Refining and Chemical NEPs, particularly elements/sub-elements/specifics of the OSHA citations, its interpretations of the PSM Standard and its aggressive use of industry standards. This information will assist you in being better prepared for a Chemical NEP inspection and in making your facility safer.

See more of this Session: OSHA's National Emphasis Programs for Refineries and Chemical Plants - Experiences and Challenges
See more of this Group/Topical: Topical 1: Global Congress on Process Safety

The Challenge to Implement and Maintain an Effective PSM Program

Ken Hanchey
ABSG Consulting Inc.
16855 Northchase Drive
Houston, TX 77060
khanchey@absconsulting.com

James R. Thompson
ABSG Consulting Inc.
16855 Northchase Drive
Houston, TX 77060
jthompson@absconsulting.com

Keywords: OSHA, NEP, MOC, corrective actions, inspections, procedures, RNEP, VPP, CNEP, RAGAGEP, PSM, National Emphasis Program

Abstract

While developing responses to help a client respond to OSHA citations, we discovered a 1997 presentation to the same industry segment and found that little or nothing has changed in 14 years at many facilities. OSHA is still finding the same gaps today. Chemical plants generally know how to complete a process hazard analysis (PHA), management of change (MOC), an incident investigation, etc. What is incomplete are the work processes to ensure continuity and completion.

It's imperative that you have a functional process safety management (PSM) program. For example, OSHA states that the best defense against a National Emphasis Program (NEP) inspection is a strong PSM program. Failures can result in citations and incidents, which can lead to repeat or egregious citations, financial losses, public disgrace, and fatalities. Once your company has been cited for minor offenses, it becomes a target for repeat inspections.

This paper will address reducing failures by designing, implementing, and using complete work processes for PSM-related activities. Many failures can be addressed by (1) improving the documentation of what you already do, (2) promptly resolving PSM-related recommendations of all types, and (3) dealing with the challenge of keeping track of the ever-changing recognized and generally accepted good engineering practices (RAGAGEP) that apply to many aspects of your PSM program. In addition, management support/involvement and systems that follow up to ensure all work activities are complete and adequately documented are essential. Instituting complete work processes will improve your PSM program, improve your efficiency, and reduce the chance of citations.

1. Introduction – The Challenge

OSHA's process safety management (PSM) standard has been in place almost 20 years, yet OSHA inspectors continue to write citations and collect fines. Less-than-robust PSM programs cost money in lost production and repairs; taint the industry's reputation; and worse yet, cause injuries to personnel.

Maybe PSM is not enough, and we need something more risk-based or a better safety culture. But, perhaps another challenge lies within the industry itself – getting back to the PSM fundamentals and then following through using a better work process.

2. Where Did We Go Wrong?

2.1 The PSM standard

The PSM standard is basically a technically sound standard that clearly states compliance requirements. You are familiar with the PSM elements. Generally, you know how to (1) manage change, (2) complete audits, (3) investigate an incident, and (4) generally meet the regulation's requirements. The PSM standard is not the problem.

2.2 The problem

The problem may be that environmental, health, and safety (EHS) is often viewed as an expense without a business driver. Activities are completed just "good enough" to get by, and no improvement is ever made toward making them better.

2.3 The solution

The solution is to fully implement the PSM work process to ensure that all work is complete. The work process needs to (1) include management, (2) ensure that the program is continued when personnel change, and (3) include steps to plan, implement, check, and correct steps as the program evolves and is implemented.

Complete work processes for PSM-related activities must be designed, implemented, and used. Many work process failures that currently exist can be addressed by (1) improving documentation already in place, (2) promptly resolving PSM-related recommendations of all types, and (3) keeping track of the ever-changing RAGAGEP that apply to many aspects of the PSM program. In addition, management support/involvement and systems to follow up on ensuring all work activities are complete and adequately documented are essential. Instituting complete work processes will improve your PSM program, improve your efficiency, and help reduce the chance of citations.

3. PSM Work Process Weaknesses

The following sections provide a few examples of some commonly found weaknesses in PSM work processes.

3.1 Management of change (MOC)

One of the most common faults with MOC programs is an over-the-top, one-size-fits-all detailed checklist. Detailed checklists are good, but every change does not have to use every checklist; changing a turnaround date requires a different checklist than changing an alarm limit. If the process is too complicated, change will naturally happen outside the MOC process. At the other extreme, an MOC process can be so vague that the analysis of potential safety and health impacts provides no information regarding potential hazards. It is very important to follow through with MOC process development and make the process appropriate for your organization.

Poor MOC communication is another common problem. If the MOC communication tool is good, it might be used for all types of communications (not just MOCs) and personnel may be overloaded with all types of notices. If that occurs, many employees may not invest any time or effort in reviewing MOCs. Again, it is essential to follow through and develop appropriate tools for each type of site communication.

3.2 Compliance audits

Repeat findings after several compliance audits have been completed are indicative that (1) something is wrong and (2) the root causes allowing the noncompliance findings have not been fixed. Investigate the root causes and eliminate the faults in the work process.

3.3 Recognized and generally accepted good engineering practices (RAGAGEP)

As previously stated, not keeping up with the changes in the applicable RAGAGEP can lead to PSM work process failures.

29 CFR 1910.119(d)(3)(i)(E)(ii) says *"The employer shall document that equipment complies with recognized and generally accepted good engineering practices."* Some companies are reluctant to even document the standards and RAGAGEP equipment is built to or even list the RAGAGEP they recognize. That means each time something is built, personnel have to decide what standards and RAGAGEP to use, which can cause deficiencies and inconsistencies. If the RAGAGEP you follow are not documented, this opens the door for OSHA (and others) to "recognize" whatever book, guidelines, or standard it wants as RAGAGEP.

The work process that covers RAGAGEP needs to include (1) a recognition process, (2) a reconciliation process, and (3) documentation. Recognition acknowledges which RAGAGEP are applicable to your systems so you can plan. Reconciliation allows changes in RAGAGEP to be tracked and determines if your systems need to be retrofitted. Documentation maintains your position and states the RAGAGEP your equipment is built to and what RAGAGEP are recognized. Follow through and add or improve your work process to keep up to date with changing RAGAGEP.

3.4 Incident investigation

Every incident does not require a full root cause analysis, and a tiered-incident investigation process should be used. Most low-potential process incidents or minor injuries require only a simple, initial report. Serious incidents (or near misses) should require (1) a PSM-type incident investigation to determine causal factors, (2) an investigation team, and (3) a formal report. Large and potentially large losses should have an extensive root cause analysis. Make sure that each incident is reviewed and evaluated but don't overwork the investigation of minor events.

Recognize that good ideas that don't address the incident or its causal factors are not corrective actions (CAs). That doesn't mean to ignore them but evaluate and resolve them outside the CA process. Recommendations from an investigation need to:

- Address a causal factor
- Prevent recurrence of the causal factor
- Correct similar existing problems
- Address the root causes of the causal factors

3.5 Corrective actions

Corrective actions are developed from several elements of PSM. When a CA is identified, promptly resolve it, track it to full and timely completion, and document what was done. Incomplete action items are the easiest noncompliances to identify and probably the easiest to fix. Develop a work process to (1) vet the CA, (2) assign a responsible person, (3) assign a reasonable due date, and (4) track and enforce compliance.

Here's an example of a less-than-full completion of a CA from a PHA that required determination if a powder was explosive. The response was "Explosion testing indicates no action required." Here are the unanswered questions from this incomplete response: (1) what data were obtained and from whom? (2) how were the data analyzed? (3) were the data added to PSI and/or the PHA? (4) will the data enable the next crew or next generation to maintain process safety? The powder was apparently not explosive, but there were no data provided to back this up and no documentation of the knowledge gained.

Often, corrective actions are not promptly resolved. If the responsible person for a CA cannot obtain authorization, the recommendation has not been resolved. Management needs to decide what, if anything, will be done within a reasonable time period (e.g., 30 to 60 days) and document the decision.

Finally, if a CA cannot be completed on time, do not just change the due date or repeatedly change the dates. Determine why it was not completed on time (e.g., the due date was too optimistic, the task is larger than anticipated, the MOC cannot be implemented until the next turnaround) and update the process. It is important to take the time to follow through and eliminate these pitfalls from your CA work process.

3.6 Conditions leading to work process weaknesses

Some conditions that can lead to PSM work process weaknesses include:

- Focusing primarily on short term results; PSM success is long term
- A lack of management/organizational commitment to solve root causes of problems (i.e., a "patch" is good enough for now)
- No champions to push ideas and systems, to make it safe, to be involved
- A "one-size-fits-all" mentality
- The right people aren't involved, so there is no ownership of problems or solutions
- There are minimal leading and lagging indicators collected/reported
- A "management is always right" attitude
- Affected personnel aren't included throughout the work process

4. Corrective Action Work Process Example

A CA work process is not a PSM element, but PSM elements generate CAs. The following example is a CA report that demonstrates what could be in a complete work process (WP) and tool. The example presents some frequently missed information and a short description of what is missing or needed follows each entry. Some of the terms may be different from what your company may use so focus on the functions not the names.

The first part of the example CA report covers the source of the CA and most of the relational data. Ask the questions (1) Was the observation from a PHA, MOC, compliance audit, incident investigation, or somewhere else? (2) Who are the responsible personnel? (3) Is it on schedule?

EXAMPLE CA WP

Corrective Action Report

OBS # 3245

CA#: 10-0025 **Status: Closed** **Priority: 3**
Assoc W/Os: 10-458921, 10-459012
Plant: Big Shiny One
Assoc HAZOP Node: #7 R-2 Reactor feed system
Source: 2010 PHA
Open Date: 02/25/10 **Due Date:** 6/30/10 **# Extensions:** 1
Owner: Bill Smith, Supt. **Assigned:** Bob Smit **Phone:** 888-555-1212
Closed Date: 06/14/10 **Closed by:** Plant Safety Committee
Effectiveness Evaluation: 01/12/11 **Evaluated by:** Plant Safety Committee

Observation:

The PHA team discovered we had no data on the explosiveness of XYZ Powder. Someone had just seen the CSB report on dust explosions.

An observation describes areas of the PSM program elements and PSM activities where an opportunity for improvement (based on good industry practice) has been identified but, based on current practice, there is no regulatory deficiency sufficient enough to warrant a finding.

Reference:

29 CFR 1910.119(e)(3) The process hazard analysis shall address: (i) the hazards of the process.
Corporate Procedure #75 – PSM
Site Policy 4A - HazCom
Directive Number CPL 03-00-008, OSHA Combustible National Emphasis Program

The reference lists the policy, standard, etc., that provide guidance in the area of concern.

Findings:

The explosive hazards of XYZ Powder are not known.

A finding is identified for issues where the design or implementation of a company's PSM (or other EHS) program does not satisfy the minimum requirements of a standard(s) listed in the reference.

Recommendation:

Determine if XYZ Powder is explosive.

A recommendation is the team's (PHA, MOC, etc.) suggested solution on how to correct the finding.

Corrective Action Plan:

Have XYZ Powder analyzed at Acme Labs for explosiveness and communicate results to PHA team.

The corrective action plan is management's resolution to the recommendation. This is a critical step and needs to be documented.

Notes:

Bob Smit requested two extra months because he just received this assignment. 4/15/10

Even the best plans may have problems; therefore, you need to document nonroutine items in a specific place.

Closing Comments

Acme labs used ASTM E 1226 test methods to determine that the mean particle size of the sample was 280 microns and the dust deflagration index, K_{ST}, for XYZ powder is 130 bar m/s in the sample provided. That puts it in the weak explosion hazard class, but a recent larger explosion at another company was caused by particles with a K_{ST} of 139 bar m/s. The mean particle size of the sample was 180 microns. If we reduce the particle size, the K_{ST} go should go up.
Acme labs suggest we test for moisture content, Minimum Ignition Energy (MIE) and Minimum Explosible Concentration (MEC).

Added items to housekeeping checklist: Look for XYZ Powder dust accumulation greater than1/16th of an inch.

Added item to PHA checklist to consider potential for dust explosions.

Results added to 2010 PHA, communicated to all Big Shiny Plant personnel at 3rd Qtr. Safety Roundup, added to operator refresher training.

Recommended a study to identify potential combustible dust problems at site, including what is appropriate analysis, what materials have potential at site and how can we mitigate the hazards. (OBS # 3524)

Closing comments are a critical part of the work process. They describe how and why the actions have completed the CA plan. Personnel have spent time and the company has spent money to complete the plan. Document the work done so that personnel have access to the information in the future.

Evaluation for Effectiveness:

The communication package to plant really opened some eyes. The dust item on the housekeeping checklist has found dust on several occasions. The recommended study of potential combustible dust problems was initiated. This CA has been effective.

Corrective actions were taken to correct a deficiency. Follow through to evaluate the corrective action for effectiveness.

Time Stamps:

02/25/10 – 14:01:23, Jim Jones opened CA 10-0025.
03/01/10 – 18:29:59, Shirley Major entered CA Plan.
03/03/10 – 07:33:33, Bill Smith approved CA Plan.
03/03/10 – 07:36:56, Bill Smith approved "Due Date"
04/14/10 – 09:45:46, Jim Jones changed "assigned to" from Shirley Major to Bob Smit.
04/21/10 – 15:22:01, Jim Jones changed "due date: from 4/31/10 to 06/30/10.
06/14/10 – 14:30:17, Plant Safety Committee Approved closure.
01/12/10 – 12:58:42, Plant Safety Committee Evaluated CA for Effectiveness

If possible, use an electronic tool to provide time stamps for events. If this is not available, items that need improvement still need to be tracked. Examples of what need to be tracked are: (1) was the project completed on schedule, (2) how many times was the schedule changed, (3) who resolved the recommendation, and (4) who evaluated the CA for effectiveness.

5. Conclusion

Designing, implementing, and using complete work processes for PSM-related activities are vital to avoiding failures in a PSM management system. Management needs to (1) acknowledge that good EHS and PSM work processes improve yields and reduce losses of materials, people, and money, (2) understand that lasting results will probably not be measurable quarterly or even yearly, and (3) be involved and be champions for changes to work processes. All personnel (i.e., all levels and all functions involved) need to be represented in the development of work processes (i.e., employee participation). The work process must be a life-cycle process and must be evaluated for efficiency and completeness for the life of the process.

How do you implement and maintain an effective PSM program? Our key recommendations are:

1. Follow through to keep improving the PSM work processes.
2. Develop tools to assist the workforce to comply with the work processes (make sure they have everything they need).
3. Track everything you want to improve (what gets measured gets done).
4. Involve management (recommendations require resolution).
5. Document all work well enough for replacements to be able to continue the process.
6. Depend on strong, rigorous systems rather than on individuals doing everything right.

6. Sources

[1] "Process Safety Management of Highly Hazardous Chemicals," 29 CFR 1910.119

[2] *Root Cause Analysis Handbook: A Guide to Efficient and Effective Incident Investigation*, Third Edition, 2008, Rothstein Associates Inc.

[3] *Guidelines for Hazard Evaluation Procedures, 3rd Edition*, 2008, Center for Chemical Process Safety, American Institute of Chemical Engineers

[4] *Guidelines for Risk Based Process Safety*, 2007, Center for Chemical Process Safety, American Institute of Chemical Engineers

[5] *Guidelines for the Management of Change for Process Safety*, 2007, Center for Chemical Process Safety, American Institute of Chemical Engineers

AUTHOR INDEX